Microbial Biotechnology- A Laboratory Manual for Bacterial Systems

Surajit Das • Hirak Ranjan Dash

Microbial Biotechnology- A Laboratory Manual for Bacterial Systems

Surajit Das
Department of Life Science
National Institute of Technology
Rourkela
Odisha
India

Hirak Ranjan Dash
Department of Life Science
National Institute of Technology
Rourkela
Odisha
India

ISBN 978-81-322-2094-7 ISBN 978-81-322-2095-4 (eBook)
DOI 10.1007/978-81-322-2095-4
Springer New Delhi Heidelberg New York Dordrecht London

Library of Congress Control Number: 2014955535

Printed on acid-free paper

Springer is part of Springer Science+Business Media (www.springer.com)

Preface

Though tiny in size, bacteria impart many useful applications for the sustainable maintenance of the ecosystem on earth. On the evolutionary lineage, they are the first to appear and had plenty of time to adapt in the environmental conditions, subsequently giving rise to numerous descendant forms. They are omnipresent in huge number and their diversity is extended from hydrothermal vents to the cold seeps. These tiny, one-celled creatures carry out many useful functions and with the advancement of science, they have been explored greatly for use in food industry, agricultural industry, clinical sectors and many others. Biotechnological industries utilise bacterial cells for the production of biological substances that are useful for human existence including foods, medicines, hormones, enzymes, proteins and nucleic acids. Despite huge benefits human beings gain out of these microscopic organisms, less attention has been paid to study these tiny creatures. Though the research on bacterial entities has gained momentum, it is estimated that only about 1 % of the microorganisms have been discovered so far. However, rapid advances in molecular biology have revolutionised the study of bacteria in the environment. It has provided new insights regarding their composition, phylogeny and physiology. New developments in biotechnology and environmental microbiology signify that microbiology will continue to be an exciting and emerging field of study in the future.

The study of bacteria dates back to 1900 AD and substantial advancement on the methodology and practices used for their study has been occurred. There are many textbooks, research and review articles dealing with state-of-art of various aspects of molecular biology of microorganisms. However, the users usually get lost in initiating an experiment due to lack of suitable easy protocols. In this regard, an assorted laboratory manual not only to motivate the researchers and students but also to enhance the acquisition of scientific knowledge as well as the scientific aptitude is the need of the hour. This laboratory manual 'Microbial biotechnology—a laboratory manual for bacterial systems' is an attempt to overcome the inherent cumbersome practices that are followed in most of the laboratories. Every effort has been made to present the protocols in a very simpler form for easy understanding of the undergraduates, graduates, postgraduates, doctoral students, active scientists and researchers. Additionally, most of the universities providing undergraduate and postgraduate courses in microbiology and biotechnology, can use for their laboratory experiments.

There is a considerable difference between a researcher and a technician. The technician can add the appropriate reagents to obtain the suitable result. However, the researcher should focus on 'how' and 'why'. Blindly following a protocol without knowing the principle and role of reagents will not be useful in a long run. Thus, an attempt has been made to make the novice students familiar with the principle of the each experimental setup and active role of each reagent to be used in each experiment. Thus, it will be helpful for the readers to modify the protocols as well as the reagents as per their requirement. The illustrative description of each experiment will be of great use in easy understanding of the readers, irrespective of their qualification and research expertise. Some specific experiments in the advanced field of environmental microbiology have been included in the last part of the manual which will increase the awareness among the students regarding the vast application of these tiny microorganisms for the sustainability of the ecosystem.

We have tried our best to incorporate all our experience and expertise to come out in the form of this manual. Throughout the writing process of this manual we have faced lots of problems and hurdles. All have been overcome due to God's grace, self-belief and people surrounding to us. We are highly thankful to each and every one for their support and encouragement in this process. We hope this manual will be of great use for the readers in their academic and research career. Wishing all the very best to the readers and their experiments!

Rourkela, Odisha, India

Surajit Das
Hirak R. Dash

Contents

About the Authors

Surajit Das is an Assistant Professor at the Department of Life Science, National Institute of Technology, Rourkela, Orissa, India since 2009. Earlier he served at Amity Institute of Biotechnology, Amity University Uttar Pradesh, Noida, India. He received his Ph.D. in Marine Biology (Microbiology) from Centre of Advanced Study in Marine Biology, Annamalai University, Tamil Nadu, India. He has been the awardee of Endeavour Research Fellowship of Australian Government for carrying out Postdoctoral research at University of Tasmania on marine microbial technology. He has multiple research interests with core research program on marine microbiology. He is currently conducting research as the group leader of Laboratory of Environmental Microbiology and Ecology (LEnME) on biofilm based bioremediation of PAHs and heavy metals by marine bacteria, metagenomic approach for drug discovery from marine microorganisms, nanoparticle-based drug delivery and bioremediation; and the metagenomic approach for exploring the diversity of catabolic gene and immunoglobulins in the Indian Major Carps, with the help of research grants from the Department of Biotechnology (DBT), Ministry of Science and Technology and the Indian Council of Agricultural Research (ICAR), Government of India. Recognizing his work, National Environmental Science Academy, New Delhi had conferred 2007 Junior Scientist of the year award on marine microbial diversity. He is the recipient of Young Scientist Award in Environmental Microbiology from Association of Microbiologists of India in 2009. Dr. Das is also the recipient of Ramasamy Padayatchiar Endowment Merit Award given by Government of Tamil Nadu for the year 2002-2003 from Annamalai University. He is the member of IUCN Commission of Ecosystem Management (CEM), South Asia and life member of the Association of Microbiologists of India, Indian Science Congress Association, National Academy of Biological Sciences and National Environmental Science Academy, New Delhi. He is also the member of the International Association for Ecology. He is the reviewer of many scientific journals published by reputed publishers. He has written three books and authored more than 40 research publications in leading national and international journals on different aspects of microbiology.

Hirak Ranjan Dash is a Senior Research Fellow at Laboratory of Environmental Microbiology and Ecology (LEnME), Department of Life Science, National Institute of Technology, Rourkela, Odisha, India. He did his M. Sc. Microbiology (2010) from Orissa University of Agriculture and Technology, Bhubaneswar, Odisha, India. Currently, he is continuing his research on diversity and genetic aspects of mercury resistant marine bacteria for enhanced bioremediation of mercury. He has also worked in the field of antibiotic resistance and genotyping of pathogenic *Vibrio* and *Staphylococcus* spp. During his research work, he has isolated many potent mercury resistant marine bacteria from Bay of Bengal, Odisha and utilised in mercury bioremediation. A number of microbiological technique has also been developed by him for monitoring the level of mercury pollution in the marine environment. A novel mechanism of mercury resistance i.e. by intracellular biosorption was reported by him in the marine bacterial isolates. He has constructed transgenic marine bacteria possessing both mercury biosorption and volatilization capability for utilization in mercury bioremediation. He has published 14 research papers, 7 book chapters and 10 conference proceedings in his credit.

Basic Molecular Microbiology of Bacteria

1

Exp. 1.1 Isolation of Genomic DNA

Objective To isolate genomic DNA from bacterial cell.

Introduction

Bacteria possess a compact genome architecture, which is distinct from eukaryotes. It shows a strong correlation between genome size and the number of functional genes, and the genes are structured in operons reflecting polycistronic transcripts. Among different species of bacteria, there is some variation in genome size which, however, is smaller than that of many eukaryotes.

DNA was first isolated during 1869 by Friedrich Miescher, which he called as nuclein, from human leukocytes. As bacteria are of much smaller size than that of eukaryotic cells, they have smaller genome contents. Most of the bacterial genome consists of single DNA molecule, and the bacterium replicates its DNA in favourable conditions of nutrition, pH and temperature. The process of bacterial cell division is much simpler than eukaryotic cells, and hence, bacteria are able to grow and divide much faster. The life styles of bacteria play an integral role in their respective genome sizes, as free living bacteria have the largest genomes, with intermediate sizes in facultative pathogens and obligate symbionts or pathogens having the smallest genomes.

Free living bacteria have the largest genomes, intermediate sizes are found in facultative pathogens and obligate symbionts or pathogens have the smallest genomes. In this context, isolation of genomic DNA from bacteria is a useful tool to determine the fate of the selected bacteria or their recombinant genes. This may also reveal genotypic diversity by determining its size and nature.

The isolation and purification of DNA from cells is one of the most common prerequisites in contemporary molecular biology that reflects a transition from cell biology to molecular biology, from in vivo to in vitro.

Principle

Many different techniques are available for isolating genomic DNA from bacterial cells; however, all follow the common steps of cell breaking, protein removal followed by release of the genetic material. The prime aim of this experiment is to yield DNA of high quality that can be stored for several years stored under proper conditions. The common steps involving DNA isolation includes lysis of cell followed by removal of proteins, carbohydrates, RNA etc. Cell walls and membrane disruptions usually are accomplished with an appropriate combination of enzymes to digest the cell wall (usually lysozyme) and detergents

S. Das, H. R. Dash, *Microbial Biotechnology- A Laboratory Manual for Bacterial Systems*,
DOI 10.1007/978-81-322-2095-4_1, © Springer India 2015

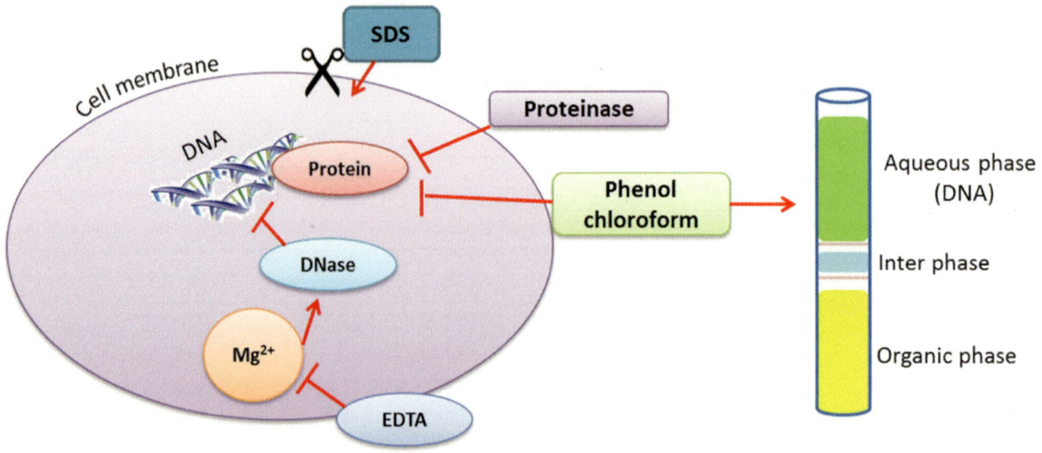

Fig. 1.1 A schematic presentation of genomic DNA isolation from bacterial cell

to disrupt membranes. Most common ionic detergent used in this step is Sodium Dodecyl Sulphate (SDS). RNA is usually degraded by the addition of DNase free RNase. The resulting oligoribonucleotides are separated from the high-molecular weight DNA on the basis of their higher solubility in nonpolar solvents (usually alcohol/water). Proteins are subjected to chemical denaturation and/or enzymatic degradation by addition of proteinase-K. The most common technique of protein removal involves denaturation and extraction into organic phase viz. phenol and chloroform (Fig. 1.1).

Reagents Required and Their Role

Luria–Bertani Broth

Luria–Bertani (LB) broth is a rich medium that permits the fast growth and better yields for many species including *Escherichia coli*. Easy to make, fast growth of the most *E. coli* strains, readily available and simple compositions contribute the popularity of LB broth. *E. coli* grow to optical density $(OD)_{600}$ 2–3 in LB broth under normal shaking incubation conditions at 24 h.

Tris EDTA Buffer

It can be prepared by mixing 50 mM Tris and 50 mM Ethylenedinitrilo tetra-acetic acid (EDTA) in water and by maintaining the pH at 8.0. As a major constituent of Tris EDTA (TE) buffer, Tris acts as a common pH buffer to control pH, while EDTA chelates cations like Mg^{2+}. Thus, TE buffer is helpful to solubilise DNA and protect it from degradation.

Sodium Dodecyl Sulphate

Ten-percent sodium dodecyl sulphate (SDS) is used for genomic DNA isolation. SDS is a strong anionic detergent that can solubilise the membrane proteins and lipids. This will help the cell membranes to break down and expose the chromosomes to release DNA.

Proteinase-K

Proteinase-K at 20 mg/ml is a very good enzyme that degrades most types of protein impurities to get a quality DNA product. It is also responsible for the inactivation of nucleases, thus preventing damage of isolated DNA.

NaCl Solution

5 M NaCl provides Na^+ ions that block negative charge of phosphates of DNA. Negatively charged phosphate in DNA causes molecules to repel each other. The Na^+ ions form an ionic bond with the negatively charged phosphates; thus, neutralise the negative charges and allowing the DNA molecules to come together.

Cetyl Trimethyl Ammonium Bromide

Cetyl Trimethyl Ammonium Bromide (CTAB) is a detergent that helps lyse the cell membrane. Apart from that, CTAB-NaCl solution binds with proteins in the digested cell lysate and helps in separation of DNA from protein making intermediate ring of protein. As a cationic detergent, CTAB is readily soluble in water as well as alcohol and can form complexes with both polysaccharide and residual protein.

Phenol:Chloroform:Isoamyl Alcohol

This is a method of liquid–liquid extraction. It separates mixtures of molecules based on differential solubility of the individual molecules in two different immiscible liquids. Chloroform mixed with phenol is more efficient at denaturing proteins than the only reagent. Chloroform isoamyl alcohol is a type of detergent that binds to protein and lipids of cell membrane and dissolves them. In this way, it disrupts the bonds that hold the cell membranes together. After dissolving the cell membrane, chloroform isoamyl alcohol forms clumps of protein-lipid complexes; thus, a precipitate is formed. The principle behind this precipitation is that, lipid–protein complex are non-aqueous compounds and DNA is an aqueous compound. Thus, the upper aqueous phase contains nucleic acid, middle phase contains lipids and the lower organic phase contains proteins.

Isopropanol

DNA is highly insoluble in isopropanol, and hence, isopropanol dissolves in water to form a solution that causes the DNA in the solution to aggregate and precipitate. Isopropanol is used as a better alternative for ethanol due to its greater potential for DNA precipitation in lower concentrations. Besides, it takes lesser time to evaporate.

Procedure

1. Grow a 5 ml bacterial culture until saturation. Centrifuge (6000 rpm for 10 min) 1.5 ml of culture for 2 min or until a compact pellet is formed.
2. Discard the supernatant and resuspend the pellet in 567 µl TE buffer.
3. Add 30 µl of 10 % SDS and 3 µl of 20 mg/ml proteinase-K, mix thoroughly, and incubate for 1 h at 37 °C.
4. Add 100 µl of 5 M NaCl and mix thoroughly. If NaCl concentration is < 0.5 M, the nucleic acid may also precipitate.
5. Add 80 µl of NaCl solution and mix thoroughly.
6. Add 1 volume (0.7–0.8 ml) of 24:1 chloroform/isoamyl alcohol, mix thoroughly, and centrifuge at 6000 rpm for 4–5 min. Transfer supernatant to a fresh tube.
7. To the supernatant, add 1 volume of 25:24:1 phenol/chloroform/isoamyl alcohol, extract thoroughly, and centrifuge at 6000 rpm for 5 min. Transfer supernatant to a fresh tube.
8. To the supernatant, add 0.6 volume isopropanol and mix gently until a stringy white DNA precipitation. Centrifuge at 10,000 rpm for 10 min briefly at room temperature, discard supernatant, and add 100 µl of 70 % ethanol to pellet.
9. Centrifuge this mixture for 5 min at room temperature, and dry the pellet by complete evaporation of ethanol.
10. Resuspend this dry pellet in 50 µl TE buffer to yield DNA. Typical yield is 5–20 µg DNA/ml starting culture (10^8–10^9 cells/ml).

11. Check the purity of the DNA by agarose gel electrophoresis and nano-drop and store at 4 °C in TE buffer till further use.

Observation

The quantity and quality of the isolated DNA can be measured by agarose gel electrophoresis and ultra-violet (UV)-visible spectrophotometer, respectively. For a 1-cm path length, the optical density at 260 nm (OD_{260}) equals 1.0 for the following solutions:

a. 50 µg/ml solution of double-stranded (ds) DNA
b. 33 µg/ml solution of single-stranded (ss)DNA
c. 20–30 µg/ml solution of oligonucleotide
d. 40 µg/ml solution of RNA

Result Table

Sample	DNA content	$OD_{260/280}$	Inference
Control			
Sample I			
Sample II			

OD optical density

Troubleshootings

Problem	Possible cause	Possible solutions
RNA contamination	If the bacterial density is too high, i.e. more than 1×10^9 cells/ml, the chances of RNA contamination becomes more	Grow the bacterial cells $\leq 10^9$ cells/ml
	RNase is not added	Add RNase (400 µg/ml) to the isolated DNA sample
Protein contamination	If the bacterial density is too high, i.e. more than 1×10^9 cells/ml, the chances of protein contamination becomes more	Grow the bacterial cells $\leq 10^9$ cells/ml
		Repeat the phenol:chloroform:isoamyl alcohol extraction step. Incubate the mixture for 10 min at −20 °C. Centrifuge, discard supernatant and add 500 µl 70 % ethanol
DNA concentration is too less	Culture volume is too less	Grow the bacterial culture upto 10^9 cells/ml or collect more pellet by repeated centrifugation
Insoluble pellet after DNA precipitation	Error in methodology and the duration of drying the pellet	Extended drying under strong vacuum may cause an overdrying of the DNA. As an acid, DNA is probably better soluble in slightly alkaline solutions such as TE or 10 mM Tris buffer with a pH of 8.0
Degraded DNA	Is the bacterial strain known as being "problematic"?	Do not let the bacterial culture grow for more than 16 h

Precautions

1. Prepare wide bore pipette tips by cutting 2–3 mm from the ends and use them. This will not allow DNA for mechanical disruption.
2. The incubation period with proteinase-K may be extended depending on the source of DNA.
3. Repetition of phenol-chloroform extraction method should be performed to obtain a pure DNA.
4. DNase-free plasticwares and reagents should be used during the entire procedure.
5. Phenol-chloroform is probably the most hazardous reagent used regularly in molecular biology laboratories. Phenol is a very strong acid that causes severe burns. Chloroform is a carcinogen. Handle these chemicals with care.
6. Wear gloves and goggles while isolating genomic DNA.

FLOW CHART

5 ml of overnight grown culture

↓

Centrifuge to collect the cell pellets

↓

Re-suspend the cell pellets in 567 µl of TE buffer

↓

Add 30 µl of 10% SDS and 3 µl of 20 mg/ml proteinase-K

↓

Mix thoroughly and incubate for 1 h at 37°C

↓

Add 100 µl of 5 M NaCl and mix thoroughly

↓

Add 1 volume of 24:1 chloroform/isoamyl alcohol, mix thoroughly; centrifuge at 6,000 rpm for 4-5 min. Transfer supernatant to a fresh tube

↓

Add 1 volume of 25:24:1 phenol/chloroform/isoamyl alcohol, extract and centrifuge at 6,000 rpm for 5 min. Transfer supernatant to a fresh tube

↓

Add 0.6 volume of isopropanol and mix gently until a stringy white DNA precipitate forms. Centrifuge at 10,000 rpm for 10 min briefly at room temperature, discard supernatant and add 100 µl of 70% ethanol to pellet

↓

Centrifuge for 5 min at room temperature and dry the pellet by evaporation of ethanol. Re-suspend in 50 µl of TE buffer

↓

Check purity of the DNA and store at 4°C in TE buffer till further use

Exp. 1.2 Preparation of Bacterial Lysates

Objective To prepare total cellular DNA of bacteria by lysis of bacterial cell from *E. coli*

Introduction

Bacteria represent a much simpler life form. They lack rigid cell wall, nuclear membrane and complex genetic organisation. This ultimately helps the researchers to carry out number of experiments by taking them as model organisms. For any molecular biology study of bacteria, extraction of its genetic material is a must prerequisite, which is much more complex procedure and lacks much handling expertise. However, for rapid extraction of total crude DNA from bacterial cells, a simple lysis of bacterial cell wall as well as cell membrane will solve the purpose (Fig. 1.2). Due to lack of the nuclear membrane, certain physical and/or chemical reagents may be used to lyse the cell to take out the cellular DNA into the aqueous medium, which can be used for

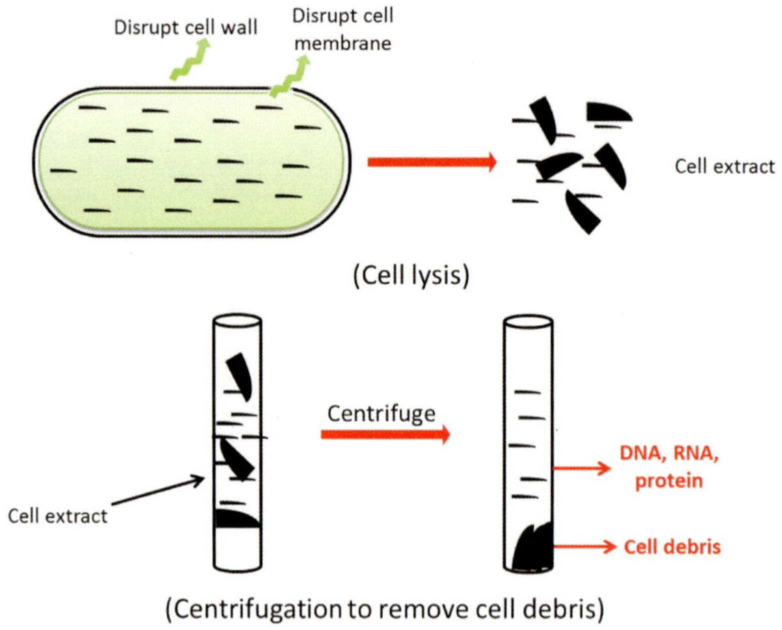

Fig. 1.2 Preparation of bacterial lysates

simple investigations like amplification of genes in PCR, detection of antibiotic resistant genes and many more.

Lysate preparation is of great use in both environmental as well as clinical microbiology. During clinical microbiological investigations, time is a crucial factor in disease diagnosis and characterisation of the potential pathogen in terms of its antibiotic resistance and pathogenicity. Hence, instead of proceeding for a much longer genomic DNA isolation or plasmid isolation, many laboratories prefer to use bacterial lysates as templates for further use in detection of genes. There are many advantages of using bacterial lysates over other conventional practices due to its less time taking steps and less expertise requirement. It can be performed without the use of any sophisticated instruments under any laboratory conditions. Till date, there are many reports of using bacterial lysates as templates for identification of the strains by 16S ribosomal (r)RNA gene amplification, analysis of the antibiotic-resistant genotype of the isolates, detection of virulence genes present or absent in the bacterial genome and restriction digestion. Thus, correct and accurate preliminary information can be obtained by using bacterial lysates rather than using the genomic DNA or plasmid DNA of bacteria, which takes a relatively longer time.

Principle

There are different methods of preparation of *E. coli* cell lysates such as boiling, sonication, homogenisation, enzymatic lysis, freezing, grinding etc. The principles of all the available practices have been provided below.

Boiling is the most common technique of preparation of bacterial cell lysates. During boiling, it requires a temperature of 100 °C. When incubated at this condition for 10 min, cell membrane of bacteria ruptures by denaturation of membrane proteins. In this process, high temperature also causes denaturation of bacterial DNA, which can be renatured by keeping the lysis product on ice for at least 5 min. When centrifuged at maximum speed, the cellular debris other than DNA and RNA precipitates at the bottom to form the pellet, and the supernatant can be used as template for further experiments.

Sonication is the most popular technique of lysing small quantity of bacterial cell. In this process, the cells are lysed by liquid shear and cavitation. DNA is also sheared by sonication; hence, there is no need of adding DNase to the cell suspension. However, the main problem associated with this practice is the temperature control. This problem can be overcome by keeping the cell suspension on ice and use of short pulses with pauses to re-establish a low temperature. There might be some additional problems with the large quantity of bacterial cultures, as it requires a long sonication time to achieve adequate lysis and in this way it becomes difficult to maintain the low temperature.

The device used for preparation of bacterial lysates is the homogeniser. In this process, the bacterial cells lyse due to the high pressure in the cell suspension followed by sudden release of the pressure. This ultimately creates the liquid shear, which is capable of lysing the bacterial cells. However, the high operating pressure in these homogenisers ultimately increases the temperature; hence, the lysed cells should be cooled to 4 °C prior to use. In addition, antifoaming agents may be used during this process, as the foam generated may inactivate many proteins.

Enzymatic lysis is based on the digestion of peptidoglycan layer of bacterial cell wall by the use of lysozyme. However, in case of Gram-negative bacteria an additional layer is present on the cell wall that needs to be permeable for the action of lysozyme on peptidoglycan composition of the cell wall. In this context, Tris, which is often used as a buffer in lysis methods, effectively increases the permeability of the outer membrane. This process can be enhanced by the addition of EDTA that chelates the magnesium ions to stabilise the cell membrane. In this process of cell lysis, a lot of DNA is liberated to the solution, and it becomes highly viscous and in order to decrease the viscosity of the solution RNase and proteinase-K may be added optionally.

The alternative lysis method of bacteria is the alternate freezing and grinding. In this method, the cells are freezed directly in liquid nitrogen, and the frozen cells are ground to a powder by the use of mortar and pestle. The obtained powder can be stored at −80 °C indefinitely and the cell lysates can be prepared by adding the powder to 5 volumes of TE buffer.

Procedure

For Boiling Lysis

1. Grow *E. coli* culture upto 0.5 McFarland suspension in LB medium. If required, dilute the grown culture to obtain 0.5 McFarland suspensions, a particular concentration.
2. Centrifuge at 6000 rpm for 5 min at room temperature.
3. Discard the supernatant and resuspend the cell pellet in 200 µl of autoclaved milli-Q water.
4. Set the water bath at 100 °C or boil water in a container.
5. Keep the centrifuge tubes containing bacterial culture in boiling water for 10 min.
6. Immediately snap the centrifuge tubes on ice for 5 min.
7. After incubation on ice, centrifuge the tube at 10,000 rpm for 5 min at 4 °C.
8. Transfer the supernatant to a fresh tube and store it at 4 °C until further use.

For Use of Sonication

1. Inoculate 4–5 fresh bacterial cultures to 2 ml of previously autoclaved LB broth and incubate at 37 °C and 180 rpm for overnight.
2. Transfer 1 ml of the bacterial suspension in a 1.5 ml of micro-centrifuge tube and centrifuge at 6000 rpm for 5 min at 4 °C.
3. Discard the supernatant and add rest of the culture to the pellet in centrifuge tube and again centrifuge at 6000 rpm for 5 min at 4 °C.
4. Discard the supernatant and add 600 µl of sterile milli-Q water and mix properly by vortexing.
5. Centrifuge at 6000 rpm for 5 min at room temperature and discard the supernatant, add 600 µl of 1 X TE buffer and mix again by vortexing.

6. Sonicate for 2 min at 22 µm amplitude with short pulses (5–10 s) and pauses (10–30 s).
7. Centrifuge at 10,000 rpm for 5 min and transfer the supernatant to a fresh micro-centrifuge tube.
8. Store the supernatant at −20 °C till further use.

By Lysozyme Digestion

1. Grow *E. coli* culture upto 0.5 McFarland suspension in LB medium. If possible, dilute the grown culture to obtain 0.5 McFarland suspensions.
2. Dissolve lysozyme in an appropriate amount of TE buffer to make a 10 mg/ml solution. Add the enzyme powder to the buffer and dissolve it slowly and keep on ice. Do not shake or mix.
3. Add 100 µl of lysozyme stock solution (10 mg/ml) to 1 ml of bacterial culture in TE buffer to obtain a final concentration of 1 mg/ml.
4. Incubate the solution at 30 °C and shake gently for 30 min to 1 h.
5. Centrifuge the solution at 10,000 rpm for 10 min at 4 °C and transfer the supernatant to a fresh vial.
6. Store the supernatant at −20 °C till further use.

By Repeated Freezing and Thawing

1. Inoculate 4–5 fresh bacterial cultures to 2 ml of previously autoclaved LB broth and incubate at 37 °C and 180 rpm for overnight.
2. Transfer 1 ml of the bacterial suspension in a 1.5 ml of micro-centrifuge tube and centrifuge at 6000 rpm for 5 min at 4 °C.
3. Discard the supernatant and add rest of the culture to the pellet in centrifuge tube and again centrifuge at 6000 rpm for 5 min at 4 °C.
4. Discard the supernatant and resuspend the cell pellet in 200 µl of sterilised milli-Q.
5. Freeze the cell pellet fairly slowly in liquid nitrogen for 3 min.
6. Place the tube in hot water bath (previously set to 80–90 °C) for 3 min.
7. Repeat the freeze-thaw cycle for three times. Make sure you mix your tube between each cycle.
8. Centrifuge the tubes at 10,000 rpm for 5 min at 4 °C.
9. Using a micro-pipette transfer the supernatant, which contains DNA, to a fresh tube and discard the pellet.
10. Store the tubes at −20 °C till further use.

By Homogenisation

1. Inoculate 4–5 isolated bacterial colonies to 2 ml of previously autoclaved LB broth and incubate at 37 °C, 180 rpm for overnight.
2. Transfer 1 ml of the bacterial suspension to a 1.5 ml of micro-centrifuge tube and centrifuge at 6000 rpm for 5 min at 4 °C.
3. Discard the supernatant, and add rest of the culture to the pellet in the centrifuge tube. Centrifuge again at 6000 rpm for 5 min at 4 °C.
4. Discard the supernatant and resuspend the cell pellet in 200 µl of sterilised TE buffer.
5. Switch the three-way valves to feed material from the tank containing the microbial cell. Set the discharge to the other tank.
6. Connect the cooling water supply to the homogeniser and ensure it is switched on.
7. Connect and switch on other utilities as required to operate the homogeniser.
8. Set operating pressure to zero and start the homogeniser. Watch the pressure rise on the instrument gauge to ensure availability of the flow path.
9. Cautiously, adjust the operating pressure to the desired value.

10. When the feed supply runs low, release the pressure back to zero and shut off the system.
11. Allow the homogenate to cool by immediately incubating the samples on ice for 5 min.
12. Centrifuge the homogenate at 10,000 rpm for 10 min at 4 °C and transfer the supernatant to a fresh vial.
13. Store the vials at −20 °C till further use.

Observation

Simple lysate is the crude extract of nucleic acid from the bacterial cell. Spectrophotometric analysis of bacterial lysate gives a clear view on the amount of DNA present in it as well as its quality. This crude extract can be used further for applications such as DNA amplification by polymerase chain reaction (PCR), restriction digestion where there is a lesser chance of interference by the RNA and protein impurities.

The absorbance at 260 nm is used to quantify the nucleic acid contents. One value of absorbance at 260 nm in 1 ml produces an OD of 1. Thus, applying the same conversion factor:

a. 1 A260 unit dsDNA = 50 μg
b. 1 A260 unit ssDNA = 33 μg
c. 1 A260 unit ssRNA = 40 μg

Result Table

Sample	DNA content	RNA content	$OD_{260/280}$ [a]	Inference
Boiling lysis				
By sonication				
By freeze-thaw				
By homogenisation				
By lysozyme digestion				

OD optical density
[a] For pure DNA $OD_{260/280}$ is 1.8 and for pure RNA it is 2.0. Thus, the inference can be drawn from $OD_{260/280}$ values <1.8 more protein contamination and >1.8 more RNA contamination

Troubleshootings

Problems	Possible cause	Possible solution
Degraded DNA	Higher temperature during boiling or sonication may degrade DNA	(1) After boiling lysis for 10 min, immediately span on ice for 5 min, so that a large fraction of denatured DNA can be renatured to ease denaturation
		(2) A lot of heat is generated during sonication. Hence, sonication may be performed at short pulses with pauses
No yield	Cell lysis is not proper	(1) Use lysozyme (for Gram-negative) or lysostaphin (for Gram-positive) at a final concentration of 1 μg/ml in addition to any lysis practices
		(2) Increase the incubation time (for boiling lysis and chemical lysis) and exposure time (for sonication and homogenisation)
No proper amplification in polymerase chain reaction	May be due to high protein contamination	(1) Phenol chloroform method may be performed after lysis of cell to get a more pure form of DNA
		(2) RNA contamination may be avoided by addition of RNase to the lysate

Precautions

1. Do not forget to make a hole at the top of the centrifuge tube before keeping it in the boiling water bath during boiling lysis technique.
2. Mark carefully the centrifuge tubes and cover them with cello-tape, otherwise they may cause confusion by erasing the mark due to steam generated in the water bath.
3. Use gloves and cryo-gloves while working with liquid nitrogen during freeze-thaw technique.
4. Carefully cover your ear during sonication, as it may damage the ear drum.
5. Always keep the samples on ice while performing sonication to minimise the chance of damage to DNA.
6. Never store the crude DNA for a longer period at 4 °C or even −20 °C, which may interfere in further experiments.

FLOW CHART

For boiling lysis

Dilute the grown *E. coli* culture up to 0.5 McFarland

↓

Take 200 µl of the culture and centrifuge at 6,000 rpm for 5 min

↓

Wash the cell pellets by re-suspending the cell pellet with 200 µl of autoclaved milli-Q water and repeat this step for two times

↓

Put the tubes in a water bath set at 100°C for 10 min

↓

Immediately snap the tubes on ice for 5 min

↓

Centrifuge the tube at 10,000 rpm for 5 min at 4°C

↓

Transfer the supernatant to a fresh tube and store it at 4°C

For Sonication

Inoculate pure culture of bacteria to 2 ml LB broth for sufficient growth

↓

Take 1 ml of bacterial suspension and centrifuge at 6,000 rpm for 5 min at 4°C

↓

Discard the supernatant and re-suspend it with 600 µl of autoclaved milli-Q

↓

Again centrifuge the culture and re-suspend the pellet with 600 µl of 1X TAE

↓

Sonicate the tubes for 2 min at 22 µm amplitude with short pulses (5-10 sec) with pauses (10-30 sec)

↓

Centrifuge at 10,000 rpm for 5 min and transfer the supernatant to a fresh centrifuge tube

↓

Store the supernatant at -20°C till further use

<u>For lysozyme digestion</u>

Grow bacterial culture up to 0.5 McFarland suspensions in LB broth and take 1 ml of culture in a centrifuge tube

↓

Add lysozyme to the bacterial suspension at a final concentration of 1 mg/ml

↓

Incubate the solution at 30°C with intermediate shaking for 30 min to 1h

↓

Centrifuge the solution at 10,000 rpm for 10 min at 4°C

↓

Transfer the supernatant to a fresh tube

↓

Store it at -20°C till further use

For repeated freezing and thawing

↓

Transfer 1 ml of culture to a 1.5 ml centrifuge tube and centrifuge at 6,000 rpm for 5 min to collect the cell pellets

↓

Re-suspend the cell pellets with 200 μl of autoclaved Milli-Q

↓

Freeze the cell pellets slowly in liquid nitrogen for 3 min

↓

Immediately incubate the tubes in hot water bath (at 80-90°C) for 3 min

↓

Mix the tubes repeatedly between each cycles and repeat the cycle of freeze-thaw for 3 times

↓

Centrifuge the tubes at 10,000 rpm for 10 min

↓

Transfer the supernatant to a fresh tube and store at -20°C

For homogenization

Transfer 1 ml of grown culture to a 1.5 ml of centrifuge tube, centrifuge at 6,000 rpm for 5 min to collect the cell pellets, re-suspend with 200 µl of TE buffer

↓

Switch the three way valves, connect the cooling water tank and switch on the other utilities

↓

Start the homogenizer with the setting the operating pressure at zero

↓

When the sample feeder runs low, release the pressure back to zero, and cool the samples immediately by keeping them on ice

↓

Centrifuge the homogenate at 10,000 rpm for 10 min, transfer the supernatant to a fresh tube and store at -20°C

Exp. 1.3 Isolation of Plasmids

Objective To isolate plasmid DNA from bacterial cell.

Introduction

Plasmid is usually a circular or sometimes linear piece of dsDNA found in bacteria. In many instances, it carries non-essential genes, which are responsible for the survival of the particular bacterium in adverse conditions. Due to their small size and versatility, bacteria plasmids have become a central part of research in biotechnology in many experiments from expressing human genes in bacterial cells to DNA sequencing.

The term 'plasmid' was introduced by an American molecular biologist Joshua Lederberg in 1952. In a single bacterial cell, the number of identical plasmids ranges from 1 to 1000 under different circumstances with a size range of 1 to over 1000 kb. Scientists have taken the advantage of plasmids to use them as tools for cloning, transferring and manipulation of genes. Plasmids used in genetic engineering are called as vectors, which are commonly used to multiply or express a particular gene. Plasmids can be introduced to bacterial cell by transformation, as bacteria divide rapidly they can be used as factories to generate DNA fragments in large numbers. There are many ways of classifying bacterial plasmids. Based on their functions they are: (i) fertility F-plasmids, (ii) resistance R-plasmids, (iii) Col plasmids, (iv) degradative plasmids and (v) virulence plasmids. Plasmids may belong to one or more than one of these functional groups. Plasmid DNA generally occurs in one of the five confirmations, i.e. nicked open circular, relaxed circular, linear, supercoiled or covalently closed-circular and supercoiled denatured DNA like supercoiled DNA.

Plasmids are the DNA molecules that are distinct from chromosome of bacterial cell and are capable of inherited stably without linking to the bacterial chromosome. It can be transferred horizontally between cells and responsible for carrying and spreading of antibiotic resistance genes among environmental and clinical strains. In addition, plasmids also carry many genes that code for wide range of metabolic activities, thus, enabling the host bacteria to degrade pollutants, production of antibacterial compounds, showing virulence and pathogenicity in bacteria. Thus, study of bacterial plasmids is of utmost importance for characterisation of a bacterial strain to explore its nature.

Principle

Plasmids need to be isolated from the bacteria to purify a specific sequence to use as vectors in molecular cloning. There are various methods and commercial kits available nowadays for the isolation of pure and desired conformation of plasmid DNA, irrespective of their copy numbers, i.e. high or low. In this section, we will discuss about the procedure that can be applied for this purpose without the use of any commercially available kits or columns.

Most of the available plasmid isolation procedures are based on the fact that plasmids generally occur in covalently closed circular configuration in bacteria. Hence, after cell lysis, most of the intra-cellular contents come out of the cell, and subsequently plasmid is enriched and purified. As plasmid DNA is highly sensitive to mechanical stress, shearing forces such as vigorous mixing or vortexing should be avoided after cell lysis. In this context, all the mixing steps should be carried out by careful inversion of the tubes several times rather than vortexing. The tip ends may be cut-off to minimise the shearing force. The trickiest stage of plasmid isolation is the lysis of bacteria, as both incomplete lysis and total dissolution of the cell may result in reduced yield of plasmid DNA. As simple lysis of the cell generates huge amount of genomic DNA from bacteria of high-molecular weight, they can be separated from the plasmid DNA by high-speed centrifugation along with other cell debris.

The most popular method of isolating plasmid DNA is the use of Birnboim and Doly (1979). This technique takes the advantage of the narrow range of pH difference (12.0–12.5), which denatures linear DNA but not covalently closed circular DNA (Fig. 1.3). Thus, on lysozyme digestion cell wall of bacteria weakens and the cellular macromolecules come out of the cell due to the treatment of SDS and sodium hydroxide. Chromosomal DNA remains in high-molecular weight form but becomes denatured. When neutralised with acidic medium, the chromosomal DNA renatures and aggregates to form an insoluble network. Additionally, high concentration of sodium acetate precipitates protein-SDS complexes and high-molecular weight RNA. As the pH of the alkaline denaturation is carefully controlled, the covalently closed circular form of plasmid DNA molecules still remain in their native form in the solution while other contaminating macromolecules coprecipitate. Thus, the precipitate can be removed by centrifugation to concentrate plasmid by ethanol precipitation. If necessary, plasmids can be purified further by gel filtration.

Reagents Required and Their Role

Luria–Bertani Broth

It is a rich medium, which permits the fast growth as well as good growth yields of many species of bacteria. It is the most commonly used growth medium for *E. coli* cell culture during molecular biology studies. LB broth can support *E. coli* to grow OD_{600} 2–3 under normal shaking incubation conditions.

Tris EDTA Buffer

TE buffer is prepared by mixing 50 mM Tris and 50 mM EDTA in water and by maintaining the pH 8.0. The major constituent of TE buffer is Tris, which acts as a common pH buffer to control pH during addition of other reagents. EDTA chelates cations like Mg^{2+}. Hence, TE buffer is helpful to solubilise DNA by protecting it from degradation.

Glucose

During isolation of plasmid DNA, glucose is added in the lysis buffer to increase the osmotic pressure outside the cells. Glucose maintains osmolarity and prevents the buffer from bursting the cells. Additionally, glucose is used to make the solution isotonic.

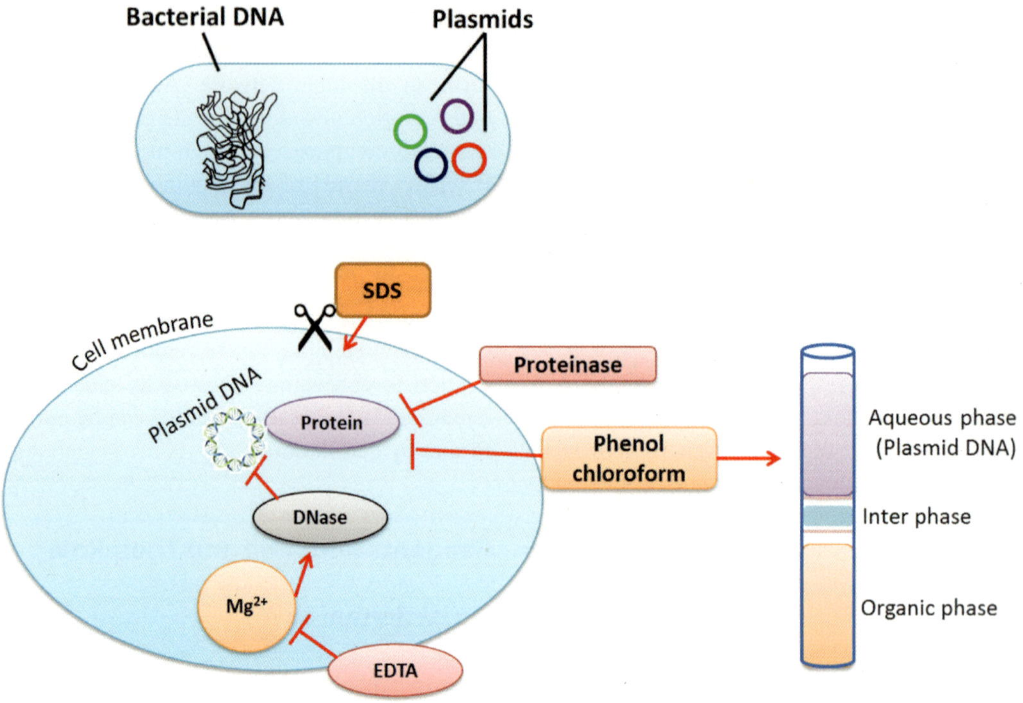

Fig. 1.3 Principle of isolation of plasmid DNA from bacteria

Ethylenedinitrilo Tetra-acetic Acid

EDTA binds with the divalent cations in the cell wall, thus weakening the cell envelope. After cell lysis, EDTA limits DNA degradation by binding Mg^{2+} ions, which are necessary cofactors for bacterial nucleases. In this way, it inhibits nucleases leading to the rupture of cell wall and cell membrane.

Sodium Hydroxide

Sodium hydroxide is used to separate bacterial chromosomal DNA from plasmid DNA. Chromosomal DNA and sheared DNA are both linear, whereas most of the plasmid DNA is circular. When the solution medium becomes basic due to addition of sodium hydroxide, dsDNA molecules are separated by denaturation and their complementary bases are no longer associated with each other. On the other hand, though plasmid DNA becomes denatured they are not separated. The circular strands can easily find their complementary strands and renature to circular ds plasmid DNA molecule once the alkaline solution is neutralised. This unique property of plasmid DNA is exploited to separate plasmid DNA from chromosomal DNA by adding NaOH.

Potassium Acetate

Potassium acetate is used to selectively precipitate the chromosomal DNA and other cellular debris away from the desired ds plasmid DNA. Potassium acetate plays three roles during plasmid DNA isolation: (i) it allows circular DNA to renature while sheared cellular DNA remains denatured as ssDNA; (ii) it allows precipitation of ssDNA as large ssDNA are insoluble in high salt concentration and (iii) when potassium acetate is added to SDS, it forms KDS, which is insoluble. This allows the easy removal of SDS contamination from the extracted plasmid DNA.

Glacial Acetic Acid

It neutralises the alkaline conditions in the solution that have been developed by addition of NaOH to solution, which helps in the rapid renaturation of the plasmid DNA. Though there is not much difference between acetic acid and glacial acetic acid, glacial acetic acid is the anhydrous acetic acid. Glacial acetic acid does not have water in it, whereas acetic acid is a weak acid which can be concentrated. Glacial acetic acid is an acetic acid of a high purity of more than 99.75 %.

Procedure

1. Prepare the following solutions with the following compositions prior to the isolation of plasmid DNA from bacteria.
   - *Solution I (Lysis buffer I):* 50 mM Tris pH 8.0 with HCl, 10 mM EDTA
     For 1 l, dissolve 6.06 g Tris base, 3.72 g EDTA.$2H_2O$ in 800 ml of milli-Q water, adjust pH to 8.0 with HCl, make up the volume to 1 l with milli-Q water, autoclave and store at 4 °C
   - *Solution II (Lysis buffer II):* 200 mM NaOH, 1 % SDS
     For 1 l, dissolve 8.0 g NaOH pellets in 950 ml of milli-Q water and 50 ml of 20 % SDS solution. Solution II should be freshly prepared just before the use.
   - *Solution III (Lysis buffer III):* 3.0 M KOAc, pH 5.5
     For 1 l, dissolve 294.5 g of potassium acetate in 500 ml of milli-Q water, adjust pH to 5.5 with glacial acetic acid (~110 ml), and make up the final volume to 1 l with addition of milli-Q water, autoclave and store at 4 °C.
2. Inoculate a single bacterial colony into 5 ml of LB broth medium and incubate the tube at 37 °C for 24 h with 180 rpm shaking.
3. Collect the bacterial cell pellet from the grown culture by centrifugation at 6000 rpm for 5 min at room temperature.
4. Discard the supernatant and resuspend the cell pellets with 600 µl of autoclaved TE buffer, again centrifuge at 6000 rpm for 5 min at room temperature and collect the cell.
5. Resuspend the cell pellet with 1 ml of ice-cold *Solution I.* Pipette up and down to completely resuspend the cell pellet.
6. Add 200 µl of *Solution II* to the suspension. Mix thoroughly by repeated gentle inversion. Avoid vortexing.
7. Add 1.5 ml ice-cold *Solution III* to the cell lysate. Do not vortex.
8. Look for the development of a white precipitate.
9. Centrifuge at 12,000 rpm for 30 min at 4 °C.
10. Transfer the supernatant to a fresh tube.
11. Add 2.5 volume of isopropanol to precipitate the plasmid DNA. Mix thoroughly by repeated inversion without vortexing.
12. Centrifuge at 12,000 rpm for 30 min at 4 °C.
13. Discard the supernatant to collect the pellet.
14. Rinse the pellet with ice-cold 70 % ethanol followed by air drying for approximately 10 min to evaporate ethanol.
15. Add 50 µl of TE buffer to dissolve the pellet.
16. Add 2 µl of RNase (10 mg/ml) and incubate for 20 min at room temperature to remove RNA contamination.
17. Store the tube at −20 °C till further use.

Observation

Run the isolated plasmid DNA on 0.8 % agarose gel and see the banding pattern, which will be visible on the following order based on their migration rate, i.e. supercoiled > linear > nicked circles > dimer > timer > others.

In addition, measure the OD at 260 and 280 to check the quantity and quality of the isolated plasmid.

Result Table

Sample	Plasmid content	$OD_{260/280}$	Inference[a]
Control			
Sample I			
Sample II			

OD optical density

[a] For pure DNA $OD_{260/280}$ is 1.8 and for pure RNA it is 2.0. Thus, the inference can be drawn from $OD_{260/280}$ values <1.8 more protein contamination and >1.8 more RNA contamination

Troubleshootings

Problems	Possible errors	Possible solution
Low yield of plasmid DNA	Growth of the culture is not adequate	Grow the culture with suitable growth medium in optimum conditions
	Lysate has not been prepared properly	Incubate for 5 min before going for final centrifugation after addition of solution III
RNA contamination	Initial centrifugation has not been performed at 20–25 °C	The residual RNA may be degraded when the initial centrifugation step of lysate is carried out at room temperature
Insoluble pellet after DNA precipitation	Pellet might have been dried excessively	As an acid, DNA is better soluble in slightly alkaline solutions such as TE or 10 mM Tris buffer with a pH of 8.0.
		Pellet may be heated for several minutes at 65 °C to enhance dissolving
Poor performance in downstream applications with the plasmid DNA	Non-complete resuspension of pellets with solution-II	Mixing must be done carefully (by inverting slowly), till a homogenous phase is obtained
Contamination with bacterial chromosomal DNA	Vortexing might have been carried out in any step	Never vortex the solution after addition of any solution, which will result in shearing of chromosomal DNA
Smear in gel/degraded plasmid	Is the bacterial strain known as 'problematic'?	Avoid the bacterial culture grown for more than 16 h
	Recommended growth time of the strain exceeded	Use the recommended growth time of the bacteria

Precautions

1. Use a fresh pipette when preparing different stock solutions to avoid cross-contamination.
2. Try to avoid touching the inner-wall of the tube while transferring the supernatant to a fresh tube.
3. Be careful not to dislodge the pellet while transferring the supernatant to a fresh tube.
4. Always wear safety goggles and gloves.
5. Never try to mix the samples by vortexing at any step of the plasmid DNA extraction procedure.
6. Use of a cut end tip will be extremely helpful during the extraction procedure.

FLOW CHART

Take 5 ml of overnight grown culture

↓

Wash the pellet with 600 µl of TE buffer

↓

Centrifuge at 6,000 rpm for 5 min to collect the cell pellet

↓

Re-suspend the cell pellets with 1 ml of ice-cooled Solution I

↓

Add 200 µl of Solution II to the suspension

↓

Add 1.5 ml of ice-cold Solution III to the cell lysate

↓

Centrifuge at 12,000 rpm for 30 min at 4°C

↓

Transfer the supernatant to a fresh tube; add 2.5 volume of isopropanol

↓

Centrifuge at 12,000 rpm for 30 min at 4°C

↓

Rinse the pellet with 70% ethanol and allow it to evaporate

↓

Add 50 µl TE buffer to dissolve the pellet and store at -20°C

Exp. 1.4 Isolation of Total RNA from Bacteria

Objective To isolate total RNA from bacterial cell.

Introduction

Central dogma of life suggests that DNA harbours all the information to code for a protein through RNA. Hence, in case of bacterial systems, RNA employs many functions, viz. (i) acts as the catalysts for most of the biochemical reactions; (ii) acts as a carrier of amino acids during protein synthesis; (iii) acts as a transmitter of genetic information to their respective function and (iv) acts as a template for protein synthesis. Thus, RNA in bacteria is the omnipresent biological macromolecule that performs many crucial roles of coding, decoding, regulation and expression of genes.

There are different types of RNA found in eukaryotic systems; however, in prokaryotic systems, transfer-messenger RNA (tmRNA) is found, which tags proteins coded by mRNAs and lack stop codons for degradation and prevents ribosome from stalling. In addition, bacteria also possess small RNAs (sRNA), which are small (50–250 nucleotide) non-coding RNA molecules which are highly structured and contain several stem loops. Though less explored, sRNA in bacteria is supposed to have their role in binding protein targets, binding to mRNA targets and thus regulating gene expression. tmRNA may form a ribonucleoprotein complex (tmRNP) together with Small protein B (SmpB), Elongation factor Tu (EF-Tu) and ribosomal protein S1. In majority of bacterial system, the desired functions have been carried out by standard one piece tmRNAs. However, in some bacterial species *ssrA* gene sometimes produces a two piece tmRNA where two separate RNA chains are joined by base-pairing.

The best example of a bacterial sRNA is the 6S RNA found in *E. coli*. 6S RNA is conserved in many bacterial species and plays an important role in gene regulation. This RNA has a major impact on the activity of RNA polymerase (RNAP), which transcribes RNA from DNA. 6S RNA inhibits its activity by binding to a subunit of polymerase to stimulate transcription during growth. This mechanism of inhibiting gene expression compels active growing cells to enter a stationary phase. Another major class of bacterial RNA is rRNA, which is generated by the endonuclease processing of a precursor transcript. Thus, the cleavage of this transcript produces 5S, 16S, 23S rRNA molecules and a tRNA molecule.

Principle

Trizol reagent has been widely used nowadays for the extraction of RNA from bacterial cell. It is the most common method developed by Chomczynski and Sacchi (1987). Though it takes a slightly longer time than the commercially available column-based methods, it has high capacity to yield more RNA. While using the chaotropic lysis buffers, this method is considered to provide the best quality of RNA.

RNA is the polymeric substance consisting of long ss chains of phosphate and ribose sugar units along with the nitrogen bases like adenine, guanine, cytosine and uracil. RNA is used in all steps of protein synthesis in all living systems; hence, its isolation and further characterisation reveals the important salient features regarding protein synthesis and gene expression analysis. Thus, isolation of RNA of high quality is the most crucial step of various molecular biology studies. In this regard, trizol is the ready to use reagent for isolation of RNA from cells and tissues. Trizol works by maintaining RNA integrity during tissue homogenisation and further extraction of the same. At the same time, it disrupts the cell membrane as well as other cell components. Addition of chloroform always separates the solution into two phases, i.e. aqueous phase and organic phase to facilitate RNA isolation in the aqueous phase.

In aqueous phase, RNA can be recovered further by precipitation with isopropyl alcohol. Additionally, DNA and protein can be recovered by sequential separation by removing the aqueous phase. Precipitation from interphase by ethanol yields DNA and an additional precipitation with isopropyl alcohol needs protein from the organic phase (Fig. 1.4). As trizol yields a pure form of RNA free from the contaminations of protein and DNA; hence, it can be used for further downstream applications like Northern blot analysis, in vitro translation, poly (A) selection, RNase protection assay as well as molecular cloning.

RNA extraction from any living system faces a huge challenge due to the ubiquitous presence of ribonuclease enzymes in the cells, which can rapidly degrade RNA. Thus, obtaining a high-quality RNA is the must prerequisite before performing other molecular biology experiments like quantitative Real Time Polymerase Chain Reaction (qRT-PCR). To generate most sensitive and biologically relevant results, RNA isolation practice must include some important steps before, during and after actual RNA extraction. Thus, three important aspects should be kept in mind for an effective extraction of RNA from bacteria, i.e. (i) treatment and handling of samples prior to RNA isolation; (ii) choice of technique used for RNA extraction and (iii) storage of the prepared RNA sample.

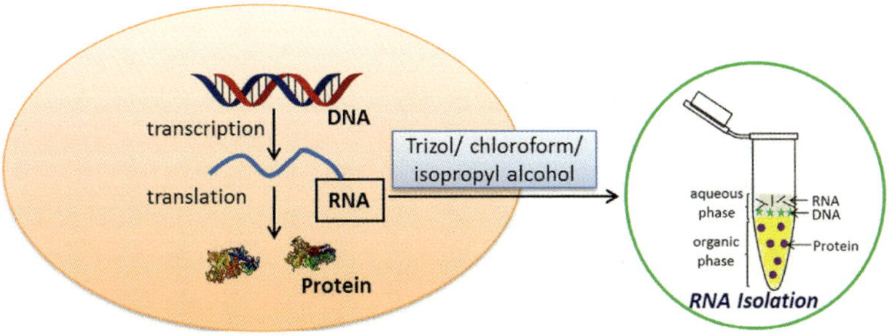

Fig. 1.4 Isolation of RNA from bacterial cell using trizol

Reagents Required and Their Role

Luria–Bertani Broth

LB broth is the rich growth medium that yields rapid good growth of many bacterial species. It is the most commonly used growth medium used for *E. coli* cell culture during most of the molecular biology studies. LB broth supports growth of *E. coli* upto 2–3 at OD_{600} under normal shaking incubation condition.

Tris Ethylenedinitrilo Tetra-acetic Acid Buffer

TE buffer can be prepared by mixing 50 mM Tris and 50 mM EDTA in water and by maintaining the pH at 8.0. As a major constituent of TE buffer, Tris acts as a common pH buffer to control pH when other reagents are added during further steps.

Trizol

Trizol is a ready to use reagent for the isolation of total RNA from bacterial cells. This reagent is a monophasic solution of phenol and guanidine isothiocyanate. Trizol generally maintains the integrity of RNA as well as disrupting cells and dissolving cell membranes. Guanidine isothiocyanate is a powerful protein denaturant that also helps in inactivation of RNases. In addition, acidic phenol partitions RNA to the aqueous supernatant for further separation in subsequent steps. Acidic pH is required for RNA isolation, as at neutral pH DNA partitions to the aqueous phase. Trizol reagent can be procured from the manufacturers in the form of TRIzol (Invitrogen brand name) or TRI (Sigma-Aldrich brand name). However, it can also be prepared in the laboratory following the methods:

Chemicals Required

The following chemicals were required: 4 M guanidinium thiocyanate, 25 mM sodium citrate (pH 7.0), 0.5 % (w/v) N-laurosylsarcosine and 0.1 M 2-mercaptoethanol.

To Prepare the Stock

Dissolve 250 g guanidinium thiocyanate. Add 17.6 ml of 0.75 M sodium citrate, pH 7.0. Add 26.4 ml of 10 % (w/v) N-laurosylsarcosine. Store for <3 months at room temperature

Chloroform

Chloroform is used to denature protein that settles in the bottom during RNA extraction. It also helps in the formation of aqueous and organic layer and in which RNA is dissolved in the aqueous layer. Chloroform, with the phenol, present in trizol reagent forms a biphasic emulsion. The hydrophobic layer of the emulsion is settled on the bottom and the hydrophilic layer remains on top after centrifugation.

Isopropyl Alcohol

RNA is insoluble in isopropyl alcohol; and hence, it aggregates and generates a pellet upon centrifugation. Addition of isopropyl alcohol also removes alcohol-soluble salts from the solution. As RNA is highly insoluble in isopropyl alcohol, it dissolves in water to form a solution that causes RNA to aggregate and precipitate. Isopropyl alcohol has been used as a better alternative than ethanol for RNA precipitation at lower concentrations. Besides, isopropyl alcohol takes much lesser time to evaporate from the solution to yield a better quality RNA.

Procedure

1. Inoculate a single bacterial colony into 5 ml of LB broth medium and incubate the tubes at 37 °C for 24 h with shaking at 180 rpm.
2. Collect the bacterial cell pellet by centrifugation at 6000 rpm for 5 min at room temperature.
3. Wash the cell pellets twice with autoclaved phosphate buffer saline.
4. Wash the pellets with autoclaved TE buffer, centrifuge at 6000 rpm for 5 min at room temperature and collect the cell pellet.
5. Resuspend the cell pellet with 1.5 ml of trizol solution.
6. Homogenate the solution by repeated pipetting or alternatively by vortexing for 1 min.
7. Alternatively, incubate the samples for 5 min at room temperature or 60 °C. A 5-min incubation at room temperature will result in the complete dissociation of nucleoprotein complexes.
8. RNA is stable in trizol as it deactivates RNases. Hence, at this step you can take a break for a shorter time or can store the samples by freezing it for a longer time.
9. Add 1/5 volume of chloroform, shake it to mix completely for 15 s.
10. Incubate the solution at room temperature for 2–5 min.
11. Centrifuge the solution at 12,000 rpm for 10 min at 4 °C. If centrifugation is not proper, DNA containing interphase will look cloudy and poorly compacted.
12. Transfer the upper aqueous layer to a fresh new tube. Take care not to aspirate the DNA containing white interface. This may lead to DNA contamination in the RNA preparation.
13. Add 1/2 initial volume of 70 % ice-cold ethanol, optionally incubate for 10 min at room temperature.
14. Centrifuge at 10,000 rpm for 15 min at 4 °C, discard the supernatant.
15. Alternatively, use RNeasy (from Qiagen) in place of ethanol for better precipitation for smaller amount of RNA and also to reduce risk of organic solvent contamination.
16. Wash the cell pellet with 500 µl of 70 % ethanol prepared with RNase-free water/ Diethylpyrocarbonate (DEPC)-treated water.
17. Dissolve pellet in 50–100 µl of RNase-free water/DEPC-treated water, mix the pellet by pipetting up and down slowly.
18. Store the tubes at − 80 °C till further use.

Observation

RNA quantitation is the important and most necessary step after completion of extraction practice. Both qualitative and quantitative analysis of RNA extraction can be predicted by UV-Vis spectrophotometry or in agarose gel electrophoresis.

The traditional method of assessment of RNA concentration and its purity is by UV-Vis spectroscopy. In this technique, the absorbance of a diluted RNA sample is measured at 260 and 280 nm, and the nucleic acid concentration is calculated using the Beer–Lambert's law.

$$\mathbf{A = \varepsilon C I}$$

Where, A = absorbance at a particular wave length

C = concentration of nucleic acid
I = path of the spectrophotometer cuvette
ε = the extinction coefficient [ε for RNA is 0.025 $(\text{mg/ml})^{-1}\text{cm}^{-1}$]

Using this equation, an A260 reading of 1.0 is equivalent to ~40 µg/ml of ssRNA. The A260/ A280 ratio is used to access RNA purity. An $A_{260/280}$ ratio of 1.8–2.1 indicates a highly purified RNA. Additionally, $A_{260/280}$ is dependent on both pH and ionic strength. The example of variation in $A_{260/280}$ ratio is as follows: [DEPC-treated water (pH 5–6) = 1.60; Nuclease-free water (pH 6–7) = 1.85; TE (pH 8.0) = 2.14].

Result Table

Sample	RNA content	$OD_{260/280}$	Inference
Control			
Sample I			
Sample II			

OD optical density

Troubleshootings

Problems	Possible errors	Possible solution
Low yield of RNA	RNA is not solubilised completely	To increase solubilisation of the pellet, pipette repeatedly in DEPC-treated water. Optionally, heat the samples at 55 °C for 10–15 min
	Residual growth medium might be present	Make sure no particulate matter remains. Be sure to take out all the supernatant after centrifugation to collect cell pellet
Degraded RNA	Sample might have been manipulated for too much time	Process the bacterial cell pellet immediately for trizol treatment
	Improper storage of RNA	Store isolated RNA at − 80 °C, not at − 20 °C
Low $A_{260/280}$	Presence of residual organic solvent in RNA	Make sure not that no organic phase with the RNA sample is present
	pH of the solution is acidic	Dissolve the sample in TE buffer instead of DEPC-treated water
	A_{260} or A_{280} outside the linear range	Dilute samples to bring absorbance into linear range
DNA contamination	Part of the interphase was removed with the aqueous phase	Ensure that no interphases was taken while transferring the upper aqueous phase to a fresh tube
	Insufficient trizol reagent was used	Use 1 ml of trizol reagent for 10^6 no. of cells
	Pellet contained organic solvent	Make sure that the original sample does not contain any organic solvent like ethanol or dimethyl sulphoxide

Precautions

1. Do not use less amount of trizol, very small volumes are hard to separate which leads to contamination.
2. Do not aspirate white interphase that contains DNA during removal of aqueous supernatant.
3. Always use acidic phenol/chloroform.
4. Always work under hood because phenol is toxic and chloroform is narcotic.
5. Always wear gloves while working, do not touch surfaces and equipment to avoid re-introduction of RNase to decontaminated material.
6. Designate a special area for RNA work only.
7. Treat surfaces of benches and glass wares with commercially available RNase inactivating agents.
8. Use sterile, disposable plasticwares.
9. Glasswares should be oven-sterilised at 180 °C for at least 2 h.
10. Use commercially available RNase-free plasticwares.
11. Use DEPC-treated water and if possible, prior to autoclaving use DEPC-treated water to wash plasticwares.

FLOW CHART

Take 5 ml of overnight grown culture, centrifuge to collect the pellet, wash twice with PBS and TE buffer

↓

Re-suspend the cell pellet with 1.5 ml of Trizol solution, homogenate the solution by vortexing for 1 min

↓

Incubate the samples for 5 min at room temperature

↓

Add 1/5 volume of chloroform, shake well to mix completely for 15 sec

↓

Incubate the solution at room temperature for 2-5 min

↓

Centrifuge at 12,000 rpm for 10 min at 4°C, check for the clear upper aqueous phase

↓

Transfer the upper aqueous phase to a fresh tube

↓

To this, add ½ volume of 70% ice-cold ethanol, incubate at room temperature for 10 min

↓

Centrifuge at 10,000 rpm for 15 min at 4°C, discard supernatant

↓

Wash the pellet with 500 μl of 70% ethanol

↓

Dissolve the pellet with 100 μl of DEPC treated water

Exp. 1.5 Amplification of 16S rRNA Gene

Objective To amplify 16S rRNA gene from bacterial genomic DNA by PCR.

Introduction

PCR is the exponentially progressing synthesis of the defined target DNA sequences in vitro. This technique was invented by Kary Mullis in 1983 for which he received Nobel Prize in chemistry in 1993. The reaction is called polymerase because the only enzyme used in this reaction is DNA polymerase. It is called as chain because the products of the first reaction become the substrate of the following one and so on. PCR relies on thermal cycling that consists of repeated heating and cooling of the reaction for denaturation of DNA followed by enzymatic replication of it. During PCR, the amplification of gene products takes place in the exponential order to leave large copies of DNA (Fig. 1.5).

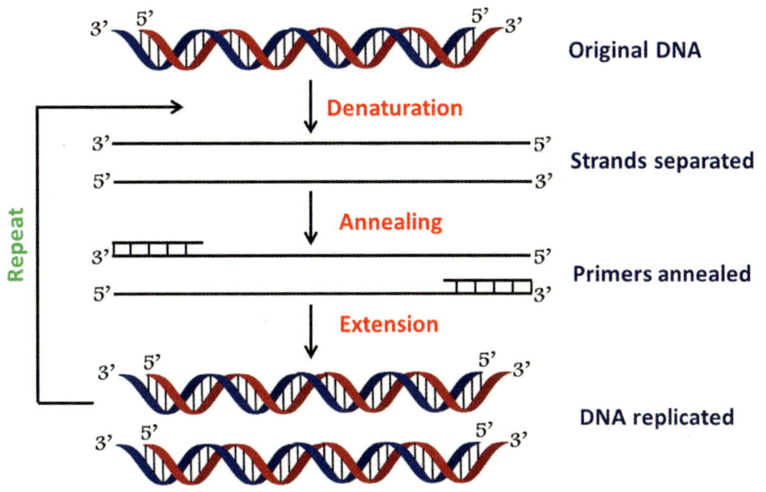

Fig. 1.5 Generation of huge copy numbers of the desired gene fragment by polymerase chain reaction

There are many widespread applications of PCR in many areas such as medical applications, infectious disease applications, forensic applications or research. PCR allows the generation of two short pieces of DNA when the two primer sequences are known. The task of DNA sequencing is also assisted by PCR. DNA cloning, genetic fingerprinting and DNA fingerprinting for forensic applications are some of the effective practical approaches of using PCR. The variations of basic PCR technique gives many advanced applications of the same, i.e. allele-specific PCR, assembly PCR, asymmetric PCR, dial-out PCR, digital PCR, hot start PCR, in-silico PCR, intersequence-specific PCR, inverse PCR, ligation-mediated PCR, multiplex PCR, nested PCR, reverse transcription PCR, touch down PCR, universal fast walking and many more.

The molecular basis of identification of bacterial species deals with the amplification and sequencing of 16S rRNA gene followed by their comparison with the existing database. Comparison of rRNA gene sequences for bacterial identification has been pioneered by Carl Woese that redefined the main lineage in the evolution of microorganisms. The major advantage of rRNA gene sequence comparison is the generation of increasingly expanding database available globally (Fig. 1.6). Nearly 60,000 16S rRNA gene sequences are currently available in the ribosomal database project (RDP II). The concept of comparing gene sequences from microbial communities has revolutionised microbial ecology. 16S ribosomal RNA is a component of the 30S small subunit of prokaryotic ribosome having a length of 1.542 kb. The reason behind the use of 16S rRNA gene amplification for identification purpose include: (i) occurrence of the gene in all organisms performing the same function; (ii) the gene sequence is conserved sufficiently containing conserved, variable and hypervariable regions, and (iv) 1500 bp of size, which is relatively easy to sequence and large enough to contain sufficient information for identification and analysis of phylogeny.

Principle

PCR is a chain reaction where a small fragment of DNA serves as template for producing the large copy numbers. One DNA molecule produces two copies, then four, then eight and so forth. This continuous doubling is accomplished by specific proteins known as polymerase. DNA polymerase also requires DNA building blocks, the nucleotide bases, i.e. adenine (A), thymine (T), cytosine (C) and guanine (G). A small fragment of DNA known as primer is also required to which the building blocks are attached and the existing DNA molecule serves as the template for constructing the new strand. When the ingredients

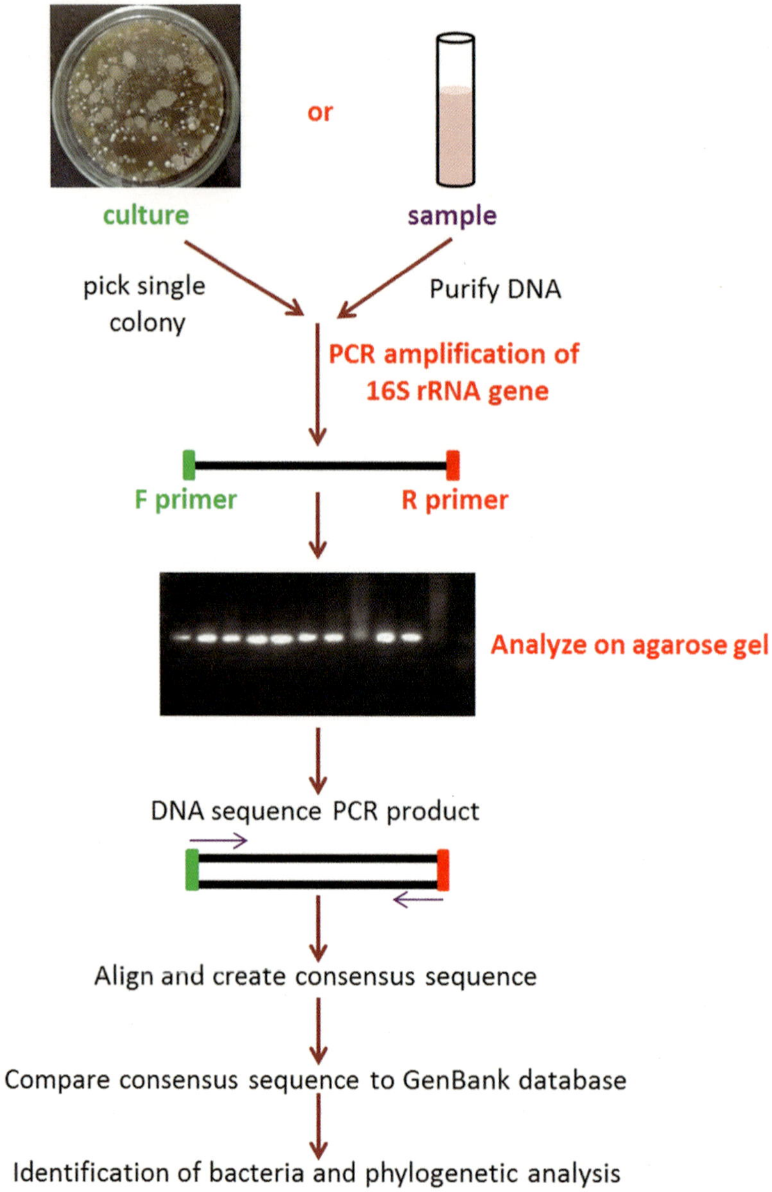

Fig. 1.6 Polymerase chain reaction amplification and sequencing of 16S rRNA gene for identification of bacterial species

are supplied, the enzyme constructs the exact copies of the template. In this way, the number of copies of DNA obtained after 'n' cycles is 2^{n+1}.

PCR can be regarded as the *in vitro* DNA synthesis that requires the same precursor molecules as in case of DNA replication in vivo. The DNA polymerase has been replaced by a thermo-stable polymerase called as Taq DNA polymerase that can withstand a temperature of >90 °C with an optimum activity of 72 °C. RNA primers in DNA replication has been replaced by oligonucleotide primers, the designing of which is the most important factor influencing the efficiency and specificity of the amplification reaction. The deoxyribonucleotides have been used as an equimolar concentration of four of them (dATP, dTTP, dGTP, dCTP). As the thermo-stable DNA polymerase requires free-divalent cations for their optimum activity, Mg^{2+} is often used in this purpose, also in certain instances Mn^{2+}. For maintaining pH level in the reaction tubes, buffer has been added to maintain pH between 8.3 and 8.8. Template DNA is the must prerequisites for a PCR, which can be added either in the form of single strand or double strand. In order to run at its best, PCR requires only a single copy of the target sequence as template.

PCR is an iterative process consisting of three steps of temperature regulation, i.e. denaturation of the template, annealing of the oligonucleotide primers to the ss target sequence and extension of the annealed primer by the thermo-stable DNA polymerase. The denaturation of the dsDNA is dependent on their G + C content. The higher the proportion of G + C, the higher the temperature required for separation of the strands of template DNA. The longer the template DNA, the higher time is required to separate two strands completely. In PCR catalysed by thermo-stable Taq polymerase, denaturation is normally carried out at 94–95 °C, which is due to the fact that the enzyme can endure for 30 or more cycles without sustaining extensive damage. The most crucial step of a PCR is annealing. If the annealing temperature is too high, the oligonucleotide primers anneal poorly to yield a poor amplified DNA. In contrary, if the annealing temperature is too low, non-specific annealing of the primers may occur resulting in the amplification of unwanted segments of template DNA. Annealing temperature is always recommended to be 3–5 °C below than the calculated melting temperature of the two oligonucleotide primers. Extension of the oligonucleotide primers is always carried out at the optimal temperature for DNA synthesis catalysed by thermo-stable DNA polymerase, i.e. 72–78 °C. Another major factor of DNA amplification by PCR is the number of cycles that depends on the number of copies of template DNA present in the beginning of the reaction. Once established in the geometric phase, the reaction proceeds until one of the components becomes limiting. This is generally the case that after ~30 cycles, the reaction contains ~10^5 copies of the target sequence and Taq polymerase (efficiency ~0.7). At least 25 cycles are required to achieve acceptable levels of amplification of a single copy target sequences of DNA templates.

Reagents Required and Their Role

DNA Template

Template contains the target DNA region that needs to be amplified. However, a proper concentration of template DNA should be used before going for PCR amplification. Approximately 104 copies of high-quality target DNA is required to detect the product in 25–30 cycles. 1 pg to 1 ng of plasmid or viral templates may be used, whereas for genomic templates 1 ng to 1 pg should be used. For 16S rRNA amplification the DNA template to be used as: purity $OD_{260}/OD_{280=}$1.8–2.0, concentration ≥50 ng/μl, DNA amount is at least 5 μg.

Forward and Reverse Primers

As DNA is a ds polynucleotide helical structure, one strand runs from 5′–3′ direction, and the other strand runs from 3′–5′ direction (complementary to the first strand), and the synthesis of the primer always takes place in the 5′–3′ direction, no matter where it is present. Hence, one primer is required in the forward direction and another is required in reverse direction. For proper amplification of the product, the final concentration of the 16S primers should be 0.05–1 μM. As primers have a greater influence on the success or failure of PCR protocols, it is ironic that primer designing is largely qualitative and are based on well-understood thermodynamic or structural principles. The set of primers to be used during this experiment are as follows:

16S Forward primer (27F)-
5′AGAGTTTGATCMTGGCTCAG3′

16S Reverse primer (1492R)-
5′ACGGCTACCTTGTTACGA3′

Taq Polymerase

Taq polymerase is a thermo-stable DNA polymerase isolated from the thermophilic bacterium *Thermus aquaticus*, which was originally isolated by Thomas D. Brock in 1965. The enzyme is able to withstand the protein denaturing conditions required during PCR. It has the optimum temperature for its activity between 75 and 80 °C, with a half-life of greater than 2 h at 92.5 °C, 40 min at 95 °C and 9 min at 97.5 °C and possesses the capability of replicating a 1000 bp of DNA sequence in less than 10 s at 72 °C. However, the major drawback of using Taq polymerase is the low replication fidelity, as it lacks 3′–5′ exonuclease proof reading activity. It also produces DNA products having 'A' overhangs at their 3′ ends, which is ultimately useful during TA cloning. In general, 0.5–2.0 units of Taq polymerase is used in a 50 µl of total reaction, but ideally 1.25 units should be used.

Deoxynucleotide Triphosphate

The dNTPs are the building blocks of new DNA strand. In most of the cases, they come as a mixture of four deoxynucleotides, i.e. dATP, dTTP, dGTP and dCTP. Approximately 100 µM of each of the dNTPs is required per PCR reaction. dNTP stocks are very sensitive to cycles of thawing and freezing and after 3–5 cycles, PCR reactions does not work well. To avoid such problems small aliquots (2–5 µl) lasting for only a couple of reactions can be made and kept frozen at −20 °C. However, during long-term freezing, small amount of water evaporate on the walls of the vial, thus, changing the concentration of dNTPs solution. Hence, before using, it is essential to centrifuge the vials, and it is always recommended to dilute the dNTPs in TE buffer, as acidic pH promotes hydrolysis of dNTPs and interfering the PCR result.

Buffer Solution

Every enzyme needs certain conditions in terms of their pH, ionic strength, cofactors etc., which is achieved by the addition of buffer to the reaction mixture. In some instances, the enzyme shift pH in non-buffered solution and stops working in this process, which can be avoided by the addition of PCR buffer. In most of the PCR buffers, the composition is almost same as: 100 mM Tris-HCl, pH 8.3, 500 mM KCl, 15 mM $MgCl_2$ and 0.01 % (w/v) gelatin. The final concentration of the PCR buffer should be 1 × concentration per reaction.

Divalent Cation

The mechanism of DNA polymerase requires the presence of divalent cations. Most essentially, they shield the negative charge of the triphosphate and allow the hydroxyl oxygen of the 3′ carbon to attack the phosphorus of the alpha phosphate group attached to the 5′ carbon of the incoming nucleotide. All enzymes that break the phosphoanhydride bonds of nucleoside di-and tri-phosphates require the presence of divalent cations. A 1.5–2.0 mM solution of $MgCl_2$ is optimal for the activity of Taq DNA polymerase. If Mg^{2+} will be too low no PCR product will be visible, whereas if Mg^{2+} is too high undesired PCR product may be obtained.

Procedure

1. Mix DNA template, primers, dNTPs and Taq polymerase in a 0.2 ml thin-walled microcentrifuge tube in the following order from a stock solution of [buffer-10 ×, $MgCl_2$−15 mM, dNTPs-10 mM, Primer-1 mM, Taq polymerase-1 U, Template-10 µg/µl]:

Sterile milli-Q water	12.5 µl
10 × Reaction buffer	2.5 µl
$MgCl_2$	2.5 µl
dNTPs mixture	0.5 µl
Primer (Forward)	1.0 µl
Primer (Reverse)	1.0 µl
Template DNA	4.0 µl
Taq DNA polymerase	1.0 µl

2. As pipetting of small volumes is difficult and often inaccurate, a master mixture is prepared where constituents common to all the reactions are combined in one tube multiplying the

Table 1.1 Preparation of master mixture (for a total volume of 25 µl)

No. of PCR tubes	5	10	15	20	25
Milli-Q water (µl)	62.5	125.5	187.5	250	312.5
10× buffer (µl)	12.5	25	37.5	50	62.5
$MgCl_2$ (µl)	12.5	25	37.5	50	62.5
dNTP mix (µl)	2.5	5.0	7.5	10	12.5
Primer (Forward) (µl)	5.0	10.0	15.0	20.0	25.0
Primer (Reverse) (µl)	5.0	10.0	15.0	20.0	25.0
Taq polymerase (µl)	5.0	10.0	15.0	20.0	25.0

PCR polymerase chain reaction, *dNTP* deoxynucleotide triphosphate

volume for one reaction with the total number of samples. Later, the appropriate amount of the master mixture is aliquoted to each tube e.g. if 1 µl of DNA template is added, 24 µl of master mixture is to be added making a total volume of 25 µl (see Table 1.1).
3. Place the tubes row wise in each well of the PCR machine.
4. Carry out the amplification with the following programme:

LID	98 °C
Initial denaturation	96 °C for 5 min
30 Cycles	
Denaturation	95 °C for 15 s
Annealing	49 °C for 30 s
Extension	72 °C for 1 min
Final extension	72 °C for 10 min
Hold	4 °C for ∞

5. At the end of the PCR cycles, take out the PCR products and run the samples by agarose gel electrophoresis in 1 % agarose gel with ethidium bromide and visualise under UV light.

Observation

Run the amplified DNA product in agarose gel to confirm the amplification along with DNA ladder. As the size of the 16S rRNA gene of bacteria is 1.5 kb, a clear, distinct band at this position will be observed when the gel will be visualised under UV light. The concentration of the amplified product can be measured by UV spectrophotometer using a nano-drop. Measurement of the OD at 260 and 280 nm will show the quality and quantity of the amplified product of 16S rRNA gene.

Troubleshootings

Problems	Possible cause	Possible solution
In correct product size	Incorrect annealing temperature	Recalculate primer Tm values by using any of the web-based software
	Mispriming	Verify the primes have no additional complementary regions within the template DNA
	Improper Mg^{2+} concentration	Optimize Mg^{2+} concentration with 0.2–1 mM increments
	Nuclease contamination	Repeat the reactions using fresh solutions
No product	Incorrect annealing temperature	Recalculate the Melting temperature (Tm) values of the primers, test the correct annealing temperature by gradient, starting at 5 °C below the lower Tm of the primer pair
	Poor primer design	Check with the literature for recommended primer design, verify that the primers are non-complementary, both internally and to each other, optionally increase length of the primer
	Poor primer specificity	Verify that the oligos are complementary to the proper target sequence
	Insufficient primer concentration	The correct range of primer concentration should be 0.05–1 μM, refer to the specific product literature for ideal conditions
	Poor template quality	Analyse DNA by agarose gel electrophoresis, check A_{260}/A_{280} ratio of DNA template
	Insufficient number of cycles	Rerun the reaction with more number of cycles
Multiple/non-specific products	Premature replication	Use hot start polymerase, set up the reaction on ice using chilled components, add samples to the PCR preheated to the denaturation temperature
	Primer annealing temperature too low	Increase the annealing temperature
	Incorrect Mg^{2+} concentration	Adjust Mg^{2+} concentration with 0.2–1.0 mM increments
	Excess primer	The correct range of primer concentration should be 0.05–1 μM, refer to the specific product literature for ideal conditions
	Incorrect template concentration	For low complexity templates (i.e. plasmid, lambda, BAC DNA), use 1 pg to 10 ng DNA per 50 μl reaction For high complexity templates (i.e. genomic DNA), use 1 ng to 1 μg template for 50 μl reaction

Precautions

1. Use pipette tips with filters.
2. Store materials and reagents properly under separate conditions and add them to the reaction mixture in a spatially separated facility.
3. Thaw all components thoroughly at room temperature before starting an assay.
4. After thawing, mix the components with brief centrifugation.
5. Work quickly on ice or in the cooling block.
6. Always wear safety goggles and gloves while performing the PCR reaction.

FLOW CHART

Thaw all the reagents on ice before set up of PCR

↓

Add all the reagents of required amount as mentioned for master mixture except template

↓

Distribute the master mixture to the individual PCR tube and add corresponding templates

↓

Set up the PCR conditions as mentioned in the procedure

↓

Take out the samples from the PCR machine and store at 4°C

↓

Run in agarose gel electrophoresis for checking the correct amplification

Exp. 1.6 To Perform Agarose Gel Electrophoresis

Objective To perform agarose gel electrophoresis for separation and visualisation of DNA.

Introduction

Agarose gel electrophoresis is the most suitable physical method of determining size of DNA. During this procedure, DNA is forced to migrate along a cross-linked agarose matrix in response to electric current. DNA contains phosphate groups that confer an overall negative charge to it; and hence, it migrates towards a positive electrode. Three factors determine the migration rate of DNA in a gel, i.e. size of DNA, conformation of DNA, agarose concentration, voltage applied, presence of ethidium bromide, type of agarose used and ionic strength of the running buffer. Electrophoresis is usually a sieving process and the higher the size of DNA it entangles in gel quiet easily and migrates slowly. On the other hand, smaller fragments move rather quickly than the larger fragments proportional to their size. The gel matrix can be adjusted by increasing or decreasing the concentration of the gel and a standard 1 % agarose gel can resolve DNA from 0.2 to 30 kb size (Fig. 1.7).

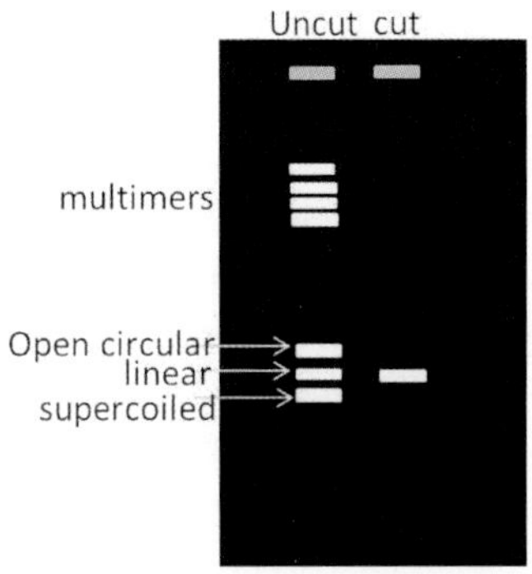

Fig. 1.7 Plasmid DNA banding pattern when subjected to agarose gel electrophoresis on 1 % agarose

Agarose is mostly isolated from the sea weed of genera *Gelidium* and *Gracilaria* and consists of repeated agarobiose, i.e. L-and D-galactose subunits. In the course of gelation, agarose polymers associate non-covalently and form a network of bundles whose pore size determines the gel's molecular sieving properties. Use of agarose gel electrophoresis has revolutionised the separation of nucleic acids as prior to the adaptation of agarose gels, DNA was primarily separated using sucrose density gradient centrifugation, and it could only provide the approximate size of DNA. Agarose gels are easy to cast and

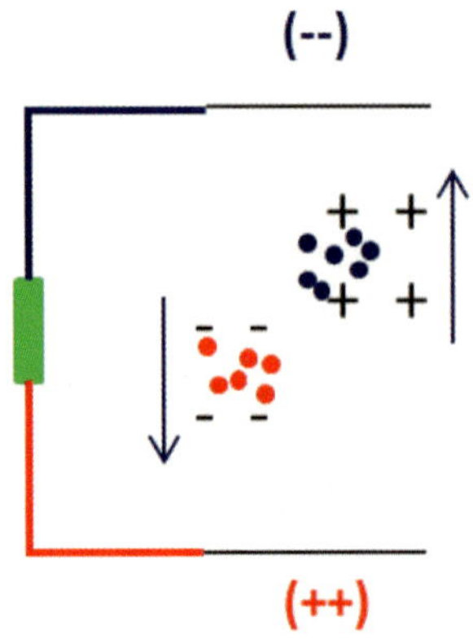

Fig. 1.8 Migration of charged molecules towards their respective electrodes in response to the difference in electric field

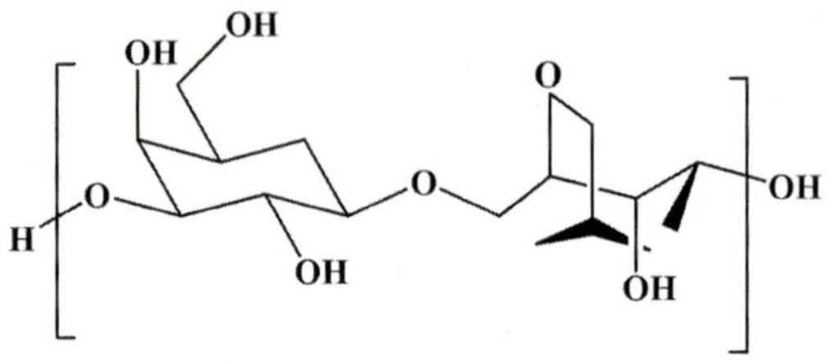

Fig. 1.9 Structure of an agarose molecule

Table 1.2 Resolution of different percentages of agarose gels for separation of DNA molecules

Agarose concentration in gel (% w/v)	Range of separation of linear DNA molecules (kb)
0.3	5–60
0.6	1–20
0.7	0.8–10
0.9	0.5–7
1.2	0.4–6
1.5	0.2–3
2.0	0.1–2

handle in comparison to the other matrices as the gel setting is a physical change rather than a chemical change as well as samples can be easily recovered from the gel, resulting gels can be stored in plastic bags in a refrigerator. DNA gel electrophoresis is usually performed for analytical purposes, mostly after amplification of DNA by PCR. In addition to that, it can be used as a preparative technique prior to the use in other methods such as mass spectroscopy, Restriction Fragment Length Polymorphism (RFLP), PCR cloning, DNA sequencing or southern blotting for further characterisation.

Principle

When charged molecules are placed in an electric field, they tend to migrate towards either the positive or negative pole as per their charge (Fig. 1.8). In contrast to proteins, which have either a net positive or negative charge, nucleic acids have a consistent negative charge imparted by their phosphate backbone and hence, migrate towards the anode. The nucleotides in a DNA molecule are linked together by negatively charged phosphodiester groups. For every base pair (average molecular weight of approximately 660), there are two charged phosphate groups. Hence, every charge in DNA molecule is accompanied by approximately the same mass, and the RNA. When heated, agarose solution becomes gel with pore size ranging from 50 to 200 nm. With the addition of fluorescent dyes like ethidium bromide or gold view, DNA can be visualised under UV detector. Most of the agarose gels are prepared between 0.7 and 2 % of agarose. A 0.7 % gel shows good separation for large DNA fragments, i.e. 5–10 kb, and a 2 % gel shows good resolution for smaller fragments with size ranging from 0.2 to 1 kb. Low percentage gels are very weak but high percentage gels are usually brittle and do not set evenly. In a gel, the distance between DNA bands of a given length is determined by the percentage of agarose. Percentage of gel is the best way to control the resolution of agarose gel electrophoresis (Table 1.2).

When a voltage is applied across the electrodes, a potential gradient 'E' is generated, which is expressed by the equation: $E = V/d$, where, 'V' is the voltage applied measured in volts and 'd' is the distance in cm between the electrodes. When, this potential gradient 'E' is applied, a force 'F' on the charged molecule is generated and expressed by the equation: $F = E.q$, where, 'q' is the charge in coulombs bearing on the molecule. This force drags the charged molecule towards the electrode. However, there is a frictional force exists that slows down the moment of charged molecules which is the function of, (1) the hydrodynamic size of the molecule, (2) the shape of the molecule, (3) the pore size of the medium where electrophoresis is taking place and (4) the viscosity of the buffer.

Reagents Required and Their Role

Agarose

It is used for the electrophoretic separation of nucleic acids. The purest form of agarose is free of DNase and RNase activities. Molecular biology grade agarose is the standard one for the resolution of DNA fragment in the range of 50 bp–50 kb, with the possibility of subsequent DNA extraction from the gel for further analysis. It possesses the following properties: gel strength (1 %)—1125 g/cm^2, gelling point (1.5 %)—36.0 °C, melting point (1.5 %)—87.7 °C, sulphate—0.098 %, moisture—2.39 %, ash—0.31 %.

Tris-acetate Buffer

The electrophoretic mobility of DNA is dependent on the composition and ionic strength of the electrophoresis buffer. During the absence of ions, there will be a minimal electrical conductance and DNA migrates slowly. The buffer of high ionic strength and high electrical conductance is efficient and additionally a significant amount of heat is generated. Thus, worsening the situation the gel melts and DNA is denatured. Several different buffers have been recommended for use in electrophoresis of native double stranded DNA. These buffers contain EDTA (pH 8.0) and Tris-acetate (TAE), Tris-borate (TBE) or Tris-phosphate (TPE) at an approximate concentration of 50 mM (pH 7.5–7.8). These buffers are generally prepared as concentrated solutions and stored at room temperature, when used the working solution is prepared as 1 ×. TAE and TBE are the most commonly used buffers and two of them have their own advantages and disadvantages. Borate has disadvantages as it polymerizes and interacts with cis diols found in RNA. On the other hand, TAE has lowest buffering capacity but it provides the best resolution for larger DNA which implies the need for the lower voltage and more time with a better product. Lithium borate is a relatively new buffer and is ineffective in resolving fragments larger than 5 kb.

Prepare a 10 × stock solution in 1 l milli-Q water: 48.4 g Tris base, 11.4 ml glacial acetic acid, 20 ml of 0.5 M EDTA or 3.7 g EDTA disodium salt. Dissolve all in 800 ml deionized water and make up the volume to 1 l. Store in room temperature and dilute it to 1 × prior use.

Ethidium Bromide

Ethidium bromide is a fluorescent dye that intercalates between nucleic acid bases and eases the detection of nucleic acid fragments in gel. When exposed to ultra violet light, DNA flourishes with an orange colour, intensifying 20-fold after binding to DNA. The absorption maxima of EtBr in aqueous solution is between 210 and 285 nm that corresponds to the UV light. Hence, as a result of this excitation EtBr emits orange light with a wave length of 605 nm. EtBr binds with DNA and slips in between its hydrophobic base pairs and stretches the DNA fragment, thus removing the water molecules from ethidium cation. This dehydrogenation results in the increase in fluorescence of the ethidium. However, EtBr is a potential mutagen, suspected carcinogen and it can irritate eyes, skin, mucous membranes and upper respiratory tract at higher concentrations. It is due to the fact that, EtBr intercalates into double stranded DNA, deform the molecule thus blocking the biological processes involving nucleic acids like DNA replication and transcription. Hence, there are many alternatives regarded as less dangerous and with better performance like Sybr dyes.

Table 1.3 Resolution of the dyes used in loading buffer during agarose gel electrophoresis

Dye	05–1.5% Agarose	2.0–3.0% Agarose
Xylene cyanol	10,000–4000	750–200 bp
Cresol red	2000–1000	200–125 bp
Bromophenol blue	500–400	150–50 bp
Orange G	<100 bp	–
Tartrazine	<20 bp	<20 bp

Gel Loading Buffer

Loading buffer is mixed with the DNA samples to be used in agarose gel electrophoresis. The dye present in the buffer is used primarily to assess how fast the samples are running during electrophoresis and to render a higher density to the samples than that of the running buffer. The increased density can be achieved by the addition of materials like ficoll, sucrose or glycerol. There are many colour combinations available to trace the migration rate of the DNA samples (Table 1.3).

To prepare a 6× gel loading buffer containing glycerol and bromophenol blue, add 3 ml glycerol (30%), 25 mg bromophenol blue (0.25%) and 10 ml of milli-Q, store at 4 °C. The working concentration of the loading buffer should be 1×.

DNA Marker

A molecular weight size marker is also called a DNA ladder is a set of standards that is used to determine the approximate size of a test molecule during agarose gel electrophoresis. DNA ladders are prepared by two ways: partial ligation, where a 100 bp of DNA piece is partially ligated that gives rise to dimers of 200 bp, trimmers of 300 bp, tetramers of 400 bp, pentamers 500 bp and so on. Secondly, restriction digestion of a known DNA sequence by a particular restriction enzyme that gives rise to DNA pieces of varied molecular masses.

DNA Sample

5 µl of amplified product is sufficient to be visualised after agarose gel electrophoresis. It can be mixed with 1 µl of 6× gel loading dye to be loaded in the wells prepared in agarose gel.

Procedure

Melting of Agarose

1. Weigh appropriate amount of agarose carefully into a conical flask. To prepare 1%, mix 30 mg agarose in 30 ml 1× TAE buffer.
2. Prepare 1× concentration of TAE buffer from the stock of 10×. Example add 1 ml of 10× buffer in 9 ml of milli-Q water to prepare a total volume of 10 ml of 1× buffer.
3. Cover the container with plastic wrap. Pierce a small hole in the plastic for ventilation.
4. Heat the solution in microwave oven on high power until it comes to boil. Watch the solution closely, agarose foams up and boils over easily.
5. Remove the container and swirl gently to resuspend any of the settled agarose.
6. Repeat this process until agarose dissolves completely.
7. Cool the agarose until you can comfortably touch the flask and add ethidium bromide solution to give a final concentration of 5 pg/ml.
8. The gel mixture is ready to be poured into the gel apparatus.

Pouring the Gel

1. Place tape across the ends of the gel casting tray and place the comb to it.
2. Pour cooled agarose into it. The agarose should come at least half way up the comb teeth.
3. Immediately rinse and fill the agar flask with hot water to dissolve any remaining agarose.
4. After solidification, carefully remove the comb.
5. Remove the tape from the ends of the gel formed.

Loading the Samples

1. Make a written record of which sample you load in each well of the gel. It may be helpful to load the samples in every other well.
2. Keep a black or dark surface under the gel for proper visualisation of the wells in the solidified agarose.
3. Mix 10 μl of the DNA sample with 2 μl of gel loading dye and load 12 μl of the sample to each well of the gel.
4. Be careful not to puncture the bottom of the wells while loading the samples.

Setting up the Gel

1. Make sure that the wells are closest to the negative (black) electrode.
2. Fill the electrophoresis tank with 1 l of 1× running buffer to just cover the wells.
3. Fill each half of the chamber adding solution until it is close to the top of the gel. Gently flood the gel from the end opposite to the wells to minimise sample diffusion.
4. Place the lid on the chamber and connect the electrode that leads to power supply.
5. Connect the black lead to the negative terminal and red lead to the positive terminal.

Running and Analysing the Gel

1. Turn on the power supply and adjust the voltage to 50–100 V.
2. Run the gel between 1 and 3 h, depending on the percentage of the gel and gel length.
3. Once the dye moves through the gel, turn off the power supply, disconnect the electrode leads and remove the chamber lid.
4. Remove the gel from the electrophoresis chamber to analyse your results.
5. View the amplified products on a UV-transilluminator or gel documentation system.

Observation

Document the amplified DNA product on agarose gel by visualising under UV light. Match the banding position with the corresponding position of the DNA ladder to get the approximate size of the amplified product.

The amplified 16S rRNA gene from bacteria is supposed to give a clear distinct banding pattern at around 1.5 kb.

Troubleshootings

Problems	Possible cause	Possible solution
Faint band or no band in the gel	Insufficient quantity/concentration of DNA	Increase the amount of DNA, but do not increase beyond 50 ng/band
	DNA degraded	Avoid nuclease contamination
	DNA electrophoresed out to the gel	Electrophorese for less time, use low voltage, higher percentage of gel
	Improper UV light source for visualisation of EtBr-stained DNA	Use shorter wave length light for greater sensitivity, i.e. 254 nm
Smeared DNA bands on the gel	Degraded DNA	Avoid nuclease contamination
	Too much DNA loaded onto the gel	Decrease the amount of DNA
	Improper electrophoresis conditions	Do not allow voltage to exceed 20 V/cm. Maintain temperature <30 °C during electrophoresis
	Small DNA bands diffused during staining	Add EtBr during electrophoresis
Anomalies in DNA band migration	Improper electrophoresis conditions used	Do not allow voltage to exceed 20 V/cm. Maintain temperature <30 °C during electrophoresis
	DNA denatured	Avoid nuclease contamination

Precautions

1. Wear proper fitting heat-resistant gloves while preparing and pouring agarose gel.
2. If UV-transilluminator is used ensure that a UV proof full face shield is on.
3. Ensure that all the glasswares are clean before preparing agarose.
4. Prepare agarose gel in a stirrer hot plate or microwave, do not over boil.
5. Take extra precaution while handing EtBr, which is highly carcinogenic.

FLOW CHART

Weigh requisite amount of agarose, mix with 1X TAE buffer

↓

Mix by boiling

↓

Add ethidium bromide at a final concentration of 5 pg/ml

↓

Pour the mixture to the gel preparation tray with suitable comb

↓

Let it be solidify

↓

Load the samples by mixing them with gel loading dye at a final concentration of 1X along with the suitable size marker

↓

Connect the gel apparatus to the power supply and adjust the voltage to 50-100 V

↓

After 1 h remove the gel and analyse it under UV light

Cloning and Transformation

2

Exp. 2.1 Preparation of Competent Cells and Heat-Shock Transformation

Objective To prepare competent cells of bacteria by chemical treatment and transformation by heat shock.

Introduction

Competent cells are bacterial cells that can accept extra-chromosomal DNA or plasmids (naked DNA) from the environment. The generation of competent cells may occur by two methods: natural competence and artificial competence. Natural competence is the genetic ability of a bacterium to receive environmental DNA under natural or in vitro conditions. Bacteria can also be made competent artificially by chemical treatment and heat shock to make them transiently permeable to DNA. Natural competence dates back to 1928, when Frederick Griffith discovered that prepared heat-killed cells of a pathogenic bacterium could transform the nonpathogenic cells into pathogenic type. Natural competence has been reported in many bacterial strains, i.e. *Bacillus subtilis, Streptococcus pneumonia, Neisseria gonorrhoeae* and *Haemophilus influenza*. The natural competence phenomenon is highly regulated in bacteria and varies across genera. In some genera, certain portions of the population are competent at a time, and in others, the whole population gains competence at the same time. When the foreign DNA enters inside the cells, it may be degraded by the cellular nucleases or may recombine with the cellular chromosome. However, natural competence and transformation is efficient for linear molecules such as chromosomal DNA but not for circular plasmid molecules Fig. 2.1.

Artificial competence is not coded by the genes of the bacterial cells. It is a laboratory procedure in which cells are passively made permeable to DNA using unnatural conditions. The procedure of artificial competence is relatively simple and easy and can be used to engineer a bacterium genetically. However, transformation efficiency is very low as only a portion of the cells become competent to successfully take up DNA.

Principle

As DNA is a highly hydrophilic molecule, normally it cannot pass through the cell membrane of bacteria. Hence, in order to make bacteria capable of internalising the genetic material, they must be made competent to take up the DNA. This can be achieved by making small holes in bacterial cells by suspending them in a solution containing a high concentration of calcium. Extra-chromosomal DNA will be forced to enter the cell by incubating the competent cells and the DNA together on ice followed by a brief heat shock that causes the bacteria to take up the DNA (Fig. 2.2). Bacteria no longer become stable when they possess holes on the cell membrane and may

S. Das, H. R. Dash, *Microbial Biotechnology- A Laboratory Manual for Bacterial Systems*,
DOI 10.1007/978-81-322-2095-4_2, © Springer India 2015

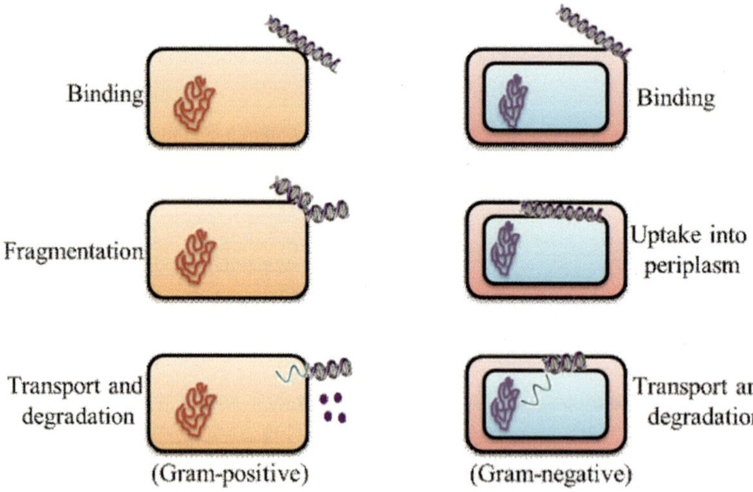

Fig. 2.1 Transformation pathways in Gram-positive and Gram-negative bacteria

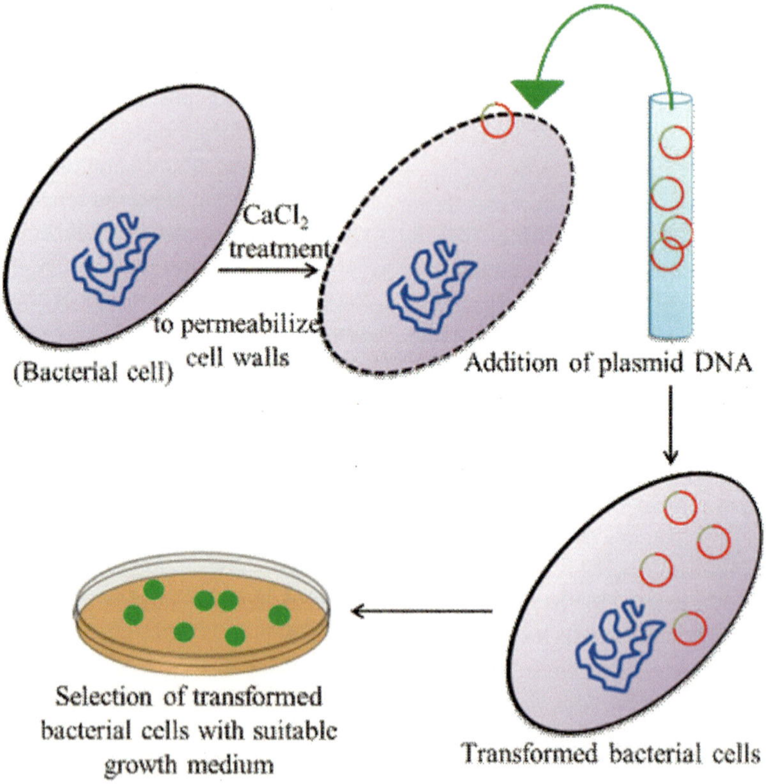

Fig. 2.2 Preparation of competent cells by $CaCl_2$ treatment and transformation

die easily. Additionally, a poorly performed procedure may lead to not enough competence cells to take up DNA. It has been reported that a naked DNA molecule is bound to the lipopolysaccharide (LPS) receptor molecules on the competent cell surface. The divalent cations generate coordination complexes with the negatively charged DNA molecules and LPS. DNA, being a larger molecule, cannot itself cross the cell membrane to enter into the cytosol. The heat shock step strongly depolarizes the cell membrane of $CaCl_2$-treated cells. Thus, the decrease in membrane potential lowers the negativity of the cell's inside potential which ultimately allows the movement of negatively charged DNA into the cell's interior. The subsequent cold shock again raises the membrane potential to its original value.

In another method, the transformation storage solution (TSS) buffer method, competence is induced by polyethylene glycol (PEG). This technique is relatively simple and does not require heat shock. Competence of bacterial cells is induced by the addition of low concentrations of divalent cations Mg^{2+} and dimethyl sulfoxide (DMSO). PEG helps in shielding the negative charges on the DNA molecule and host cell membrane; thus, repulsion between them is reduced. The pH of the buffer is maintained at slightly acidic conditions to increase the cell's viability as well as transformation efficiency up to 10^7–10^8.

Reagents Required and Their Role

Luria-Bertani Broth

Luria-Bertani (LB) broth is a rich medium that permits fast growth and good growth yields for many species including *E. coli*. It is the most commonly used medium in microbiology and molecular biology studies for *E. coli* cell cultures. Easy preparation, fast growth of most *E. coli* strains, ready availability and simple compositions contribute to the popularity of LB broth. LB can support *E. coli* growth (OD_{600}=2–3) under normal shaking incubation conditions.

Calcium Chloride

Calcium chloride transformation technique is the most efficient technique among the competent cell preparation protocols. It increases the bacterial cell's ability to incorporate plasmid DNA, facilitating genetic transformation. Addition of calcium chloride to the cell suspension allows the binding of plasmid DNA to LPS. Thus, both the negatively charged DNA backbone and LPS come together and when heat shock is provided, plasmid DNA passes into the bacterial cell. Prepare 2000 ml of 50 mM Calcium chloride stock solution by adding 14.701 g of $CaCl_2 \cdot 2H_2O$ in 2 l of milli-Q water, autoclave and store at 4 °C.

Polyethylene Glycol

PEG is a polyether compound having many functions. In this case, it helps in shielding the negative charges present on the DNA and host cell membrane. This results in lowering of repulsion. In addition, PEG, being a larger molecule, coordinates with the water molecules present in the bacterial suspension. This results in the increased concentration and bioavailability of the plasmid DNA to pass through the membrane of a bacterial competent cell. In other words, a highly effective plasmid concentration results in a more effective DNA transformation into bacteria.

Dimethyl Sulfoxide

DMSO is an organosulfur compound and is a polar aprotic solvent. It can be dissolved both in polar and nonpolar compounds and is miscible with a wide range of organic solvents as well as water. DMSO brings together the reagents that have been added during the reaction. It also acts as a preserving agent as the prepared competent cells need to be stored at −80 °C for a longer period without losing their viability.

$MgCl_2$

$MgCl_2$ acts in the same way as does $CaCl_2$. It induces the ability of the cells to take up DNA by altering the permeability of the membranes. The negatively charged incoming DNA is repelled by the negatively charged macromolecules present on the bacterium's outer surface which is neutralized by the addition of $MgCl_2$ to neutralize the unfavourable interactions.

Procedure

By $CaCl_2$ Treatment

1. Inoculate the *E. coli* culture into the LB medium and incubate at 37 °C for 24 h with vigorous shaking at 180 rpm.
2. Aliquot 0.5 ml of the grown culture into 50 ml of LB in a 200-ml conical flask. Pre-warm the broth to 37 °C.
3. Incubate at 37 °C with shaking at 180 rpm.
4. Monitor the growth regularly till the OD_{600} reaches to 0.35–0.4.
5. When suitable growth has been reached, chill the culture on ice.
6. Transfer the culture to an autoclaved centrifuge tube and collect the cell pellets by centrifugation at 6000 rpm for 5 min at 4 °C. Discard the supernatant.
7. Resuspend the cell pellets in 20 ml of an ice-cold 50-mM $CaCl_2$ solution. Incubate the resuspended cells on ice for 20 min.
8. Collect the cell pellets by centrifugation at 6000 rpm for 5 min at 4 °C.
9. Resuspend the cells with 2.5 ml of ice-cold 50-mM $CaCl_2$. Optionally, if required to store the competent cells for a longer period, resuspend the cells with 2.5 ml ice-cold 50-mM $CaCl_2$ containing 10 % glycerol.
10. Use 100 µl of the prepared competent cells for transformation.
11. Dispense the competent cells into aliquots of 100 µl and store them at − 80 °C for further use.

By Using Transformation Storage Solution Buffer

1. Grow a 5-ml overnight culture of *E. coli* in LB broth medium. In the morning, dilute the culture into 25–50 ml of fresh LB medium in a 200-ml conical flask to dilute the culture at least by 1/100.
2. Grow the diluted culture to an OD_{600} of 0.2–0.5.
3. Put the micro-centrifuge tubes on ice so that they are cold when cells are aliquoted. If the culture is 'X' ml, 'X' number of tubes will be required. At this point, be sure that the TSS buffer is chilled. It should be stored at 4 °C, but if freshly prepared, put it on an ice bath.
4. Split the culture into two 50-ml falcon tubes and incubate on ice for 10 min. All subsequent steps should be carried out at 4 °C and the cells should be kept on ice wherever possible.
5. Centrifuge the cells for 10 min at 3000 rpm at 4 °C temperature.
6. Discard the supernatant. The cell pellets should be sufficiently solid to pour off the supernatant and pipette out the remaining medium.
7. Resuspend the pellet in chilled TSS buffer. The volume of the TSS to use is 10 % of the culture volume that was spun down.
8. Vortex gently to fully resuspend the culture, keep an eye out for the small cell aggregates even after the pellet is completely off the wall.
9. Add 100 µl aliquots to the chilled micro-centrifuge tubes and store at − 80 °C till further use.

Preparation of Transformation Storage Solution Buffer

To prepare 50 ml of TSS buffer, add 5 g PEG 8000, 1.5 ml 1 M $MgCl_2$ or 0.30 g $MgCl_2.6H_2O$, 2.5 ml DMSO and add LB medium to 50 ml followed by filter sterilization with a 0.22-µm filter.

For Heat-Shock Transformation

1. Thaw 50–100 µl of competent cells carefully on ice.
2. Add 1 µl of plasmid solution (concentration 1 µg/µl) to 100 µl of competent cells.
3. Incubate the cells on ice for 30 min.
4. Quickly transfer the tubes to a water bath previously set at 42 °C. Incubate for 1 min, and then quickly transfer to ice.
5. Add 1 ml LB broth medium to the tube.
6. Incubate at 37 °C for 30 min to 1 h.
7. Streak out 50–500 µl of culture onto plates containing suitable antibiotic markers.

Observation

Observe the number of colonies grown on the plates after successful transformation. Transformation efficiency (Transformant/μg of plasmid) may be calculated as the number of colony-forming units (CFU) produced by 1 μg of DNA and is measured by performing a control set of transformation reaction using a known quantity of DNA and then calculating the number of CFU formed per μg of DNA.

$$\text{Transformation efficiency} = \frac{\text{No. of transformants (colonies)} \times \text{Final volume at recovery (ml)}}{\mu\text{g of plasmid DNA} \times \text{Volume plated (ml)}}$$

Troubleshooting

Problems	Possible cause	Possible solutions
Low transformation efficiency with chemically competent cells	Impurities in DNA	Remove phenol, protein, detergents and ethanol by ethanol precipitation
	Excess DNA	Use not more than 1–10 μg of DNA for transformation purpose
	Cells handled improperly	Thaw cells on ice and use immediately; refreezing decreases the efficiency; do not vortex the cells
	Poor cell growth	Incubate cells for a minimum of 90 min during recovery and incubate the transformed colonies for a longer period
	Calculation errors	Ensure to use the correct dilution factor and DNA concentration to calculate efficiency
Few or no colonies	Too little DNA	Maintain the DNA concentration at 1–10 μg
	Wrong antibiotic concentration	Check the optimum antibiotic concentration of the vector
Satellite colonies	Degraded antibiotic	Check expiry date of the antibiotic, avoid repeated freeze-thaw cycle
	Too many colonies on the plate	Plate the transformants from a higher dilution
	Ampicillin use	You may use carbenicillin rather than ampicillin to reduce satellite colonies

Precautions

1. $CaCl_2$ is a hazardous material for skin, eyes and the respiratory system and may cause burns. Hence, use gloves while using the same.
2. Avoid thawing of cells before use.
3. Keep cells on ice not longer than 3 h; do not use cells again that have been on ice.
4. You may stock-freeze the competent cells in liquid nitrogen. The stock-freezing might keep cells viable for a longer period, but it decreases the transformation efficiency by at least a factor of ten.
5. During preparation of TSS buffer, if non-chemically resistant filters (e.g. cellulose nitrate) are used, add DMSO after sterilization.

FLOW CHART

By $CaCl_2$ treatment

Grow the *E. coli* cultures by monitoring the cell growth till OD_{600} reaches to 0.35-0.4.

↓

Chill the culture on ice

↓

Transfer the culture to a centrifuge tube and collect the cell pellet by centrifugation at 6,000 rpm for 5 min at 4°C

↓

Re-suspend the cell pellets in 20 ml of ice cold 50 mM $CaCl_2$

↓

Use 100 µl of the prepared competent cells for transformation

↓

Dispense the competent cells into aliquots of 100 µl and store them at -80°C till further use

Using TSS buffer

Grow *E. coli* culture to OD_{600} of 0.2 to 0.5

↓

Incubate micro-centrifuge tubes on ice for 30 min to 1 h

↓

Pre-chill the TSS buffer on ice or use the 4°C stored TSS buffer directly

↓

Centrifuge the cells for 10 min at 3,000 rpm and 4°C to collect the cell pellets

↓

Re-suspend the cell pellets in chilled TSS buffer; the volume of TSS buffer should be 10% of the culture volume that have been spun down

↓

Vortex gently to fully re-suspend the culture, make sure for the proper mixing of the small cell aggregates

↓

Add 100 µl aliquots to the chilled micro-centrifuge tubes

↓

Store at -80°C till further use

For transformation

Thaw previously prepared competent cells on ice

↓

Add 1 μl of 1 μg/μl plasmid solutions to 100 μl of competent cells

↓

Incubate on ice for 30 min

↓

Heat shock at 42°C for 1 min and quickly transfer them to ice

↓

Add 1 ml LB broth medium to the tubes

↓

Incubate at 37°C for 1 h

↓

Confirm the transformation by spreading 50-500μl of culture on plates through Blue-White screening

Exp. 2.2 Electroporation

Objective To perform transformation in bacteria by electroporation.

Introduction

Electroporation is the significant increase in electrical conductivity of cell membrane for subsequent increase in its permeability. In most cases, electroporation is used to introduce certain foreign substances into the bacterial cells as a piece of coding DNA and plasmid. An electric pulse of high intensity in kilovolts per centimetre for few microseconds to milliseconds causes a temporary loss of the semipermeable nature of the cell membrane. This phenomenon increases the uptake of drugs, molecular probes and DNA into the cell. Electroporation has many applications such as introduction of plasmids or foreign DNA into living cells for gene transfections, fusion of cells to prepare heterokaryons and hybridomas, insertion of proteins into cell membranes, improvement of drug delivery, increased effectiveness of chemotherapy of cancerous cells, activation of membrane transporters and enzymes, and alteration of gene expression in living cells.

The electrical device used to porate cells for transformation and gene expression analysis relies upon the discharge of the capacitor by cellular suspension to generate the required electric field. In all the commercially available devices, a capacitor between 2 and 1000 μF is charged to a voltage between 200 and 2000 V and subsequently discharged through the cell suspension by using an electronic or mechanical switch. This ultimately results in a voltage pulse with a rise time of less than 10 μs. Because of their smaller size, bacteria require a much higher electric field to induce poration than mammalian cells or even larger plant cells.

Thus, electroporation is the physical mechanism of allowing cellular introduction of highly charged molecules like DNA through cell membrane. This process is approximately ten times

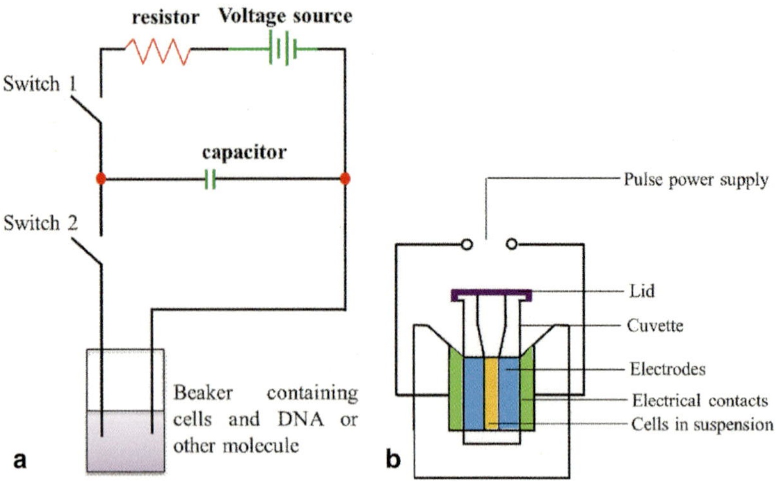

Fig. 2.3 **a** Diagram of the basic circuit setup of an electroporation apparatus, **b** specially designed elelctroporatic cuvette

more effective than that of chemical transformation. During electroporation, the natural barrier function of the membrane is overcome so that ions and water soluble molecules can cross the membrane. Though the microscopic mechanism by which molecular transport occurs is not yet established, significant progress has been made regarding their electrical as well as mechanical behaviour. However, little is known regarding membrane recovery and the ultimate fate of the transformed cells.

Principle

Electroporation capitalizes the weak nature of the cellular membrane and its ability to spontaneously reassemble after disturbance. Ultimately, the quick voltage shock disrupts the membrane temporarily, allowing the polar DNA molecules to pass, and then the membranes reseal quickly to leave the cells intact. During electroporation, the host cell suspension and the DNA molecules to be inserted are kept in the same suspension and the electric field is applied in the electroporation apparatus. Though the electroporation apparatus is commercially produced, the basic process inside this apparatus is quite the same as represented in the diagram in Fig. 2.3.

When the first switch is closed, the capacitor charges up, and a high voltage is stored. When the second switch is closed, the voltage gets discharged through the cell suspension. However, typically 10,000–100,000 V/cm in a pulse lasting for microseconds to milliseconds are necessary for proper electroporation purpose. This, in turn, generates an electric potential across the cell membrane so that the charged molecules such as DNA are driven across the membrane through the pores in a similar fashion to that of electrophoresis. After the charged ions and molecules pass through the pores, the cell membrane discharges, and the pores are closed quickly to reassemble the cell membrane and the intended molecules remain inside the cell for further use (Fig. 2.4).

There are many advantages of using electroporation technique over conventional transformation techniques—technical simplicity, ease of operation, rapidity and reproducibility, greater transformation efficiencies, avoidance of deleterious toxic side effects of chemicals like PEG, no need for pre-incubation of cells and DNA, better control of size and position of the electropores and many others. In order to achieve a good transformation yield, this technique is dependent on several critical factors and parameters that have been categorised into three determinants: (1) cellular factors—growth phase at time of

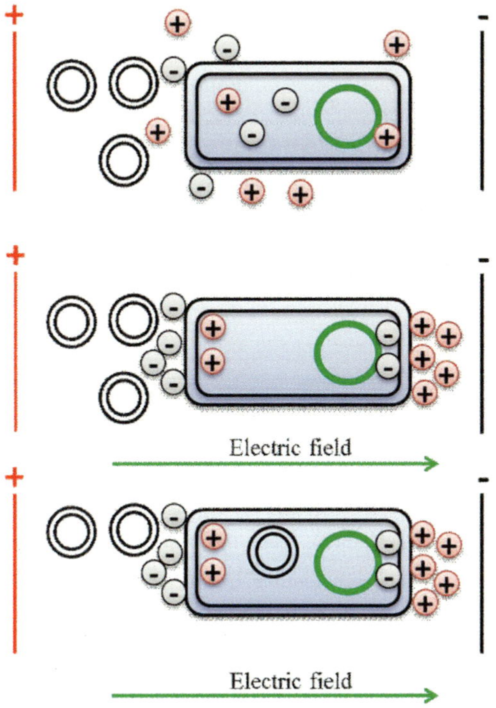

Fig. 2.4 Transfer of foreign genetic material into the cell when electric field is applied, and pathways are formed across the cell membrane allowing DNA to enter the cell

harvesting, cell density, cell diameter, cell wall rigidity and its susceptibility to electroporation; (2) physico-chemical factors—temperature, pH, osmolarity, ionic concentration of electroporation buffer, DNA concentration, etc.; and (3) electrical parameters—optimum field strength, critical voltage, pulse length, number of repetitive pulses, uniform or non-uniform electric fields, etc.

There are many techniques available nowadays for the introduction of foreign DNA into a cell. However, electroporation is the most extensively used technique with huge advantages over other practices such as its versatility, i.e. it is effective with nearly all cell and species types; efficiency, i.e. up to 80 % of the foreign DNA can be taken up by the host cell; small scale, i.e. a smaller amount of DNA is required than that of the other techniques; in vivo, i.e. this technique can be performed with intact tissue. Additionally, this method also has certain disadvantages, i.e. cell damage. Similarly, wrong pulse length may rupture the cell membrane which ultimately fails to close even after discharging the membrane potential. The non-specific transport of materials into the cytoplasm during electroporation may result in ion imbalance and lead to improper cell function and cell death.

Reagents Required and Their Role

Luria-Bertani Broth

Luria-Bertani (LB) broth is a rich medium that permits fast growth and good growth yields for many species including *E. coli*. It is the most commonly used medium in microbiology and molecular biology studies for *E. coli* cell cultures. Easy preparation, fast growth of most *E. coli* strains, ready availability and simple compositions contribute to the popularity of LB broth. LB can support *E. coli* growth (OD_{600}=2–3) under normal shaking incubation conditions.

Antibiotics

Prepare antibiotics by dissolving in the appropriate solvents followed by filter sterilization. All the antibiotic solutions can be stored at −20 °C till further use. Antibiotics act as the markers for the correct transformation of the plasmid DNA into the bacterial cell. The final concentration of the antibiotics depends on the plasmid and the host. In most of the cases, concentrations of stock solutions for each antibiotic as shown in Table 2.1 are used.

Procedure

Preparation of Electro-competent Cells

1. Streak out *E. coli* cells from the stock culture on fresh LB plates. Incubate at 37 °C for 24 h.
2. Use a single colony and inoculate into 10 ml of LB broth. Incubate at 37 °C for 24 h with shaking at 250 rpm.
3. Dilute the culture to an OD_{600} of 0.5–1.0.

Table 2.1 List of antibiotics and concentration of final and stock solutions

Antibiotic	Solvent	Stock solution (mg/ml)	Final concentration in medium (μg/ml)
Ampicillin	H_2O	50	100
Chloramphenicol	C_2H_5OH	34	34
Gentamycin	H_2O	50	50
Kanamycin	H_2O	50	50
Rifampicin	CH_3OH	50	As required
Streptomycin	H_2O	300	10
Tetracycline	C_2H_5OH	5	300
Timenton	H_2O	300	300

4. Harvest the cells by centrifuging at 4000 rpm for 15 min at 4 °C.
5. Remove supernatant and resuspend the cell mass with cold sterile milli-Q.
6. Divide the cell suspension to aliquots of 100 μl per tube. Store the cells at −80 °C till further use.

Pre-electroporation Procedure

1. Gently thaw the cells on ice.
2. Add 0.5–2.0 μl of cold plasmid DNA to the cells. Mix well and incubate on ice for 1 min.
3. Transfer the cell/DNA mixture into an ice-cold 1-mm electroporation cuvette. Make sure to avoid bubbles or gaps in the cuvette's electrode gap.

Gene Pulser (Bio-Rad, USA) Apparatus Setup

1. Turn the gene pulser apparatus on, make sure the display is illuminated and read '0.00'.
2. Use the corresponding buttons to set the voltage to 1.8 kV.
3. Adjust the capacitor and set the capacitance at 25 μF.
4. Adjust the parallel resistor to a resistance of 200 Ω on the gene pulser control panel.

Electroporation

1. Wipe the cuvette with tissue paper so that no water bubbles remain on the cuvette.
2. Insert the cuvette into the white slide; push the slide into the chamber till the cuvette makes firm contact with the chamber electrodes.
3. Charge the capacitor and deliver a pulse, press and hold both the red pulse buttons until a continuous tone sounds. At this point the display should flash 'Chg', indicating that the capacitor is being charged.
4. Release the pulse buttons once the display signals the delivery of the pulse.
5. Remove the cuvettes from the chamber and add 1 ml of LB broth medium to the cuvette.
6. The cells are supposed to be fragile at this stage, transfer the cell culture carefully to a sterile micro-centrifuge tube.
7. Turn off the gene pulser apparatus.

Plating of Cells

1. Incubate the cells at 37 °C on a heat block for 60 min.
2. From each transformation, plate 50 μl and 100 μl of cell suspension to the two LB plates with suitable antibiotics.
3. Incubate the plates at 37 °C for 24 h and observe for the growth of the transformants.

Observation

Transformation efficiency can be calculated after observing the actual number of observed colonies using different amounts of plasmid concentrations. Transformation efficiency is the number of transformants per microgram of supplied plasmid.

Result Table

The result table showing transformation efficiency by electroporation is as follows:

Transformation condition	Observed colony numbers with different amounts of plasmid (μg)		
	0	5	10
Control	0		
Set 1	0		
Set 2	0		
Set 3	0		

Troubleshooting

Problems	Possible cause	Possible solutions
When the chamber is opened, the sample is splattered over the inside of the chamber	Sample conductivity is too high	Reduce the ionic conductivity of the sample
Cells are slightly foamy after being pulsed	Though this condition is normal for *E. coli*; field strength may be too high	Reduce the voltage range setting, compare the cell viability and obtained transformation
Cells are lysed after being pulsed	Field strength may be substantially too high	Reduce the voltage range setting, reduce conductivity of the sample
No transformants when cells are pulsed with a known DNA	Electrical connections may not be proper	Check all the connections properly, test with a batch of cells with known efficiency
Transformation efficiency is less than expected	Voltage range may not be optimum	If voltage range is set to HIGH or LOW, reset to MEDIUM and repeat the experiment; if voltage range is set to MEDIUM, compare the results with LOW or HIGH
Mutations produced in chromosomal and plasmid DNA	High-voltage pulses producing indirect mutations	Reduce the voltage till an efficient transformation is achieved
Alteration in membrane structure and function	Possible membrane damage allowing leakage of cytosolic components and entry of medium and buffer components into cells, thus loss of metabolic or transport functions	Reduce the voltage to an optimum level

Precautions

1. Avoid spilling of any liquid on to the electroporation apparatus.
2. Always use a paper towel or a cloth wet with either water or alcohol to clean the surface.
3. Read the instruction manual of the electroporation instrument carefully before performing the experiment.
4. Mix the samples carefully to avoid any bubbles in the sample.
5. Never used cracked cuvettes; carefully check any used cuvettes for cracks.
6. Do not use DNA samples with excess salt content—it may hinder the electroporation process.
7. Wash the cuvettes properly with 0.1 N NaOH, distilled water and 95–100 % ethanol before use.
8. Always use gloves throughout the experiment.

FLOW CHART

Prepare OD_{600} of 0.5 to 1.0 culture of *E. coli* from the over-night grown cultures

↓

Harvest the cells by centrifugation at 4,000 rpm for 15 min at 4°C

↓

Re-suspend the cell mass with cold sterile milli-Q

↓

Divide the cell suspensions to 100 µl per tube and store them at -80°C till further use. Thaw the cells on ice

↓

Add 0.5-2.0 µl of cold plasmid to the cells, mix well, and incubate on ice for 1 min

↓

Transfer the cell/DNA mixture to ice cold 1 mm electroporation cuvette

↓

Set the voltage of electroporation apparatus to 1.8 kV, adjust the capacitance to 25 µF and the parallel resistance to 200 Ω.

↓

Switch on the electroporator after placing the cuvette at appropriate place

↓

Release the pulse button once the display signals the delivery of the pulse

↓

Remove the cuvettes from the chamber, add 1 ml of LB broth to the cuvette and incubate at 37°C on a heat block for 60 min

↓

Plate 50 and 100 µl of cell suspension to two LB plates with suitable antibiotics

↓

Incubate plates at 37°C for 24 h to observe the growth of the transformants

Exp. 2.3 Restriction Digestion and Ligation

Objective To perform restriction digestion and ligation.

Introduction

Restriction digestion is the enzymatic technique used for the cleavage of DNA molecules at specific sites so that all the DNA fragments containing a particular sequence have the same size. This cleavage method is performed by the use of DNA

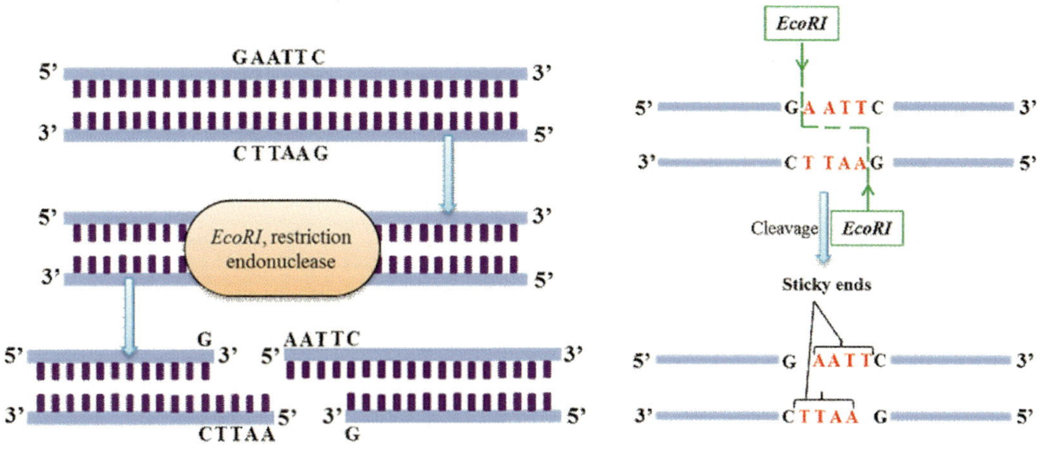

Fig. 2.5 Schematic representation of restriction digestion of *Eco*RI at specific site on the DNA fragment

cleaving enzymes of bacteria. This group of enzymes called restriction enzymes are capable of cleaving DNA molecules at particular short-base sequences located at a particular position on the DNA. This technique is sometimes also referred to as DNA fragmentation.

DNA ligation is the molecular method of joining two DNA strands together by the formation of a phosphodiester bond by the use of the DNA ligase enzyme. DNA ligase forms two covalent phosphodiester bonds between 3′-hydroxyl ends of one nucleotide with 5′-phosphate end of the other and this process is energy dependent. DNA ligase has many applications during DNA repair and replication in vivo.

Nowadays each molecular biology experiment is coupled with restriction digestion for wider applications in the field of cloning as well as other analytical techniques. The most common use of restriction digestion is the cloning of a DNA fragment into a suitable vector such as a cloning and/or expression vector. Other important applications of restriction digestion include restriction mapping, analysing population dynamics, rearranging DNA molecules, preparation of molecular probes, creating mutants and many others. Restriction enzymes are known to make double-strand cuts in DNA molecules. Under natural conditions, bacteria secrete these enzymes to defend themselves against the foreign invading DNA from other bacteria and viruses. Now restriction enzymes have become an important part of the molecular biology tool box for wide applications.

Principle

Restriction enzymes fall under the group of nucleases that cleave the sugar-phosphate back bone of a DNA molecule. The restriction enzymes which cut within the DNA molecule are called restriction endonucleases. They can cleave the DNA molecules at specific points called restriction sites to generate a set of smaller fragments (Fig. 2.5).

There are three types of restriction enzymes based upon the restriction and modification systems.

Type I Enzymes They exhibit both restriction digestion and modification which require cofactors such as Mg^{2+} ions, S-adenosylmethionine (SAM) and adenosine triphosphate (ATP) for their activity. Type I restriction enzymes cleave DNA at non-specific sites which may be 1000 bp or more. As methylation reaction is also performed by the same enzyme and the target DNA is modified prior to cutting, this type of restriction enzyme is of less value in gene manipulation practices.

Table 2.2 List of some commonly used restriction enzymes

Restriction enzyme	Organism	Sequence specification	Cut site	Nature of generated ends
EcoRI	*E. coli*	5'GAATTC 3'CTTAAG	5'-----G AATTC------3' 3'-----CTTAA G-------5'	Sticky
BamHI	*B. amyloliquefaciens*	5'GGATCC 3'CCTAGG	5'-----G GATCC------3' 3'-----CCTAG G-------5'	Sticky
BglII	*B. globigii*	5'AGATCT 3'TCTAGA	5'-----A GATCT------3' 3'-----TCTAG A-------5'	Sticky
PvuII	*Proteus vulgaris*	5'CAGCTG 3'GTCGAC	5'-----CAG CTG------3' 3'-----GTC GAC-------5'	Blunt
HindIII	*H. influenza*	5'AAGCTT 3'TTCGAA	5'-----A AGCTT------3' 3'-----TTCGA A-------5'	Sticky
Sau3A	*S. aureus*	5'GATC 3'CTAG	5'----- GATC------3' 3'-----CTAG -------5'	Sticky
AluI	*Arthrobacter luteus*	5'AGCT 3'TCGA	5'-----AG CT------3' 3'-----TC GA------5'	Blunt
TaqI	*Thermus aquaticus*	5'TCGA 3'AGCT	5'-----T CGA------3' 3'-----AGC T-------5'	Sticky
HaeIII	*Haemophilus aegyptius*	5'GGCC 3'CCGG	5'-----GG CC------3' 3'-----CC GG-------5'	Blunt
NotI	*Nocardia otitidis*	5'GCGGCCGC 3'CGCCGGCG	5'-----GC GGCCGC------3' 3'-----CGCCGG CG-------5'	Sticky

Type II Enzymes Type II enzymes possess two separate proteins for restriction cleavage and modification. In this case, the restriction activity is not dependent on cofactors such as ATP or SAM. The only requirement as cofactors in this case is Mg^{2+} ions. The major advantage of this group of enzymes is their site-specific nature to hydrolyze specific phosphodiester bonds in both DNA strands. Hence, this group of restriction enzymes is used for the purpose of restriction digestion and recombinant technology in most cases which also includes genome mapping, restriction fragment length polymorphism (RFLP), sequencing and cloning.

Type III Enzymes These enzymes are similar to that of type I systems, possessing both restriction digestion and modification activities simultaneously. They can recognise and cleave 25–27 bp outside the recognition sequence downstream to 3′ direction and also require Mg^{2+} for their activity.

Type II restriction enzymes generate three types of DNA ends possessing a 5′-phosphodiester bond and 3′-hydroxyl groups, i.e. cohesive 5' ends (e.g. *Eco*RI), cohesive 3′ ends (e.g. *PstI*) and blunt ends (*HaeIII*). Some commonly used restriction enzymes and their restriction sites have been provided in Table 2.2.

Plasmids act as useful vectors for transferring the genetic material from one organism to another by recombinant DNA technology. Restriction endonuclease is used in this process

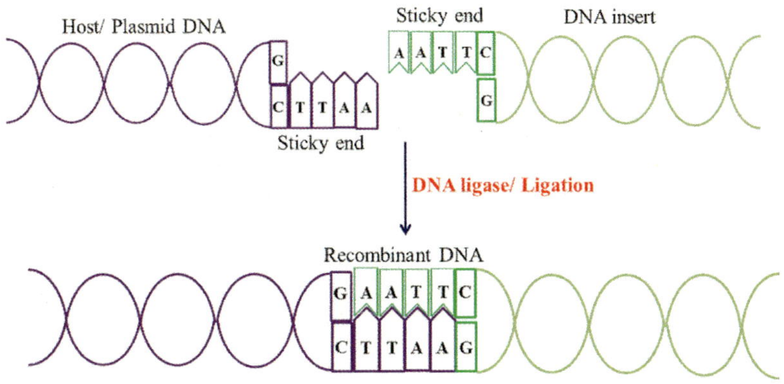

Fig. 2.6 Ligation of two sticky ends generated by *Eco*RI digestion

to cut and insert foreign DNA pieces into plasmid vectors. Ligation involves the formation of phosphodiester bonds between DNA molecules of the insert DNA and plasmid vector. DNA ligase catalyses this function by the formation of phosphodiester bonds between the adjacent 3′-hydroxyl and 5′-phosphoryl terminus of the nucleic acids (Fig. 2.6). Restriction digestion-generated cohesive ends are more efficient for ligation than blunt ends. Salt and phosphate concentrations play an important role in efficient ligation.

In most cases, two different enzymes are used to add an insert into a suitable vector, one enzyme from the 5′ end and another from the 3′ end. The advantage of using two different enzymes is that they ensure the correct orientation of the insert and prevent the vector from self-ligation. Self-ligation of the sticky ends of the vector can be minimized by treatment of the digested vector backbone with a phosphate before performing ligation reaction. Phosphatase removes the 5′ phosphate, thus preventing the ligase from fusing the two ends of the vector. During the process of nick sealing, DNA ligase uses the nicked double-stranded DNA, ATP and Mg^{2+}. Ligation practice involves three steps. In step 1, adenosine monophosphate (AMP) is linked to the amino group of a lysine molecule present at position 159 of the active site of DNA ligase and releases pyrophosphates (PPi) from ATP. In step 2, the ligase-adenylate develops a transient complex which searches the 5′-phosphorylated end through successive transient complexes. When it finds the suitable position for transfer of the adenylate group to the 5′-phosphorylated site, a stable complex develops and the 3′ end becomes available for sealing reaction. In step 3, ligase catalyses an attack on the pyrophosphate bond of the OH group at the 3′ end releasing the free enzyme and AMP (Fig. 2.7).

There are many factors that affect the activity of restriction enzymes in vitro which include temperature, buffer systems, ionic conditions and methylation of DNA. Most digestions are generally carried out at 37 °C. However, certain restriction enzymes like *SmaI* require lower temperatures (25 °C), whereas some require higher temperatures (TaqI: 65 °C). A buffer system is mandatory for the restriction action of the enzymes as most of them have the optimum activity at a range of pH 7.0–8.0. Most restriction enzymes also require Mg^{2+} for their activity; however, in some cases other ions like Na^+ and K^+ are also required, which depends on the nature of the enzyme. Methylation of specific adenine or cytidine residues affects the action of the restriction enzymes in an adverse manner.

In this experiment, pUC18 and lambda DNA (λDNA) will be double digested with *Eco*RI and *Hin*dIII followed by the insertion of λDNA into a linearized pUC18 vector by ligation.

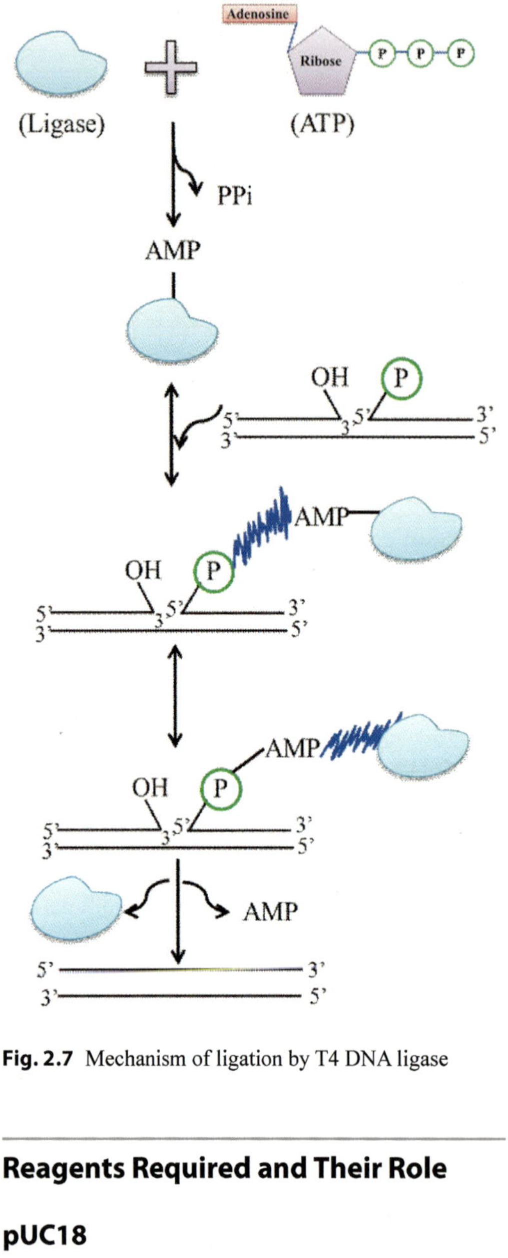

Fig. 2.7 Mechanism of ligation by T4 DNA ligase

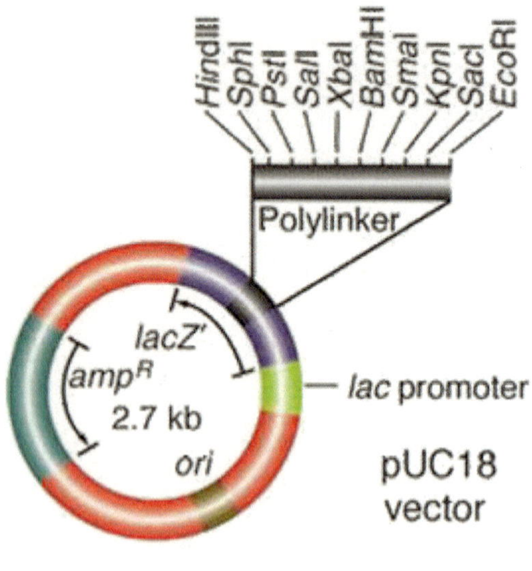

Fig. 2.8 pUC18 plasmid vector

Reagents Required and Their Role

pUC18

pUC18 is an artificial plasmid that has been genetically engineered and accommodates genes of antibiotic resistance (*amp*^{R}) and promoters for the enzyme beta-galactosidase (*lacZ*; Fig. 2.8). The *lacZ* gene contains a polylinker region with a series of unique restriction sites which is not present anywhere else in the plasmid. Digestion with any of the restriction enzymes makes a single cut and linearizes the circular plasmid DNA. This allows the recombination with the foreign DNA which has been cut with the same endonuclease.

λDNA

λDNA is the DNA molecule isolated from an *E. coli* bacteriophage or lambda phage. The genetic material of the phage is linear, double-stranded with 12 bp single-stranded complementary 5′ ends. Phage λDNA has become a common substrate for restriction digestion and generating smaller DNA fragments. It has wide applications in the field of studying activity and specificity assay of restriction enzymes, preparation of DNA molecular weight standards as well as cloning.

HindIII

*Hin*dIII is the type II restriction enzyme that cleaves the palindromic sequences AAGCTT in presence of Mg^{2+} by hydrolysis isolated from *Haemophilus influenza*. After digestion, it produces 5′ overhangs on DNA and sticky ends. It is a homodimer comprising four β-sheets and a single α-helix. It has wide applications in the field of DNA sequencing and mapping.

EcoRI

*Eco*RI isolated from *E. coli* produces sticky ends with 5′ overhangs, and the enzyme is a homodimer of 31 kDa-subunits. However, *Eco*RI can exhibit non-site-specific cutting called star activity on varied conditions in the solution. Low salt concentration, high glycerol concentration, excessive enzymes, high pH and presence of organic solvents may induce star activity of *Eco*RI. *Eco*RI has wide applications in the field of cloning, DNA screening, deleting sections of DNA and others.

Assay Buffer

The assay buffers include: for *Hin*dIII, 10 mM Tris-HCl, 50 mM NaCl, 10 mM $MgCl_2$ and 1 mM dichlorodiphenyltrichloroethane (DTT); for *Eco*RI, 50 mM Tris-HCl, 100 mM NaCl, 10 mM $MgCl_2$ and 5 mM mercaptoethanol. As all the enzymes work at their optimum environmental conditions, specific buffers provide such optimum conditions. The assay buffer is available commercially along with the respective restriction enzyme.

0.5 M Ethylenediaminetetra-acetic Acid

In order to stop further digestion of the DNA fragment, a stop solution is added to the reaction that contains ethylenediaminetetra-acetic acid (EDTA). However, heat inactivation is the simplest technique of stopping the restriction reaction. Addition of EDTA chelates Mg^{2+} ions from the solution which is the necessary prerequisite for the restriction enzyme to perform its action. Hence, due to unavailable Mg^{2+} ions, the reaction stops.

T4 DNA Ligase

DNA ligase catalyses the formation of a phosphodiester bond between the 5′-phosphate group and the 3′-hydroxyl group of the DNA molecule. T4 DNA ligase can join both the blunt and cohesive ends produced by restriction digestion using any restriction enzyme. The advantage of using T4 DNA ligase is that it can repair nicks developed in double-stranded DNA. During ligation, the double-stranded DNA is produced in three steps: (1) release of the phosphate group by adenylation of a lysine residue of the enzyme, (2) formation of the phosphodiester bond by transferring AMP to 5′ phosphate and (3) formation of the phosphodiester bond.

5 X Ligation Buffer

The composition of a 5X ligation buffer for the optimum activity includes: 250 mM tris-HCl (pH 7.6), 50 mM $MgCl_2$, 5 mM ATP, 5 mM DTT and 25 % (w/v) PEG 8000. Ligation buffer provides the optimum reaction conditions and energy sources for enhanced activity of DNA ligase.

Procedure

For Restriction Digestion

1. Take two clean autoclaved micro-centrifuge tubes and add the reagents listed in Table 2.3 accordingly.
2. After adding the reagents, mix them well gently and centrifuge briefly.
3. Incubate the tubes for 4 h at 37 °C.
4. After incubation, stop the enzyme reaction by adding 1 μl of 0.5 M EDTA solution.
5. To confirm restriction digestion, load 5 μl of the digested sample in 1 % agarose gel.
6. Store the digested product in −20 °C till further use.

Table 2.3 Reaction mixture for restriction digestion

pUC18 digestion		λDNA digestion	
Components	Volume (μl)	Components	Volume (μl)
pUC18 plasmid (1 μg/μl)	1	λDNA	1
*Hin*dIII assay buffer (10X)	2	*Hin*dIII assay buffer (10X)	2
*Hin*dIII (10 U/μl)	0.5	*Hin*dIII (10 U/μl)	0.5
*Eco*RI (10 U/μl)	0.5	*Eco*RI (10 U/μl)	0.5
Sterile milli-Q	16	Sterile milli-Q	16
Total	*20*	*Total*	*20*

Table 2.4 Reaction mixture for ligation

Components	Volume (μl)
pUC18/*Eco*RI/*Hin*dIII (100 ng/μl)	9
λDNA/*Eco*RI/*Hin*dIII (100 ng/μl)	3
5X ligase buffer	4
T4 DNA ligase (0.1 U/μl)	1
Sterile milli-Q	4
Total	*20*

For Ligation

1. Add the components given in Table 2.4 in a 0.5-ml micro-centrifuge tube.
2. Mix the components gently and centrifuge briefly.
3. Incubate the mixture for 2 h at 16 °C.
4. Store the samples at −20 °C till further use.

Observation

Observe the gel electrophoresis result carefully for the digested products. The orientation of the insert can also be confirmed by determining the approximate size of the DNA fragment resulting from restriction digest.

Troubleshooting

Problems	Possible cause	Possible solutions
Incompletely digested or undigested DNA	Nature and confirmation of DNA may not be suitable for enzyme digestion	Verify the activity of the enzyme using substrate DNA of linear or super-coiled nature
	Impure DNA	Remove impurities such as phenol, chloroform, alcohol, detergents, EDTA or salts that interfere with the restriction endonuclease activity
	Lack of recognition sequence	Check for the presence of DNA sequences that can be recognised by the particular enzyme
	Reaction conditions not suitable	Use the reaction buffer as recommended by the manufacturer. Use freshly prepared buffer
	Improper dilution of the enzyme	Use appropriate enzyme concentration for an efficient result
	Partially or completely inactivated enzyme	Check for the expiry of the reagents; use fresh stock
Atypical banding pattern	Incomplete digestion	Use fivefold to tenfold excess restriction enzyme for proper digestion of the DNA
	Enzyme star activity	To avoid enzyme star activity, use enzyme buffer and reaction conditions recommended by the manufacturer
	DNA contamination	Digest the DNA with other restriction enzymes

Problems	Possible cause	Possible solutions
Incompletely digested or undigested DNA	Nature and confirmation of DNA may not be suitable for enzyme digestion	Verify the activity of the enzyme using substrate DNA of linear or super-coiled nature
	Impure DNA	Remove impurities such as phenol, chloroform, alcohol, detergents, EDTA or salts that interfere with the restriction endonuclease activity
	Lack of recognition sequence	Check for the presence of DNA sequences that can be recognised by the particular enzyme
	Reaction conditions not suitable	Use the reaction buffer as recommended by the manufacturer. Use freshly prepared buffer
	Improper dilution of the enzyme	Use appropriate enzyme concentration for an efficient result
	Partially or completely inactivated enzyme	Check for the expiry of the reagents; use fresh stock
Smear on gel after digestion	Impure substrate	Run a control set of reaction with DNA without enzyme and with other enzymes
	Impure enzyme	Use fresh stock of enzyme
	Protein contamination	Incubate the samples with 1 % SDS at 65 °C for 10 min before gel electrophoresis
Poor ligation efficiency	Degraded buffer components	Repeat ligation with fresh ligation buffer
	Restriction enzyme active in ligation mixture	Stop restriction digestion using 0.5 % EDTA or incubating at 65 °C for 10 min
	Low ligase concentration	Use high amount of ligase, i.e. 0.3–0.6 Wu/1 μg of fragments

SDS sodium lauryl sulphate, *EDTA* ethylenediaminetetra-acetic acid

Precaution

1. Always keep restriction enzymes and buffer in ice and put them back to −20 °C as soon as possible.
2. Always spin the tubes gently to bottom down the contents prior to opening.
3. Use fresh autoclaved microtips.
4. Check for pipetting errors; a higher or lower value may alter the reaction.
5. Use gloves throughout the reaction as the buffers and enzymes are highly sensitive.
6. Mix the reaction components thoroughly.
7. When setting up a large number of reactions, make a master mixture containing water, buffer and enzyme, aliquot into tubes containing the DNA to be digested.
8. Do not forget to inactivate the enzyme for proper downstream applications.

FLOW CHART

For restriction digestion

Add the reaction components as mentioned in the table into two separate tubes, one for pUC18 and another for λ DNA

↓

Mix gently and centrifuge briefly

↓

Incubate for 4 h at 37°C

↓

Add 1 µl of 0.5% EDTA to stop the reaction

↓

Confirm digestion by gel electrophoresis, store at -20°C till further use

For ligation

Add the components as mentioned in the table in a 0.5 ml microcentrifuge tube

↓

Mix the components gently and centrifuge briefly

↓

Incubate the mixture at 16°C for 2 h

↓

Store the sample at -20°C till further use

Exp. 2.4 Selection of a Suitable Vector System for Cloning

Introduction

Nowadays most molecular biology experiments deal with the transfer of genetic material from one organism to another. In a bacterial system, genetic material can be transferred naturally in three ways, i.e. transformation, transduction and conjugation. However, for the transfer of any particular gene sequence to another host system a carrier or vector is required. There are three pre-requisites for cloning genes to a new host. Firstly, a method should be standardized to introduce the gene of interest into the potential host. Secondly, the introduced DNA should be maintained in the new host either as a replicon or by integrating into the chromosome or pre-existing plasmid. The uptaken gene should be maintained and expressed in the new host system.

A vector is used to amplify a single molecule of DNA into many copies. The DNA fragment should be inserted into the cloning vector which will be carrying the gene of interest to the new host organism. The cloning vector is a DNA molecule that must possess an origin of replication (ORI) which is capable of replicating in the bacterial cell. Most of the commercially available vectors nowadays are genetically engineered plasmids or phages. Other available vectors include cosmid vectors, bacterial artificial chromosomes (BACs) and yeast artificial chromosomes (YACs).

In the 1970s, construction of first-generation general cloning vectors started, and till now numerous numbers and varieties of cloning vectors have been generated and are available in the market. Thus, it has become a critical decision to select the suitable plasmid or vector for a particular function. Though numerous varieties of vectors are available commercially, the choice of selection of the suitable system includes consideration of a few criteria, i.e. insert size, copy number, compatibility, selectable marker, cloning sites and specialized vector functions. Though many plasmid vectors are available, all of them are derived from the plasmids that were originally isolated from the bacterial cells in nature. Commercial cloning vectors are designed in such a way that they become quite useful during the recombinant DNA technology practice. All the commercially available plasmids for cloning use have three things common in them:

An Origin of Replication (ORI) This is the DNA sequence which makes sure that the plasmid will be recognised by the bacterial replication machinery and can be replicated along with the bacterial genome. The plasmid will remain as an autonomous unit and can replicate to a very high copy number within the single bacterial cell.

A Selectable Marker Though many techniques have been discovered so far, the efficiency of transformation is too low. Because of this limitation, the plasmid to be used as vector must possess certain selectable markers which allow the transformed cells to grow, and those which are not transformed will not grow. The most commonly used selective marker are the antibiotic resistance genes, e.g. the *amp*^{R} gene. After transformation, positively transformed cells will grow in the presence of media containing ampicillin.

The Multiple Cloning Sites (MCSs) Restriction enzymes cut at specific DNA sequences. Thus, the plasmid containing multiple restriction sites will be very useful as there will be flexibility in choosing different restriction enzymes to design an experimental strategy.

An exhaustive list of the commercially available vectors and their use has been listed in Table 2.5.

Different Types of Cloning Vectors

There are different types of cloning vectors which can be chosen based upon the type of cloning experiments. The cloning vector is the small piece of DNA into which a foreign piece of a DNA fragment can be inserted. This insertion can be achieved by treating the vector as well as the insert with a restriction enzyme, creating the same overhang and subsequently ligating the fragments together. Many types of available vectors include genetically engineered plasmids, bacteriophages, BACs and YACs.

Plasmid Vectors

Plasmid vectors are used for the cloning of DNA fragments of the size of a few base pairs to several thousand base pairs i.e. 100 bp–10 kb. The cloning strategy involving plasmids depends upon the starting information and the desired endpoints that include protein sequence, positional cloning information, mRNA sequence, cDNA libraries, known or unknown DNA sequences, genomic DNA libraries and the nature of the polymerase chain reaction (PCR) product. There are many advantages of using plasmids as cloning vectors due to their smaller size, circular nature, replication independent of host cell, presence of several copies and frequently available antibiotic resistance. Hence, they are easy to manipulate, more stable, facilitating replication, and it becomes easy to detect their presence and location after cloning. Additionally, there are also certain disadvantages of using plasmids as vector systems as they cannot accept large fragments, the suitable

Table 2.5 Commonly used cloning vectors

Plasmid/vector	Features	Commercial source
pUC18, pUC19	Small size (2.7 kb)	New England Biolabs, UK
	High copy number	
	Multiple copy number	
	Resistant marker ampicillin	
	Blue-white selection	
pBluescript vectors	Originated from pUC	Stratagene, USA
	Single-stranded ORI	
	T7 and SP6 promoters flanking MCS	
pACYC vectors	Low copy number (15 copies per cell)	New England Biolabs, UK
	p15A ORI	
Supercos	Cosmid vectors	Stratagene, USA
	Two *cos* sites	
	Insert size 30–42 kb	
	Selectable marker ampicillin	
	T3 and T7 promoter flanking cloning sites	
EMBL3	λ replacement vector	Promega, USA
	MCS sites: *Sal*I, *Bam*HI, *Eco*RI	
λ ZAP	λ vector	Stratagene, USA
	In vivo excision into pBluescript phagemid vector	
	Cloning capacity 10 kb	
	Blue-white selection	
pBeloBAC11	BAC vector	New England Biolabs, UK
	Insert up to 1 Mb	
	T7 and SP6 promoters flank insertion site	
	Blue-white selection	
	Cos site	
	*Lox*P site	
pALTER-Ex1, pALTER-Ex2	*E. coli* vector	Promega, USA
	T7 promoter	
	Selection marker tetracycline	
pBAD/His	*E. coli* vector	Invitrogen
	araBAD promoter	
	Selection marker ampicillin	
	Originated from pUC	
pCal-n	*E. coli* vector	Stratagene, USA
	T7-lac promoter	
	Selection marker ampicillin	
	Protease cleavage site *Thr*	
	Originated from ColE1	
pcDNA 2.1	*E. coli* vector	Invitrogen, USA
	T7 promoter	
	Selection marker ampicillin	
	Originated from pUC	
pDUAL	*E. coli* vector	Stratagene, USA
	T7-lac promoter	
	Selection marker kanamycin	
	Protease cleavage site *Thr*	
	Originated from ColE1	

Table 2.5 (continued)

Plasmid/vector	Features	Commercial source
pGEX-2T	*E. coli* vector	Pharmacia, Sweden
	tac promoter	
	Selection marker ampicillin	
	Protease cleavage site *Thr*	
	Originated from pBR322	
pHAT20	*E. coli* vector	Clontech, USA
	Lac promoter	
	Selection marker ampicillin	
	Protease cleavage site EK	
	Originated from pUC	
pLEX	*E. coli* vector	Invitrogen, USA
	P_L promoter	
	Selection marker ampicillin	
	Originated from pUC	
pQE-60	*E. coli* vector	Qiagen, Netherlands
	T5-lac promoter	
	Selection marker ampicillin	
	Originated from ColE1	
pRSET	*E. coli* vector	Invitrogen, USA
	T7 promoter	
	Selection marker ampicillin	
	Protease cleavage site EK	
	Originated from pUC	

MCS multiple cloning site, *BAC* bacterial artificial chromosome, *kb* kilobase, *ORI* origin of replication

range of DNA fragment is 0–10 kb and the available standard methods of transformation are highly inefficient. The peculiar characteristic features of plasmids making them suitable cloning vectors are discussed below:

Extra-Chromosomal Self-Replicating DNA Molecules Plasmids are circular, double-stranded DNA molecules that are different from the bacteria's chromosomal DNA. The range of the size of plasmids is from a few thousand base pairs to more than 100 kilobases (kb). In addition to the host cell's chromosomal DNA, plasmid DNA is replicated prior to each cell division and at least one copy of the same is segregated to each daughter cell. Plasmids harbour many genes responsible for antibiotic resistance that can spread the drug resistant plasmids, thus expanding the drug resistant bacteria in the environment. This drug resistant phenotype can be treated as marker to be used in cloning experiments.

Plasmids Can Be Engineered to Be Used as Cloning Vectors Most of the plasmids used during recombinant DNA technology can replicate in *E. coli*. These plasmids can be manipulated easily to be used as suitable vectors during cloning. For example, the length of the naturally occurring plasmids can be reduced to ~3 kb length as most of them contain many non-essential nucleotide sequences required for DNA cloning. The desired nucleotide sequences in plasmids to be used for cloning should include the ORI, a drug resistant gene and the region where foreign DNA can be inserted (Fig. 2.9).

Simplified Plasmid DNA Replication ORI is the specific DNA sequence of 50–100 bp present on the plasmid DNA and is responsible for its replication. Host enzymes responsible for its replication also bind with the ORI region of the plasmid to initiate its replication. Once initiated, the whole plasmid can be replicated independent of

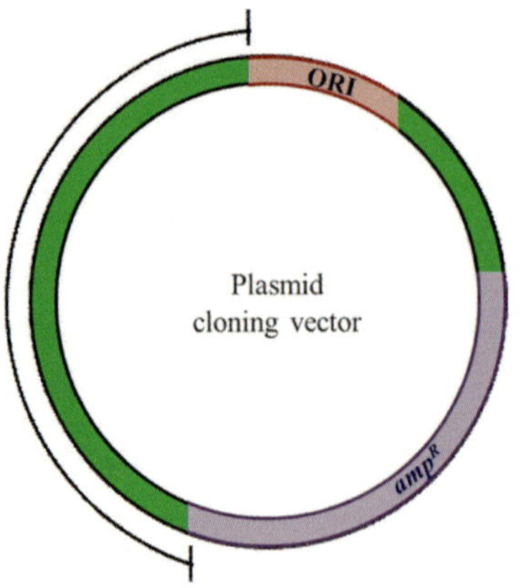

Fig. 2.9 A simple cloning vector derived from a plasmid, the circular double-stranded DNA molecule capable of replicating in *E. coli*

its nucleotide sequences. Thus, any foreign DNA sequence present on the plasmid gets replicated along with the plasmid which is one of the prime requisites for molecular cloning.

Better Isolation of DNA Fragments from Complex Mixtures The foreign DNA bases of ~20 kb which have been inserted into the plasmid vector are responsible for developing the antibiotic resistant progeny cells that can be distinguished clearly from the rest of the cells. As all the cells grown as a single colony arise from a single transformed cell constituting a clone of cells, the cloned DNA fragment inserted into the plasmid DNA can be isolated readily from the developed colony for further use.

Better Action of Restriction Enzymes As most of the restriction enzymes are derived or synthesized from the bacteria themselves, they are more prone to the same restriction enzyme if not methylated properly. In this regard, the plasmid vector is quite useful in restriction digestion to be used in subsequent steps of cloning and expression analysis of the target genes.

Presence of Polylinkers Plasmid vectors can be manipulated easily to insert a polylinker site, i.e. a synthetic MCS sequence (Fig. 2.10) containing one copy of several restriction sites in vitro. When such a vector is targeted with any restriction enzyme, it recognises the sequences in the polylinker, cut at that point to generate sticky ends for further insertion into the plasmid.

Bacteriophage Lambda

Phage lambda is the bacterial virus that uses *E. coli* as a host having the typical phage structure of head, tail and tail fibres. The lambda viral genome comprises a 48.5 kb linear DNA with 12 bases of single-stranded DNA (ssDNA) sticky ends at both sides which are complementary to each other and can hybridize to each other possessing the cohesive ends. When infected in an *E. coli* host, lambda begins its life cycle by circularizing at the *cos* site. A cosmid can be defined as a type of hybrid plasmid containing lambda phage *cos* sequences. The major advantage of using cosmids over plasmids is that they can transfer about 37–52 kb of DNA in contrast to 1–20 kb by the plasmids alone. They possess a strong selection for the cloning of large inserts, and cosmids can be maintained as phage particles in solutions. A generalized structure of a cosmid has been provided in Fig. 2.11.

Yeast Artificial Chromosome

YACs are genetically engineered plasmid shuttle vectors capable of replicating in common bacterial and yeast hosts. The most promising flexibility of using such a vector system is that when they are amplified and manipulated in bacteria, they render to be circular and of smaller size, i.e. approximately 12 kb, whereas when introduced to a yeast host, they become linear with large sizes, i.e. few hundred kilobases.

YACs are the cloning vehicles that propagate in prokaryotic cells as eukaryotic chromosomes being capable of cloning large inserts of DNA in the range of 100 kb–10 Mb. The final chimeric DNA of YAC is a linear DNA molecule with telomeric ends. Additionally, they have a bacterial ORI, a selection marker for the propagation of YAC through bacteria as well as yeast as its host system.

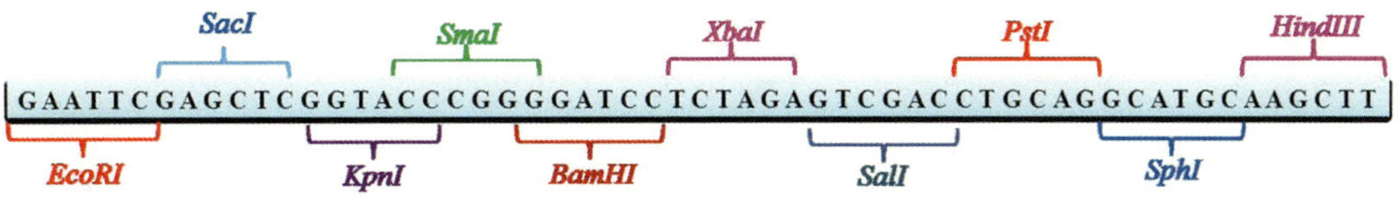

Fig. 2.10 Schematic presentation of a polylinker sequence

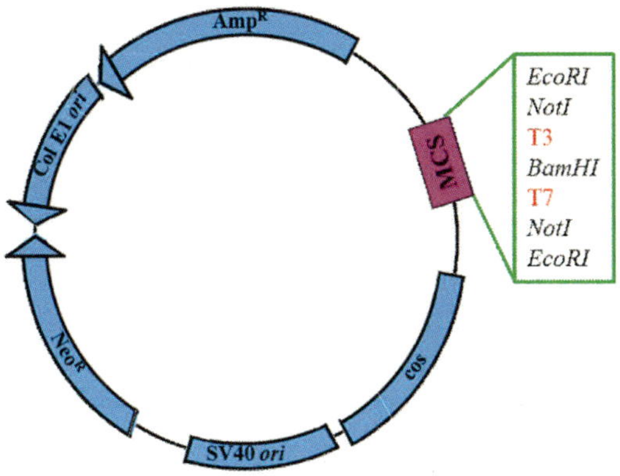

Fig. 2.11 A generalized structure of cosmid pWE15 of size 8.2 kb

Retroviral Vectors

Retroviral vectors are used to introduce new or altered genes into the genomes of human as well as animal cells. The viral RNA is converted to DNA by the viral reverse transcriptase followed by efficient integration into the host genome. Any foreign or mutated host gene introduced into the retroviral genome can be integrated into the host chromosome to reside indefinitely. Most of the applications of retroviral vectors include the study of oncogenes and human genes.

Expression Vectors

Expression vectors are used to produce a large amount of a specific protein which permits the study of the structure and function of the desired proteins. They are extremely useful when the proteins consist of rare cellular components and are difficult to isolate. Expression vectors may be of three types: expression vectors with a strong promoter which codes for more proteins from more mRNA; expression vectors with an inducible promoter, e.g. drug inducible (isopropyl β-D-1-thiogalactopyranoside, IPTG, or arabinose) and heat inducible; and expression vectors with a fusion tag for affinity purification which facilitate the purification of the expressed protein, e.g. 6-histidine tag, glutathione transferase tag (GST) or maltose-binding protein tag.

However, there are many disadvantages of using bacterial expression systems which include:

Low Expression Levels This problem can be solved by changing the promoter, plasmid, cell type or by addition of rare tRNAs for a rare codon on a second plasmid.

Severe Protein Degradation Use of proteasome inhibitors and other protease inhibitors and induction at low temperature may solve this problem.

Mis-Folded Proteins Co-expression with a chaperone GroEL and using different refolding buffers may solve such problems.

Specialized Cloning Vectors

Nowadays, TA cloning has become extremely useful in the field of molecular biology and cloning. It has been established that most of the DNA polymerases, including Taq DNA polymerase, add additional non-template-directed nucleotides to the 3′ end of the amplified DNA. Taq polymerase adds a single 3′ A to the blunt end of double-stranded DNA by terminal transferase-like activity. Hence, most of the PCR products amplified by Taq polymerase possess the 3′-A overhangs at both ends (Fig. 2.12). Thus, the target DNA vector tailed with dideoxythymidine triphosphate (ddTTP) adds only T residues on each blunt end. The TA cloning kit is commercially

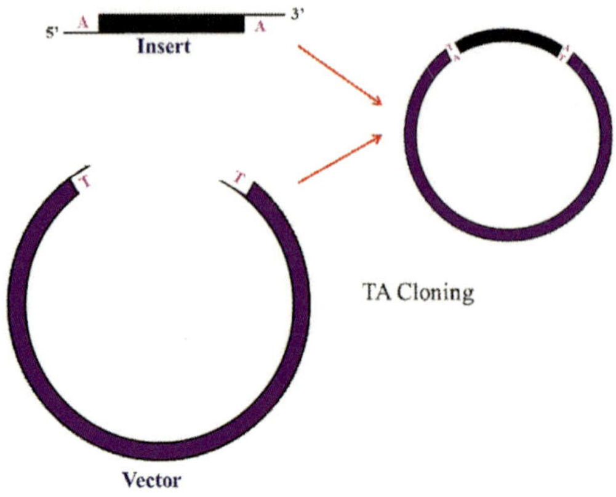

Fig. 2.12 Schematic presentation of TA cloning procedure

available for proper instant cloning reactions without the use of restriction enzymes.

There are many advantages of using TA cloning vectors over other available practices of cloning which include no requirement of restriction enzymes to generate linearized vectors. In addition to that, this procedure is much simpler and faster to perform. It is not necessary to add restriction sites while designing primers, and at places where no viable cloning sites are found, this technique becomes extremely useful. However, the major drawback of this technique is that there is no possibility of directional cloning. Hence, the genes have a 50% chance of getting cloned in reverse direction.

Criteria for Choosing a Suitable Cloning Vector

Insert Size

The size of the DNA to be cloned to the host system is the critical factor in choosing the vector system. Most of the available plasmids can carry around 15 kb of DNA fragment. A higher length of insert size creates problems with the replication and stability of the plasmid in the host system. Hence, several other vectors are found capable of transferring higher sizes of gene sequences. These vectors capable of carrying higher DNA fragments are mostly used to construct libraries representative of the entire genome, and the clones can be screened further to detect the clone carrying the DNA of interest.

Lambda Vectors The genome of the λ bacteriophage is of 48,502 bp, and inside the host system it can adopt either of the two life cycles, i.e. lytic or lysogenic cycle. The lytic life cycle of the phage can be used for cloning purposes. The ability of replacing the genes responsible for the lysogenic life cycle with foreign DNA without affecting the lytic life cycle makes the λ bacteriophage a suitable cloning vector. Lambda vectors can accommodate inserts of a size of 5–11 kb.

Cosmids Cosmids are the conventional vector systems harbouring the cohesive end site, i.e. *cos* in the lambda bacteriophage system. In this case, the linear DNA fragment is ligated to the vector DNA in vitro and then transferred to the bacteriophage. Cosmids are well known for generating large insert libraries. However, the vector plus insert length in case of cosmids should ideally range between 28 and 45 kb.

Bacterial Artificial Chromosomes BACs are circular DNA molecules containing replicons consisting of *oriS* and *repE*. In this case, the average insert size is 120 kb; however, it can increase up to 350 kb. In some new BACs, certain sites have been introduced for recovery of cloned DNA. BACs are capable of cloning in similar processes by ligation of DNA to the linearized vector and introducing it to the host system by electroporation.

Copy Number

Depending on the replicon of the plasmid, different vectors have different copy numbers. In certain cases like maintenance and subsequent manipulation greater yield of insert DNA is required. Thus, in this case, vectors with high copy numbers can be used. A high copy number in the vector system can be achieved where replication is dependent on the ColE1 replicon. The wild ColE1 replicon is responsible for producing 15–20 copy numbers; however, genetic engineering causing point mutation within the RNAII regulatory increases the copy number of the mu-

Table 2.6 Commonly used vectors and their copy numbers

Vector	Replicon	Copy number	Nature
pBR322	ColE1	15–20	High copy number
pACYC	p15A	18–22	High copy number
pSC101	–	Around 5	Low copy number
BACs	–	One per cell	Low copy number
pUC19	pMB1	More than 100	High copy number

BAC bacterial artificial chromosome

tant to 500–700. In contrast to that, in certain instances, a high copy number may render problems in cloning. At high copy number, certain membrane proteins produce toxic effects on the cell. Even when the protein is poorly expressed, the presence of a high copy number renders the toxic effects. Thus, a vector with low copy number is required. Certain examples of the commonly used vectors and their copy numbers have been provided in Table 2.6.

Compatibility

Different plasmids present in the same cell sometimes become incompatible to each other to share the same genetic machinery. This situation occurs primarily when the different plasmids have common functions in replication and portioning to the daughter cells. Hence, there is the chance of losing one plasmid due to such competition during cell growth. In another aspect, one smaller plasmid vector and one larger vector may not be sustained simultaneously in one host system as the larger one requires more time for its replication which is outcompeted by faster replicating smaller plasmids. For example, vectors with a ColE1-based replicon cannot be sustained with vectors containing R6K or p15A replicons. However, the issue of compatibility comes into the picture only when there is the possibility of using two vectors in the same host system.

Selectable Markers

Introduction of an extra-chromosomal DNA to a bacterial system decreases the growth rate of the bacteria. This is due to the fact that bacteria need to replicate their own genetic system in addition to the vector containing inserted DNA. In this process, the bacteria may tend to lose the inserted plasmid from their cytoplasm. This problem arises critically with plasmids of high copy number and large-sized plasmids. However, selection pressure located in the genetic machinery of the plasmid can be induced for proper maintenance of the vector inside the bacterial system. Most of the naturally occurring plasmids carry certain antibiotic resistance genes. Hence, the selection marker which should not be present in the host system de novo should be chosen carefully. Thus, the genotype of the host system should be checked and known correctly prior to inserting the plasmid or vector into it.

Certain vector systems possess more than one antibiotic resistant cassette. For example, pACYC177 contains both ampicillin and kanamycin resistant markers. Most of the new vectors carry special cloning sites, i.e. polylinkers or MCSs. In such instances, cloning of insert DNA does not interfere with the subsequent vector functions and stability. The most common antibiotic resistance markers include ampicillin, kanamycin, tetracycline and chloramphenicol.

Ampicillin Ampicillin inhibits bacterial transpeptidase responsible for bacterial peptidoglycan synthesis. Thus, it kills the bacterial cells at log phase but not in the stationary phase. β-lactamase renders resistance to this antibiotic by destroying the β-lactam ring of the antibiotic. In most of the cloning vectors, the ampicillin-resistant genotype is coded by the *bla* gene.

Kanamycin Kanamycin interacts with the three ribosomal proteins and rRNA in the 30S ribosomal subunit, and hence prevents the formation of chain-elongating complex inhibiting protein synthesis. Aminophosphotransferase confers resistance against kanamycin coded by *aph* (3′)

I and *aph* (3′) II derived from Tn903 and Tn5, respectively, by transferring phosphate from ATP to kanamycin and thus inactivating it. These two resistant genotypes have different gene sequences and different restriction maps.

Chloramphenicol Chloramphenicol is a broad-spectrum antibiotic which inhibits the activity of ribosomal peptidyl transferase to inhibit protein synthesis. However, the resistance to chloramphenicol is conferred by chloramphenicol acetyl transferase (CAT) by transferring an acetyl group from acetyl-CoA to chloramphenicol and inactivating it.

Tetracycline Tetracycline binds to the 30S ribosomal subunit and hence blocks the attachment of aminoacyl tRNA to the acceptor site, inhibiting protein synthesis. Certain efflux proteins in bacteria confer resistance to them by catalysing the energy-dependent export of tetracycline outside the cell, e.g. *tetA*, *tetB*, *tetC*, *tetD* and *tetE*.

Cloning Sites

In general, cloning involves the restriction digestion of the vector and the insert DNA sequences by the same restriction enzyme to generate the compatible ends which are further ligated by DNA ligase. To achieve compatible cohesive ends of insert and vector DNA fragments, both the vector and insert DNA sequence should possess the same restriction site. However, due to advancement of molecular biology techniques any blunt end fragment can be ligated to any other blunt end fragment, and overhangs generated by restriction digestion can be made blunt ended. In this regard, MCSs or polylinkers present in the vector system increase the wide choice of using restriction enzymes. This also limits the cloning site to one small region allowing the specific positioning of the insert DNA. MCSs also allow the amplification and detection of the insert DNA regardless of its position by using the universal primer.

Conclusion

The investigators face a huge challenge in selecting a proper vector system with enormous available choice for a particular work. Thus, the application of quite a small number of factors as discussed above can guide them quickly for selecting a suitable vector. Due to advancement of technologies and resources, a wide number of vectors are available commercially containing sufficient features to be suitable for many applications. Hence, the vector targeted should cover most of the needs of the researchers.

Exp. 2.5 Confirmation of Transformation by Blue-White Selection

Objective To confirm successful transformation in a bacterial system by using the blue-white selection method.

Introduction

Blue-white selection is a technique utilizing the *E. coli lacZ* gene expression for visual detection of bacterial colonies containing plasmids with DNA inserts. When a DNA insert is present in the vector after transformation, there are chances of getting two types of colonies, i.e. colonies with vector alone and colonies with recombinant vector. However, it is time-consuming to check individual colonies for detection of the presence of the desired DNA insert. Hence, the blue-white screening technique has become useful because of its less time- and labour-intensive methodology. This technique relies on the distinguished colony colour on selective plates for screening of the correct transformant-harbouring recombinant vector.

Though antibiotic stress is used as a marker for selection of positive transformation, it allows the growth of transformants harbouring both vector as well as recombinant one. Hence, some additional methodology is required to distinguish between a positive recombinant vector and a

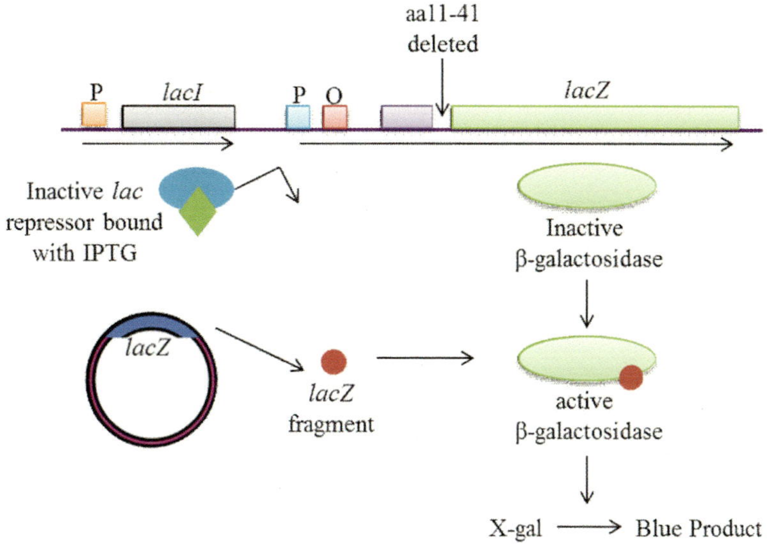

Fig. 2.13 Schematic presentation of blue-white screening assay

non-recombinant one. After many years of application of antibiotic stress for transformant selection, nowadays the blue-white selection strategy is used for the same. In blue-white selection, DNA is cloned into the restriction site of the *lacZ* gene. *lacZ* encodes β-galactosidase, the enzyme that degrades the disaccharide lactose into the monosaccharides glucose and galactose. However, when the function of the gene is interrupted by the insert gene fragment, *lacZ* is incapable of producing functional *lacZ* to deliver its function. X-gal is the lactose analogue that has been specially designed to produce a colour change when interacting with the β-galactosidase enzyme.

Principle

The protein β-galactosidase encoded by *lacZ* in the lac operon is a homotetramer. β-galacosidase catalyses β-galactosides into monosaccharides by hydrolysing the β-glycosidic bond formed between galactose and its organic moiety. α-complementation exhibited by this enzyme is responsible for the basis of the blue-white selection. The enzyme splits into the two peptides *LacZα* and *LacZΩ* which, when resembled together, form the functional enzyme (Fig. 2.13).

The blue-white screening technique takes advantage of the disruption of the α-complementation process. The plasmids carrying the *lacZα* sequence internal to MCSs are digested by the restriction enzyme to accommodate the insert DNA. Thus, the gene gets disrupted and the production of α-peptide gets disrupted. Finally, the cells containing the plasmid with the insert cannot form a functional β-galactosidase. The presence of the active β-glactosidase is detected in vitro by a colourless analogue of lactose, the X-gal which is cleaved to form 5-bromo-4-chloro-indoxyl. 5-bromo-4-chloro-indoxyl dimerizes and oxidises to form 5,5′-dibromo-4,4′-dichloro-indigo a bright blue insoluble pigment. Thus, the blue-coloured colonies harbour the functional β-galactosidase gene confirming the presence of the vector with the uninterrupted *lacZα*. In contrary to that, the white colonies confirm that the X-gal is not hydrolysed, and thus the presence of the insert sequence in *lacZα* disrupts the formation of the active β-galactosidase (Fig. 2.14).

The greatest advantage of using the pUC series of plasmid vectors is the presence of concentrated restriction sites in one region called multiple cloning sites (MCSs). Additionally, a part of the MCSs code for the β-galactosidase polypeptide. During the use of pUC plasmid in host *E. coli*, the gene is switched on by the addition of IPTG that acts as an inducer. Subsequently, the functional enzyme hydrolyses the colourless substrate called X-gal to blue insoluble material to develop

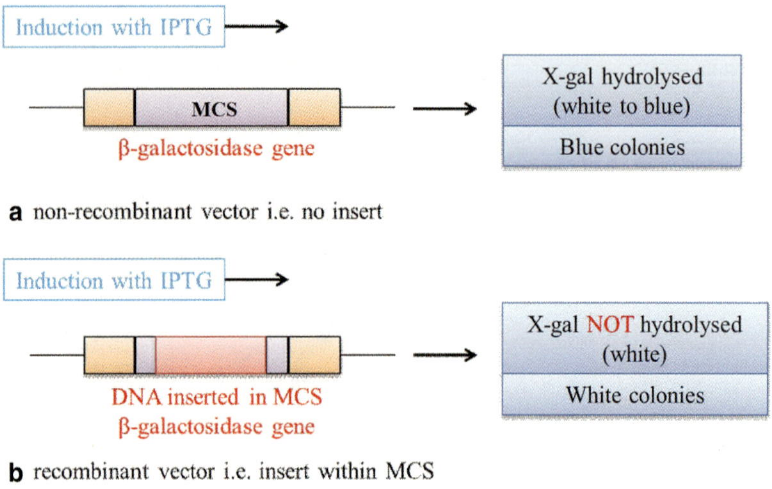

Fig. 2.14 Principle of blue-white selection for the detection of recombinant vectors

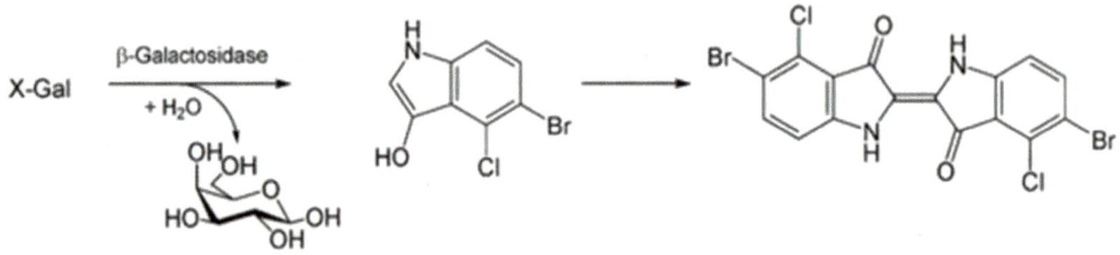

Fig. 2.15 Structure of X-gal and the end products of β-galactosidase enzyme action

blue-coloured colonies. This elegant system allows the initial identification of recombinants quickly present in a number of vector systems. However, this selection method and insertional inactivation of antibiotic resistant genes do not provide information on the character and proper orientation of the DNA insert; it only provides information on the status of the vector. The limitation of this screening technique is that it can be applied to certain selected host systems, e.g. XL1-Blue, XL2-Blue, DH5α F′, DH10B, JM101, JM109 and STBL4.

Reagents Required and Their Role

X-gal

X-gal, the 5-bromo-4-chloro-3-indolyl-β-D-galactopyranoside (BCIG), consists of galactose and indole and yields insoluble blue compounds due to enzyme-catalysed hydrolysis. It is an analogue of lactose and hence can be cleaved by β-galactosidase to yield galactose and 5-bromo-4-chloro-3-hydroxyindole. It is used in most of the molecular cloning experiments for easy identification of the active enzyme coded by the *lacZ* gene. The reaction during reporter gene assay for *lacZ* using X-gal is given in Fig. 2.15.

X-gal is light sensitive and is not soluble in water. It should be wrapped in aluminium foil to protect it from light and can be dissolved in DMSO. The stock solution of X-gal required for the purpose of blue-white selection is 20 mg/ml.

IPTG

Isopropyl β-D-1-thiogalactopyranoside is the molecule that mimics allolactose which is capable of inducing the transcription of the *lac* operon in *E. coli*. As IPTG is not hydrolysed by β-galactosidase, its concentration throughout the experiment remains constant. The effective concentration of IPTG to be used in vitro for the *lac* operon induction ranges from 100 μm to 1.5 mM. However, the concentration of IPTG to be used depends upon the intensity of induction required during the process.

Antibiotics

A suitable antibiotic selection marker should be used to select the appropriate colony of transformants. In addition to IPTG and X-gal, antibiotics allow the correct selection of a transformant by limiting the growth of non-transformants under selection pressure. Use of antibiotics recognises the protein produced by the particular gene. When the petri plate contains antibiotics like ampicillin or kanamycin, the bacteria which are resistant to the same can grow, and those which do not harbour the resistance genotype cannot grow.

pBluescript

pBluescript is a commercially available vector plasmid that includes an MCS and ampicillin resistance marker. The MCS of the vector is present within the *lacZ* gene and hence develops blue colourization when expressed in bacteria. This can be achieved by the induction from IPTG and X-gal. During blue-white selection, the plasmid can be used as a control set of the experiment that gives blue colourization in plates supplemented with ampicillin, IPTG and X-gal.

Transformation Reaction Product

The transformation reaction product from Exp. 2.3 can be used for this experiment. The product includes the competent cell, the competent cell with vector alone and the competent cell with vector and the insert sequence. This screening technique allows the confirmation of transformation and correct cloning practice.

Procedure

1. Prepare LB agar medium plates using ampicillin as selection marker.
2. The final concentration of ampicillin to be added in media should be 50 μg/ml. In order to prepare 100 ml of medium, 50 μl of 100-mg/ml stock solution of ampicillin can be added to the same.
3. Spread 50 μl of 10-mM IPTG stock solution on the solidified medium. The final concentration of IPTG in the medium should range from 0.1 to 0.5 mM.
4. Spread 100 μl of 20-mg/ml stock solution of X-gal on the solidified medium. The final concentration of X-gal in the medium should be 20 μg/ml.
5. Incubate the plates in laminar air flow for a few minutes for proper drying of the plates.
6. Store the plates at 4 °C by aluminium foil wrapping as X-gal is light sensitive.
7. Spread 50–100 μl of the transformation reaction product onto each plate.
8. In the control plate, spread 50 μl of the pBluescript vector alone.
9. Incubate the plates for 24 h at 37 °C and observe for the development of blue- or white-coloured colonies.

Observation

The blue-white screening technique is not a selection technique as it does not kill the unwanted bacteria; hence, it is a screening technique. Thus, the following observation should be made from the colour of grown bacterial cultures on Luria agar supplemented with IPTG and X-gal.

Conditions	Observations
Untransformed bacteria	No growth on ampicillin plates
Bacteria transformed with vector only	Blue colonies
Bacteria transformed with vector + insert DNA	White colonies
Control set	Blue colonies

Troubleshooting

Problems	Possible cause	Possible solution
All white colonies but no insert	IPTG not used properly	Spread 30 μl of 100-mM IPTG solution on surface of plates
	X-gal does not diffuse the plates properly	Spread 50 μl of 2% X-gal solution on the surface of the plate
	Plates not stored properly	Store plates with X-gal at 4 °C in dark
		Do not use plates with X-gal in agar which are more than 4 months old
	Colour may not have developed fully	Incubate the plates for a longer period of time
	Vector may not allow for α-complementation	Check the vector for α-peptide of the *lacZ* gene
All blue colonies with recombinant DNA	Insert cloned in frame with α-peptide	Perform other screening techniques like PCR to confirm transformation
	E. coli used possess intact β-*gal* gene	Use a proper *E. coli* host that contains *lacZ6M15*, partial deletion of *lacZ* that allows α-complementation
Any problem with blue-white screening	Incorrect amount of X-gal and/or IPTG in agar plates	Check for the addition of correct amount of IPTG and X-gal to the plates
Satellite colonies	Degraded or expired antibiotic	Check for the antibiotic expiry; avoid multiple freeze–thaw cycle
	Too many colonies on plates	Plate from a higher dilution
	Plates were incubated for a much longer time	Observe results after overnight at 37°C
	Ampicillin used	You can use carbenicillin instead of ampicillin to reduce satellite colonies

Precautions

1. Store all the reagents at −20 °C and thaw on ice before starting the assay.
2. Do not store the reagents for more than 6 months at −20 °C.
3. Always use DMSO or dimethylformamide (DMF) for dissolving X-gal. Do not mix X-gal in water to prepare the stock solution.
4. Avoid the exposure of blue-white screening reagents to light as it may cause degradation of X-gal to form some yellow impurities.
5. In-frame insertion of small DNA fragments into the β-galactosidase site does not interfere with its activity that results in the formation of blue or light blue recombinant plasmids bearing bacterial colonies.
6. Always use gloves throughout the experiment.
7. Do not get exposed to IPTG and X-gal as they are harmful.

FLOW CHART

Prepare IPTG, X-gal, ampicillin plates by adding at a final concentration of 20 μg/ml, 0.1-0.5 mM and 50 μg/ml respectively

↓

Store the plates at 4°C under dark conditions till further use

↓

Spread 100 μl of the transformant solution to the plates

↓

Incubate the plates at 37°C for overnight

↓

Check for the presence of blue and/or white colonies

↓

Confirm the presence of target gene fragment in the white coloured colonies by PCR

↓

Use these white colonies for further down-stream applications as positive transformants with insert DNA

Exp. 2.6 Confirmation of Cloning by PCR

Objective To confirm cloning in the positive transformants by polymerase chain reaction (PCR).

Introduction

During cloning and transformation techniques, it is of upmost importance to select the correct cloning product harbouring the target gene of the insert for further downstream applications. There are many techniques available nowadays for the screening and selection of positive transformants like blue-white selections, selection under antibiotic stress and many more; however, all of them are loaded with lacunae and each of them have their own flaws. There are many disadvantages of using the blue-white screening technique such as the vector used for generating recombinant strains may not contain a functional *lacZ* gene which in turn results in a non-functional β-galactosidase. Thus, the cells are not able to produce blue colour and many times give false positive results. These false positive clones are not due to the recombinants but may be due to just the background vector. In some cases, white colonies do not contain the exact DNA fragment but another small piece of DNA, generated during improper restriction digestion, ligates into the MCSs of the vector and disrupts the *lacZα* fragment. Ultimately, it prevents the expression of the *lacZ* gene and gives rise to false positive results.

In addition, few linearized vectors after transformation into the host strain may self-ligate together. Thus, no *lacZ* is produced, and ultimately, they cannot convert X-gal to the blue end product. On the contrary, certain blue-coloured colonies may also contain the insert DNA fragment. When the insert is in frame with the *lacZα* gene and is devoid of a stop codon, it leads to the expression of a fusion protein and is still functional as *LacZα*. Lastly, the major disadvantage of this technique is the use of X-gal which is highly expensive, unstable and cumbersome to use. Hence, the PCR cloning method is regarded as a suitable

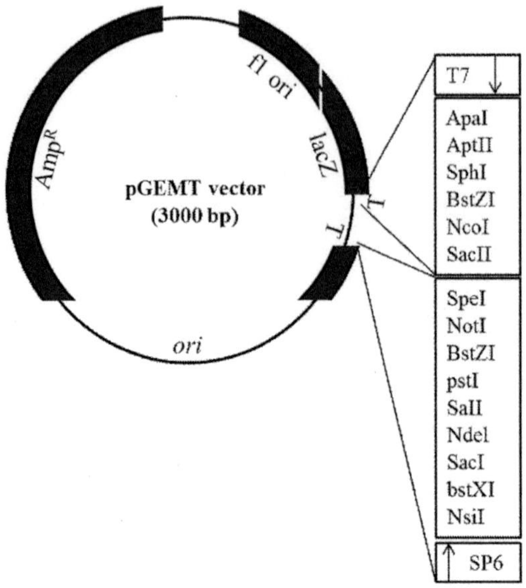

Fig. 2.16 pGEMT vector map and sequence reference points for use of universal primers

efficient alternative for identification of positive clones with the desired DNA insert.

Principle

The insert DNA sequences after suitable restriction digestion and/or simple PCR cloning bind to the MCSs of the designed vector. The vectors have been designed in such a way that they contain certain universal primer recognition sequences to amplify the insert sequence. Hence, it is quite useful for the confirmation of the presence of the target gene sequence in the vector. For example, in the TOPO TA cloning vector, T7 and M13 primers are used as designed by the manufacturer for amplification of the insert DNA in the vector. Similarly, in case of the pGEMT vector, T7 and SP6 primer sets are used for amplification of the insert sequence.

The target region of the designed primers is present in the upstream and downstream sequences of the MCSs of the vector. Thus, whatever may be inserted in the MCSs of the vector DNA, they give rise to the amplified product with the almost exact size of the target DNA insert (Fig. 2.16). Thus, the amplified product may be further sequenced, and, based upon the size, correct cloning can be confirmed after transformation of the exact piece of the DNA fragment into the host cells.

Nowadays many vectors have been designed with suitable primer sequences for amplification of cloned fragments for confirmation of cloning experiments. An extensive list of the available DNA vectors and their possible universal primers for confirmation of cloning by PCR is provided in Table 2.7.

In addition to the universal primers, gene-specific primers can be used for the amplification of the target gene sequences that have been cloned into the vector system. This technique requires the purification of plasmids from the host cells, and the plasmids can be used as the template DNA for the amplification of the DNA fragment of our interest by using the gene-specific primers. For example, if we have cloned the *merA* gene fragment in the TOPO vector in *E. coli* cells, the plasmid/vector can be isolated from the host cell and subsequently the *merA* gene can be amplified using the gene-specific primers, i.e. merAF (5′ TCGTGATGTTCGACCGCT 3′) and merAR (5′ TACTCCCGCCGTTTCCAAT 3′). The DNA fragment amplified with the gene-specific primers confirms the gene product to be cloned successfully into the host cell to carry out further downstream applications.

Reagents Required and Their Role

Plasmid DNA Template

The template contains the plasmid DNA that can be isolated from the host cells by using a protocol as described in Exp. 1.3. However, a proper concentration of template DNA should be used before going for PCR amplification. Approximately 10^4 copies of high-quality target DNA are required to detect the product in 25–30 cycles. Of plasmid or viral templates, 1 pg–1 ng may be used.

Forward and Reverse Primers

As required, the forward and reverse primers can be used for the amplification of the cloned gene

Table 2.7 List of universal primers and their sequences to be used for confirmation of cloning

Vector	Primers	Primer sequences (5′-3′)	Manufacturer
pGEMT	T7	TAATACGACTCACTATAGGG	Promega, USA
	SP6	ATTTAGGTGACACTATAG	
TOPO cloning vector	T7	TAATACGACTCACTATAGGG	Invitrogen, USA
	M13	CAGGAAACAGCTATGACC	
pETM14-ccdB	LP1rev	GGGCCCCTGGAACAGAACTT	European Molecular Biology Laboratory, Germany
	LP2for	CGCCATTAACCTGATGTTCTGGGG	
pDONR P4-P1r	B4	GGGGACAACTTTGTATAGAAAA GTTG	Invitrogen, USA
	B1r	GGGACTGCTTTTTTGTACAAACT TG	
pDONR 221/207	B1	GGGGACAAGTTTGTACAAAAAA GCAGGC T	Invitrogen, USA
	B2	GGGGACCACTTTGTACAAGAAAGC TGGGT	
Pichia	3′ AOX1	GCAAATGGCATTCTGACATCC	Invitrogen, USA
	5′ AOX1	GACTGGTTCCAATTGACAAGC	
pAc5.1/V5-His A, B, and C	Ac5 forward	ACACAAAGCCGCTCCATCAG	
	a-Factor	TACTATTGCCAGCATTGCTGC	
Pichia methanolica expression vector	AUG1 forward	CAATTTACATCTTTATTTATTAACG	
	AUG1 reverse	GAAGAGAAAAACATTAGTTGGC	
pBlueBac4.5	cI forward	GGATAGCGGTCAGGTGTT	
	BGH reverse	TAGAAGGCACAGTCGAGG	
pMT/V5-His	MT forward	CATCTCAGTGCAACTAAA	
	M13/pUC reverse	AGCGGATAACAATTTCACACAAGG	
pUni	pUni forward	CTATCAACAGGTTGAACTG	
	pUni reverse	CAGTCGAGGCTGATAGCGAGCT	

fragment. The target gene can be amplified either by using universal primers which are specific for the vector system used or the gene-specific primer against the target gene.

Taq Polymerase

This enzyme is able to withstand the protein denaturing conditions required during PCR besides carrying out the extension of DNA fragments. It has the optimum temperature for its activity between 75 and 80 °C, with a half-life >2 h at 92.5 °C, 40 min at 95 °C and 9 min at 97.5 °C and possesses the capability of replicating a 1000 bp of DNA sequence in less than 10s at 72 °C. However, the major drawback of using Taq polymerase is the low replication fidelity as it lacks 3′-5′ exonuclease proofreading activity. It also produces DNA products having ‘A’ overhangs at their 3′ ends which is ultimately useful during TA cloning. In general, 0.5–2.0 units of Taq polymerase are used in 50 µl of total reaction, but ideally 1.25 units should be used.

Deoxynucleotide Triphosphate

Deoxynucleotide triphosphates (dNTPs) are the building blocks of new DNA strands. In most of the cases, they come as a mixture of four deoxynucleotides, i.e. dATP, dTTP, dGTP and dCTP. Around 100 µM of each of the dNTPs is required per PCR reaction. dNTP stocks are very sensitive to cycles of thawing and freezing, and after 3–5 cycles, PCR reactions do not work well. To avoid such problems, small aliquots (2–5 µl) lasting for only a couple of reactions can be made and kept frozen at −20 °C. However, during long-term freezing, small amounts of water evaporate on the walls of the vial, thus changing the concentration of the dNTP solution. Hence, before using, it is essential to centrifuge the vials, and it is always recommended to dilute the dNTPs in TE buffer as acidic pH promotes hydrolysis of dNTPs and interfering with the PCR result.

Buffer Solution

Every enzyme needs certain conditions in terms of their pH, ionic strength, cofactors, etc., which is achieved by the addition of buffer to the reaction mixture. In some instances, the enzyme shift pH in non-buffered solution and stops working in this process which can be avoided by the addition of a PCR buffer. In most of the PCR buffers the composition is almost the same as 100 mM Tris-HCl, pH 8.3, 500 mM KCl, 15 mM $MgCl_2$ and 0.01 % (w/v) gelatine. The final concentration of the PCR buffer should be 1X concentration per reaction.

Divalent Cation

The mechanism of DNA polymerase requires the presence of divalent cations. Most essentially, they shield the negative charge of the triphosphate and allow the hydroxyl oxygen of the 3′ carbon to attack the phosphorus of the alpha phosphate group attached to the 5′ carbon of the incoming nucleotide. All enzymes that break the phosphoanhydride bonds of nucleoside di- and tri-phosphates require the presence of divalent cations. Of $MgCl_2$, 1.5–2.0 mM are optimal for the activity of Taq DNA polymerase. If Mg^{2+} will be too low, no PCR product will be visible whereas if Mg^{2+} is too high, an undesired PCR product may be obtained.

Procedure

1. Isolate the plasmid DNA from the white-coloured colonies according to the protocol described in Exp. 1.3.
2. Optionally you can perform the PCR by taking the template DNA from bacterial lysates.
3. Mix the template DNA, primers, dNTPs and Taq polymerase in a 0.2-ml thin-walled micro-centrifuge tube in the following order from a stock solution of [Buffer: 10X, $MgCl_2$: 15 mM, dNTPs: 10 mM, Primer: 1 mM, Taq polymerase: 1 U/μl, Template: 10 μg/μl]

Sterile milli-Q water	12.5 μl
10X reaction buffer	2.5 μl
$MgCl_2$	0.5 μl
dNTPs mixture	2.0 μl
Primer (forward)	2.0 μl
Primer (reverse)	2.0 μl
Template DNA	2.5 μl
Taq DNA polymerase	1.0 μl

4. As pipetting of small volumes is difficult and sometimes may be inaccurate, a master mixture can be prepared instead where constituents common to all the reactions are combined in one tube multiplying the volume for one reaction with the total number of samples. Later, the appropriate amount of the master mixture is aliquoted to each tube, and finally template DNA is added to each tube individually.
5. Place the tubes row wise in each well of the PCR machine.
6. Carry out the amplification with the following programme:

LID	98 °C
Initial denaturation	94 °C for 5 min
30 cycles	
Denaturation	94 °C for 1 min
Annealing	54 °C for 1 min
Extension	72 °C for 1 min
Final extension	72 °C for 4 min
Hold	4 °C for ∞

7. At the end of the PCR cycle, take out the PCR products and run the samples by agarose gel electrophoresis in 1 % agarose gel with ethidium bromide and visualise under UV light.

Observation

Run the amplified DNA product in agarose gel to confirm the amplification along with DNA ladder. As the size of the insert DNA fragment is known, it can be confirmed for the presence of the suitable fragment at the desired position on the gel. The presence of the same size of DNA fragment in the gel confirms the cloning of the desired piece of DNA into the host system. The concentration of the amplified product can be measured by UV spectrophotometer using a nano-drop. Measurement of the OD at 260 and 280 nm will show the quality and quantity of the amplified product of the 16S rRNA gene.

Troubleshooting

Problems	Possible cause	Possible solutions
Size of the amplified product of a higher size	Amplified with the universal primer	Amplify the template with the gene-specific primer
The sequencing product gives a lesser sequence size than the exact size of the gene fragment	Sequencing reaction is carried out with the gene-specific primers	Repeat the sequencing reaction with universal primer; you will get the exact length of sequences as of the target gene
No PCR product	Incorrect primer	Use appropriate primer, either the universal primer specific for the vector or the gene-specific primer
	Incorrect annealing temperature	Follow the manufacturer's instruction for the correct annealing temperature
Multiple or non-specific bands	Incorrect annealing temperature	Increase the annealing temperature or follow the manufacturer's instruction
Incorrect product size	Incorrect annealing temperature	Use the appropriate annealing temperature for both universal primers as well as gene-specific primers
	Improper concentrations of the PCR reagents	Optimize the concentrations of PCR constituents by following the manufacturer's guidelines

PCR polymerase chain reaction

Precautions

1. Use pipette tips with filters.
2. Store materials and reagents properly under separate conditions and add them to the reaction mixture in a spatially separated facility.
3. Thaw all components thoroughly at room temperature before starting an assay.
4. After thawing, mix the components with brief centrifugation.
5. Work quickly on ice or in the cooling block.
6. Always wear safety goggles and gloves while performing the PCR reaction.

FLOW CHART

Take the white colonies from the IPTG+X-gal+Amp plates and isolate plasmid from the colonies

↓

Thaw all the reagents on ice before set up of PCR

↓

Add all the reagents of required amount as mentioned for master mixture except template

↓

Distribute the master mixture to the individual PCR tube and add corresponding templates

↓

Set up the PCR conditions as mentioned in the procedure

↓

Take out the samples from the PCR machine and store at 4°C

↓

Run in agarose gel electrophoresis for checking the correct amplification

Advanced Molecular Microbiology Techniques

3

Exp. 3.1. Synthesis of cDNA

Objective To synthesise cDNA from the isolated total RNA of *Escherichia coli.*

Introduction

Complimentary DNA (cDNA) is the DNA synthesised from messenger RNA (mRNA). The reaction is catalysed by two enzymes; reverse transcriptase as well as DNA polymerase. As per the central dogma of life, DNA is transcribed to mRNA, which is subsequently translated to proteins to carryout cell's functions. The major difference between the prokaryotic and eukaryotic mRNA is the presence or absence of intron sequences among the exons. During transcription of eukaryotic mRNA, intron sequences are spliced-off, which can further be translated to the amino acids to carry out their desired function. As prokaryotic genes do not contain introns, their RNA is not subjected to cutting and splicing. Thus, complementary DNA is used for gene cloning or as gene probes or in the creation of cDNA libraries. The widest application of cDNA is to monitor the expression level of specific functional genes by using quantitative real-time PCR (qRT-PCR) with cDNA as template.

cDNA can be defined as the double-stranded DNA (dsDNA) version of an mRNA molecule. In most of the cases, cDNA is prepared from mRNA by the use of a special enzyme called reverse transcriptase, which was originally isolated from retrovirus. Reverse transcriptase synthesises single-stranded DNA templates from mRNA, which can be further used as template for dsDNA synthesis. There are many uses of cDNA in molecular biology studies which include gene cloning, gene probes, cDNA libraries or gene expression. Sequence-specific primers can be designed for hybridisation in 5′-3′ ends of cDNA fragment that code for a protein. After amplification, both ends of the amplified product can be cut using nucleases and can be inserted into the expression vectors. As the expression vectors can readily go for self-replication inside the host cell, proteins will be translated and the expression pattern of the gene of interest can be studied.

Principle

With advancement in the field of genetic engineering, gene expression pattern analysis has become an indispensable tool to find out whether the gene of interest is turned on or off. For this purpose, mRNA is located, quantified and can be coded by DNA for translation to produce respective proteins. Due to the unstable and fragile nature, RNA is readily degraded by omnipresent RNase. To overcome such problems, cDNA is prepared from RNA that contains the entire sequence of mRNA, which is further used for subsequent studies. However, many primers can be used as per the requirement of cDNA synthesis,

S. Das, H. R. Dash, *Microbial Biotechnology- A Laboratory Manual for Bacterial Systems,*
DOI 10.1007/978-81-322-2095-4_3, © Springer India 2015

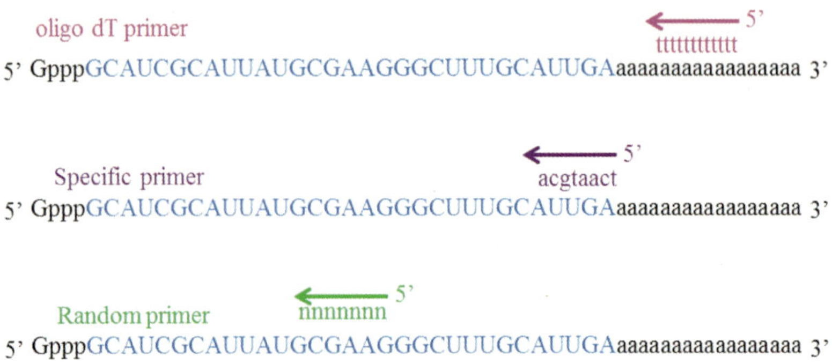

Fig. 3.1 Use of three different types of primers for the preparation of cDNA from RNA

i.e. oligo-dT, sequence-specific primer or random primer. Oligo-dT primers are used when the mRNAs have a poly-A tail mostly in case of eukaryotic RNAs. In contrast, if it is required to generate a particular cDNA fragment from a pool of RNAs, sequence-specific primers can be used that binds to the specific RNA sequence only. To design sequence specific primers, the sequence of mRNA of interest should be known; most preferably the 3′-terminal sequence. For shorter DNA fragments with unknown sequence, a random cocktail of primers can be used. The advantage of using random primers is the increased probability of converting the entire 5′ end of the RNA to cDNA. Hence, random primers are found to be extremely useful in this regard (Fig. 3.1).

cDNA synthesis relies on the rapid and reliable synthesis of dsDNA from mRNA. In many cases, cloned Moloney murine leukemia virus (M-MLV) reverse transcriptase is used to synthesise first strand cDNA from the mRNA (Zhu et al. 2001). The single reaction system allows the synthesis of double-stranded cDNA immediately after the synthesis of the first strand which can minimise the extraction and precipitation between steps (Fig. 3.2).

The enzyme reverse transcriptase plays a key role in the synthesis of cDNA from the mRNA pool. Under in vivo conditions, this enzyme is responsible for the transcription of viral RNA to form dsDNA to be inserted into the host genome. In this way it is vital for viral replication, however, when used in bacterial and other eukaryotic systems reverse transcriptase is capable of producing cDNA. The major functions of reverse transcriptase include its complex and multifunctional roles. It acts as an RNA-dependent DNA polymerase that transcribes single-stranded DNA from RNA (Fig. 3.3). In this way it demonstrates the activity of ribonuclease H (RNaseH), a subunit of reverse transcriptase enzyme. Further, it acts as DNA-dependent polymerase to transcribe the second DNA strand complementary to the first strand of DNA. However, reverse transcriptase is central to the infectious nature of many viruses that include HIV and HTLV-1. The most important application of reverse transcriptase includes the reverse transcription polymerase chain reaction (RT-PCR), a powerful tool for the detection of diseases such as cancer and to study the expression level of many other genes in both prokaryotic and eukaryotic systems under various environmental conditions.

To produce DNA copies from RNA templates by using reverse transcriptase enzyme is well-documented. This method involves the synthesis of cDNA from mRNA by forming a short double-stranded region which is provided by an oligo (dT) primer against poly (A) RNA. Though this enzyme always does not produce full-length transcripts but all the complimentary strands are finished-off by a short hair pin loop. This ultimately provides a ready-made primer for second strand synthesis, which can be performed by reverse transcriptase or by DNA polymerase. Prior to application in other downstream applications the hair pin loops can be removed using single-strand-specific nuclease S1.

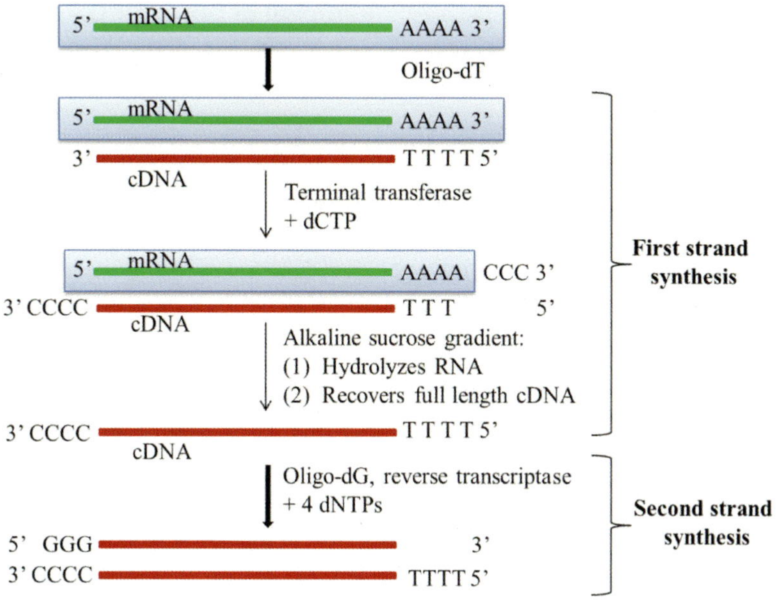

Fig. 3.2 First and second strand cDNA synthesis from RNA by reverse transcriptase

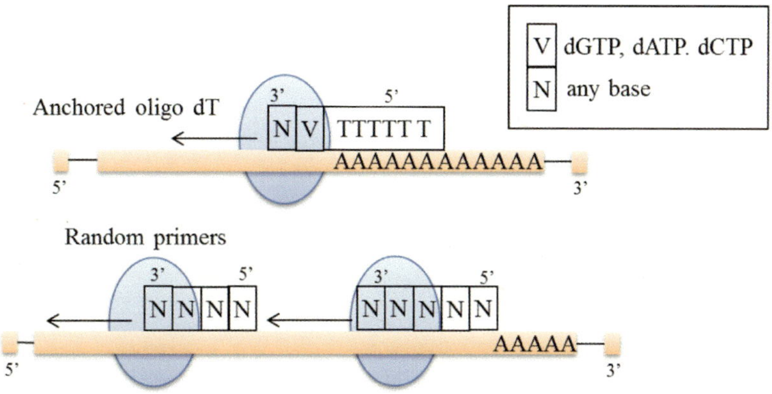

Fig. 3.3 Synthesis of cDNA from mRNA by reverse transcriptase

Reagents Required and Their Role

The current experiment aimed to prepare cDNA for *czrCBA* gene from isolated total RNA. The protocol describes the use of Thermo Scientific RevertAid first strand cDNA synthesis kit.

Reverse Transcriptase

It is the enzyme used to synthesise cDNA from mRNA templates through reverse transcription. Reverse transcriptase isolated from retroviruses may play as RNA-dependent DNA polymerase, ribonuclease H and DNA-dependent DNA polymerase. The most widely studied and used reverse transcriptases have been isolated from

HIV1, M-MLV and AMV. In contrast to the classical PCR techniques, reverse transcriptase can be used to synthesise DNA from the RNA template.

RiboLock RNase Inhibitor

Ribonuclease inhibitor (RI) are the 50 kDa cytosolic proteins rendering them to be inactive. The complex formed between RI and target ribonucleases are one of the strongest known biological interactions. These compounds have high-cysteine content and are sensitive to oxidation which consists of an alternating α-helix and β-helices as its backbone. Despite having low-sequence affinity, RI is able to bind wide varieties of RNases. However, their affinity to RNA renders them highly cytotoxic and cytostatic effect and hence should be handled carefully under laboratory conditions.

Random Hexamers/Oligo dT

These are the oligonucleotide sequences of six bases that are synthesised randomly to give rise to numerous ranges of sequences having potential to anneal at random points of DNA sequence. This acts as a primer for first strand cDNA synthesis. This primer is mainly the synthetic single-stranded 18-mer oligonucleotide with 5′ and 3′ hydroxyl ends.

Reaction Buffer

The buffer keeps all the components in the reaction mixture in their stable form to perform correctly. The buffer contains dNTPs as well as the cations responsible for DNA synthesis under optimum conditions, i.e. K^+ and NH_4^+. However, the exact composition of the buffer differs from one manufacturer to another.

Procedure

1. Thaw all the reagents on ice. After proper thawing, briefly centrifuge the components.

Removal of Genomic DNA from RNA

1. Add the reagents in the following order into an RNase-free tube:

RNA	1 l (1 µg/µl)
10X reaction buffer with $MgCl^2$	1 µl
DNase I	1 µl
Nuclease-free water	7 µl

2. Incubate at 37 °C for 30 min.
3. Add 1 µl of 50 mM ethylenediaminetetra-acetic acid (EDTA) (mix 0.731 g of EDTA in 50 ml of milli-Q water to prepare 50 mM of EDTA) and incubate at 65 °C for 10 min. Alternatively, phenol: chloroform extraction method can be used (see the Exp. 1.4).
4. Use this prepared RNA as template for the action of reverse transcriptase.

First Strand cDNA Synthesis

1. After thawing, mix and briefly centrifuge the components and keep them on ice.
2. Add the following reagents to a sterile, nuclease-free 200 µl tube in the following order:

Total RNA	1 µl (0.1–5.0 ng/µl)
Random hexamer primer	1 µl
Nuclease-free water	10 µl

3. If the RNA is GC rich, mix gently, briefly centrifuge and incubate at 65 °C for 5 min.
4. Add the following components in the order given below:

5X reaction buffer	4 µl
RNase inhibitor (20 U/µl)	1 µl
10 mM dNTPs mix	2 µl
Reverse transcriptase (200 U/µl)	1 µl
Nuclease-free water	12 µl

5. Mix gently and centrifuge.
6. For oligodT or gene-specific primers for cDNA synthesis, incubate at 42 °C for 60 min.
7. Terminate the reaction by heating at 70 °C for 5 min.

8. The reaction product can be directly used for PCR amplifications or can be stored at −20 °C for less than 1 week or for long-term storage keep at −80 °C.

PCR Amplification of First Strand cDNA

1. Use the first strand cDNA synthesis product directly for PCR or qRT-PCR purpose.
2. Use the reaction as given below for the amplification of *czrCBA* gene from the synthesised cDNA.
3. Add the following components as indicated below:

Sterile milli-Q water	12.5 µl
10X reaction buffer	2.5 µl
$MgCl^2$	2.5 µl
dNTPs mixture	0.5 µl
czrCBA F	1.0 µl
czrCBA R	1.0 µl
Template DNA	4.0 µl
Taq DNA polymerase	1.0 µl

4. Carry out the amplification with the following programme:

LID	98 °C
Initial denaturation	96 °C for 5 min
30 cycles	
Denaturation	95 °C for 15 s
Annealing	49 °C for 30 s
Extension	72 °C for 1 min
Final extension	72 °C for 10 min
Hold	4 °C for ∞

5. At the end of the PCR cycle, take out the PCR product and run in agarose gel electrophoresis in 1 % agarose gel with ethidium bromide and visualise under UV light.
6. Store the product at −20 °C till further use of quantitative analysis of the gene by qRT-PCR.

Observation

Observe carefully the gel profile of double-stranded cDNA synthesised during the experiment. The number and position of cDNA on the gel differ with respect to the RNA used during the study. When gene-specific primers are used for cDNA preparation, observe carefully for the exact amplicon size that corresponds to the size of the gene of interest.

Trouble-Shootings

Problems	Possible cause	Possible solutions
Poor yield, truncated PCR product from RNA	Organism specific band may be present	Electrophorese the samples on a formaldehyde/agarose/EtBr gel to determine the concentration and analyse its quality
	Concentration of RNA is low, but the quality is good	Repeat the experiment with more RNA and more PCR cycles
	RNA is degraded partially before or during first strand cDNA synthesis	Repeat the experiment with fresh RNA
	Quality of RNA is not good, contains impurities that inhibit cDNA synthesis	To remove impurities, wash twice with EtOH or isolate the RNA using different technique
No amplification product	Error in procedure	Include positive control with appropriate primers
	Degraded RNA sample	Check the RNA integrity by formaldehyde agarose gel electrophoresis
	RNA concentration too low	Increase the amount of RNA sample
	Too high ratio of random primer: RNA	Optimise the ratio
Unexpected bands	Genomic DNA contamination	Treat the prepared RNA with DNase
	Non-specific primer	Increase annealing temperature, Decrease primer and RNA concentration, Optionally you can check non-specific priming by 3′ primer without addition of 5′ primer
	Degraded RNA	Check the quality of RNA on gel, if possible prepare a fresh RNA sample
	Carry over from the other reactions	Be careful to avoid any chance of contaminations

Precautions

1. As RNA is very unstable in nature, always keep the samples on ice.
2. Keep the samples out of direct sunlight always.
3. Never vortex RNA.
4. Always use gloves while handling RNA, change gloves frequently and spray the gloves with 70 % alcohol.
5. Always work quickly and keep RNA on ice to avoid degradation.
6. Use RNaseZap, the solution that removes RNase contamination completely by cleaning the equipment and the work bench.
7. Follow proper microbiological aseptic techniques.
8. Handle tube as less as possible and always avoid touching the inside of the lid and keep tubes closed when not in use.
9. Always use RNase-free tubes and tips, be cautious to keep them sterile once the package is opened.
10. Store the RNA samples away from amplified PCR products and plasmids.

FLOW CHART

Thaw all the reagents on ice, briefly centrifuge the components

↓

Remove as much genomic DNA contamination as possible by adding 1µg of RNA, 1 µl of 10X buffer, 1 µl of DNaseI, and nuclease free water, incubate at 37°C 30 min

↓

Use this prepared RNA as template for the action of reverse transcriptase

↓

Add 0.1-5.0 ng total RNA, 1 µl of random hexamer primer and nuclease free water to a total volume of 12 µl.

↓

Mix gently and centrifuge, incubate at 42°C for 60 min, terminate the reaction by heating at 70°C for 5 min

↓

Use first strand cDNA synthesis product for second strand synthesis of *czrCBA* gene

↓

Add the following components: sterile milli-Q 12.5 µl, 10X buffer- 2.5 µl, MgCl2- 2.5 µl, dNTPs mixture- 0.5 µl, czrCBA F and R primer- 1.0µl, template DNA- 4.0 µl and 1.0 µl of Taq polymerase

↓

Use PCR programme: 96°C for 5 min, 30 cycles at 95°C for 15 s, 49°C for 30 sec, 72°C for 1 min followed by Final extension of 72°C for 10 min and hold at 4°C for ∞

↓

Run the PCR product in 1% agarose gel to confirm the amplification of the desired gene of interest

↓

Store the product at -20°C till further analysis of qRT-PCR

Exp. 3.2. Gene Expression Analysis by qRT-PCR

Objective To detect the level of *czrCBA* gene expression from isolated total RNA from bacterial cell.

Introduction

The technique used for simultaneous amplification and quantification of target DNA is termed as quantitative polymerase reaction or real-time PCR. It is based upon the principle of PCR and the quantification may be the absolute number of copies or the relative amount after normalisation with a standard DNA copy. There has been much advancement since its discovery in terms of using DNA-intercalating fluorescent dyes or use of sequence-specific DNA probes labelled with fluorescent reporter. Real-time PCR is also termed as kinetic PCR as this technique is regarded as the most sensitive technique to quantify the gene targets compared to other commonly available techniques like northern and southern blotting techniques. This technique is very sensitive to detect the level of RNA from a single cell.

Cells transcribe to RNA and regulate the level of gene expression by gene transcripts. Hence, the number of expressed genes in a cell can be measured by the number of copies of the mRNA transcript of that particular gene present in the sample. PCR employs the formation of cDNA from RNA by reverse transcriptase, the details of which has been described in the previous chapter (Exp. 3.1). The target DNA can be amplified from the synthesised cDNA and the quantitative detection of DNA samples can be done after each cycle.

qRT-PCR and DNA microarray are the most advanced methodologies for gene expression analysis. They have many advantages over other older methods of differential display, RNase protection assay and northern blot. It can be used to quantify DNA in both relative and absolute terms. Hence, it can also be applied to quantify the expressed mRNA level from endogenous genes as well as transfected genes from either stable or transient transfection. To date, it is the most sensitive technique for quantitative analysis of the expression level of mRNA. There are numerous applications of qRT-PCR in laboratory for diagnostic as well as research purposes that include diagnosis of infectious diseases, cancer and abnormalities in genes. In microbiology experiments, qRT-PCR is mainly used in the field of food safety, food spoilage and fermentation as well as for the microbiological risk assessment of drinking water quality. In bioremediation studies, qRT-PCR is a useful technique to detect the expression level of the genes responsible for the bioremediation of the toxic pollutants.

Principle

In quantitative PCR, the amplified DNA product of the target sequence is linked to the fluorescent intensity using a fluorescent reporter molecule. There are two ways to calculate the quantity of initial template, i.e. at the end of each reaction (end-point qPCR) or while the amplification is still progressing (real-time qPCR). In this regard, the use of a reporter molecule is of utmost importance which may be (i) the sequence-specific probe composed of oligonucleotide labelled with a fluorescent dye such as TaqMan probe, or (ii) the non-specific DNA binding dye such as SYBR green. Following considerations should be made while selecting a probe for detection of DNA quantity: the level of sensitivity and accuracy, budget available, available skill and expertise for designing and optimising qPCR assay.

DNA binding dyes such as SYBR green are easy to use as well as cost-effective. SYBR green possesses a unique property of displaying low intensity of fluorescence when free in solution. However, when binds to a dsDNA, its fluorescence level increases by over 1000-fold. Hence, fluorescence measurement is performed at the end of each PCR cycle to monitor an increased amount of amplified DNA. As it can bind and give an increased fluorescence level with any dsDNA, the specificity of this assay is greatly decreased by amplification of non-specific PCR products and formation of primer-dimers. However, the characteristic melting peak at melting point (Tm) of the amplicon distinguishes amplification artefacts from the desired amplicon that melt at lower temperature, thus, giving rise to broad peaks. In addition, for proper detection of accurate DNA fragment, hydrolysis probes like 6-carboxyfluorescein (FAM), tetrachloro-6-carboxy-fluorescein (TET), hexachloro-6-carboxyfluorescein (HEX), 6-carboxytetramethyl-rhodamine (TAMARA), 4-(dimethylaminoazo) benzene-4-carboxylic acid (DABCYL) can be used. During PCR process, probes anneal to the target DNA and facilitate Taq polymerase to cleave the probe, thus allowing the increased fluorescent emission. Thus, the increased fluorescent intensity is directly proportional to the amount of amplicon produced (Fig. 3.4).

During real-time PCR experiments, specific errors get introduced due to minor differences between the initial amount of RNA, difference in efficiency of cDNA synthesis and PCR amplification. Thus, in order to avoid such problem, cellular RNA is simultaneously amplified with a reference gene for normalisation of the obtained result. The most common genes used for normalisation are mostly termed as housekeeping genes that include β-actin, GAPDH and 16S

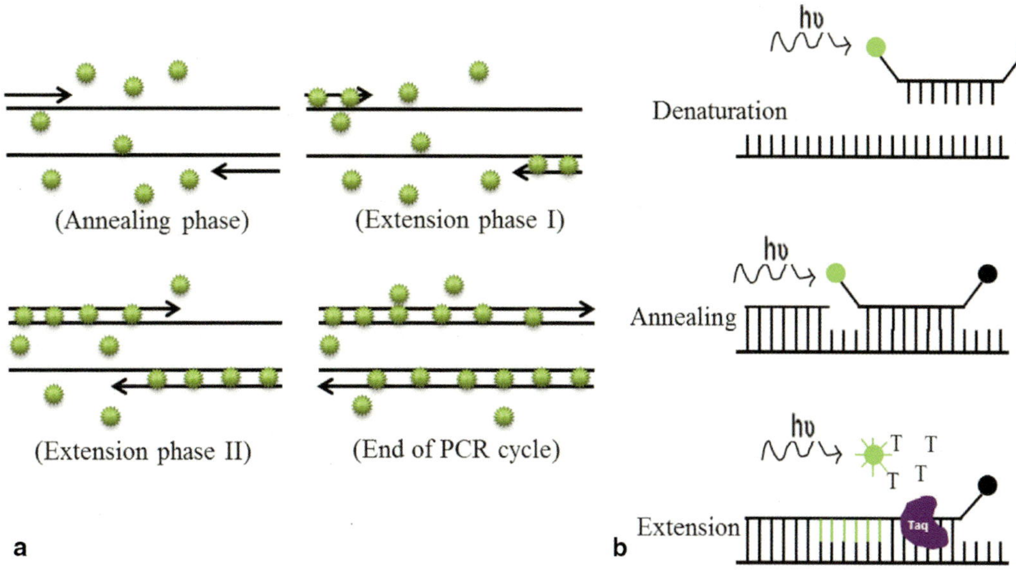

Fig. 3.4 Role of fluorescent probes during DNA replication in qRT-PCR for detection of level of gene expression. **a** DNA-intercalating probe SYBR green. **b** Hydrolysis probe TaqMan

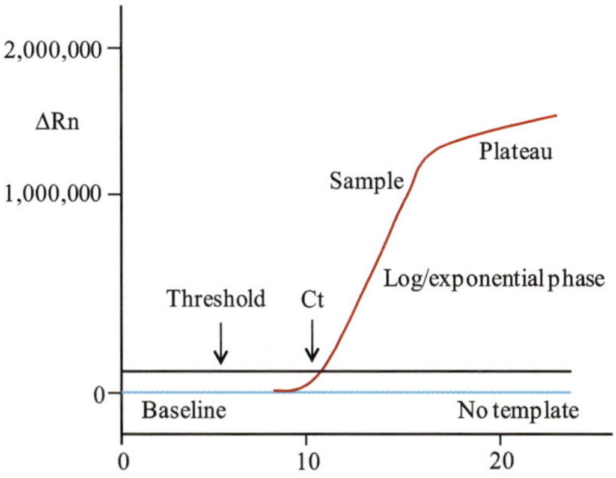

Fig. 3.5 Model of the single-amplification plot with the commonly used nomenclatures in qRT-PCR. Δ*Rn* is the fluorescence emission by the product at each time point—fluorescence emission by the base line. C_t is the threshold cycle

rRNA. Theoretically, these genes are expressed at a constant level at all stages and should remain constant under different experimental conditions, hence, some important terms associated with qRT-PCR analysis have been discussed below (Fig. 3.5):

Base Line Base line of a qRT-PCR is the PCR cycles in which the reporter fluorescent signal is accumulated but not detected by the instrument. By usual practice, the software of the instrument sets the base line from 3 to 15 cycles; however, it can be modified as per the requirement of the user.

ΔR_n It can be calculated by the designed software using the formula of $R_{nf}-R_{nb}$, where R_{nf} is the fluorescent emission of the product at each time point and R_{nb} is the fluorescent emission from the base line correction. These values are generally plotted against the cycle number and it should be taken care that the ΔR_n value should not exceed the base line.

Threshold Based upon the variability of the base line, an arbitrary threshold is chosen by the software programme. Mostly it is calculated as ten times to the standard deviation of the average signal between three to fifteen cycles. The fluorescent signal detected above the threshold point is considered to be a real signal, which can be used to know the threshold cycle (Cr) of the sample. The threshold can be changed manually to obtain the region of exponential amplification across the whole amplification plot.

C_t C_t is the fractional PCR cycle number at which the reporter fluorescence is greater than the threshold detection level. C_t value in RT-PCR is responsible for producing accurate and reproducible data. During the exponential phase, reaction components become limiting and thus C_t values become reproducible to replicate reactions with the same starting copy number.

General PCR amplification reaction follows the below formula, i.e. $A = B(1+e)^n$ where, A represents amplified products, B is input templates, n is number of cycles, e is amplification efficiency. However, the amplification efficiency in RT-PCR depends upon many factors—efficiency of reverse transcription, Mg^{2+}, dNTPs/primer concentrations, enzyme activity, pH, annealing temperature, number of cycles, variation in temperature and tube to tube variation. As PCR product results in the million fold amplification, minute variation in these above-mentioned factors affect significantly the final output. Thus, an appropriate control is required for the proper analysis of RT-PCR results. In this regard, the qRT-PCR relies on the following formula, i.e. $A / A' = B(1+e)n / B'(1+e)n$ where, A is the amplified product, B is the input templates, A' is amplified control products, B' is input control templates, e is amplification efficiency, n is number of cycles. Thus, any effect on the amplification efficiency equally affects both the templates. It provides a linear relationship between both the wild and control templates during the amplification process, i.e. $A / A' = B / B'$.

Reagents Required and Their Role

SYBR Green

SYBR green is the asymmetrical cyanine dye that results in DNA-dye complex which absorbs blue light at λ_{max} of 497 nm and emits green light at λ_{max} of 520 nm. The dye binds with dsDNA more preferentially than that of single-stranded DNA

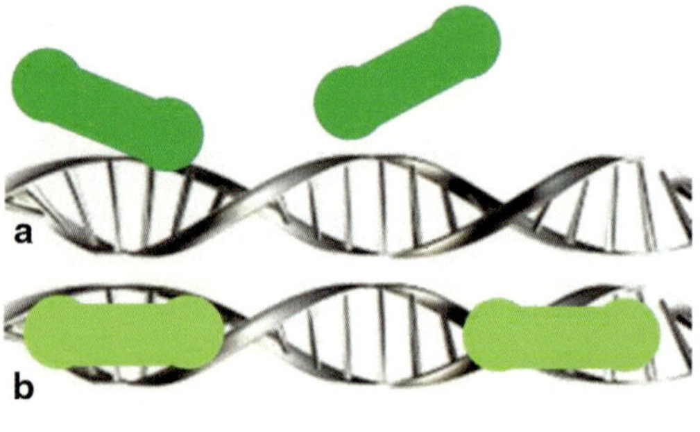

Fig. 3.6 Increase in intensity of the SYBR green after binding with dsDNA. **a** Unbound SYBR green, **b** fluorescent bound SYBR green

(Fig. 3.6). In most of the qRT-PCR experiments, SYBR green is used to quantify the dsDNA. The other applications of SYBR green includes the visualisation of DNA in agarose gel electrophoresis and labelling of DNA within the cells in flow cytometry and fluorescence microscope.

Forward and Reverse Primer

Forward primer is used to amplify the sense strand of the target DNA from 5′ to 3′ direction in both target and house-keeping gene. In a similar fashion, reverse primer amplifies the antisense strand of the target gene from 5′ to 3′ direction. In this experiment, primers for both 16S rRNA gene (16S Forward 5′ AGAGTTTGATCMTGGCTCAG 3′ and 16S reverse 5′ ACGGCTACCTTGTTACGA 3′) and czrCBA (czrCBA Forward 5′ TCCTCAAATCCGAACTGGGC 3′ and czrCBA Reverse 5′ GCTCGATGGCGAATTGGATG 3′) will be used.

Nuclease-Free Water

Nuclease-free water is used to prepare the same reaction volume and concentration to compensate differences among the volume of the ingredients. The water used for this purpose should be nuclease free in order to prevent the degradation of DNA and RNA in the reaction mixture.

Table 3.1 Components and amount of the constituents of the master mixture

Components of master mix	Volume (μl)	Final concentration
SYBR green RT-PCR (2X) master mix	35	1X
Reference dye	0.7	–
10 μM forward primer	3.5	0.5 μM
10 μM reverse primer	3.5	0.5 μM
cDNA	13	–
MiliQ water	14.3	–
Total	70	–

Template DNA/cDNA

The DNA sample should be the extracted genomic sample from the targeted organism or the RNA based complementary DNA whose cDNA is required to be amplified to check the expression level under various conditions.

Procedure

Pre-experiment Preparations

1. Before starting the RT-PCR reaction, RNA isolation and cDNA preparation should be carried out following the protocol described in the previous chapters. Prepare cDNA from the mRNA isolated from the bacterial cells grown at Luria Bertani broth and Luria Bertani broth supplemented with 1250 ppm of Cd as $CdCl_2$ separately.
2. All the required reagents should be thawed properly by incubating them on ice.

Experimental Setup for RT-PCR Reaction

1. Make sure the working area is clean and free from any distractions.
2. Prepare the PCR master mixture in a 1.5 ml micro centrifuge tube (Table 3.1) to maximise uniformity and minimise labour when working with multiple samples.
3. Prepare each sample in triplicates, i.e. both for the gene of interest and house-keeping gene. Always prepare master mixture in excess that may be needed to account the pipetting variations.
4. For each reaction, the total volume should be a minimum of 5 μl.
5. In this experiment, gene expression level of two sets of genes (gene of interest *czrCBA* and house-keeping gene 16S rRNA) in two conditions (control and in presence of 1250 ppm of Cd as CdC12) will be measured.
6. Mix the SYBR green, water and reference dye thoroughly and dispense equal aliquots to 0.5 ml PCR tubes.
7. Add equal volume (3.5 μl each) of target and house-keeping primers in both the dispensed tubes, mix thoroughly and divide equally into two tubes.
8. Add 13 μl of template cDNA to each reaction tube as needed and aliquot them equally in triplicates into standard optically cleared strips.
9. Tightly seal the strips with the strip caps and short spin the contents to bring them at the bottom of the tube.
10. Set up the machine as per the instruction of the manufacturer. A typical PCR programme consists of the DNA melting curve for quality control analysis.
11. Place the strips on the RT-PCR system (Mastercycler® ep realplex, Eppendorf, Germany) and start the program.
12. Set the PCR programme with an initial denaturation of 95 °C for 2 min followed by 40 cycles of 95 °C for 15 s, 53 °C for 30 s, 60 °C for 45 s and the melting curve (Fig. 3.7).
13. After completion of the program, save the assay and start the analysis.

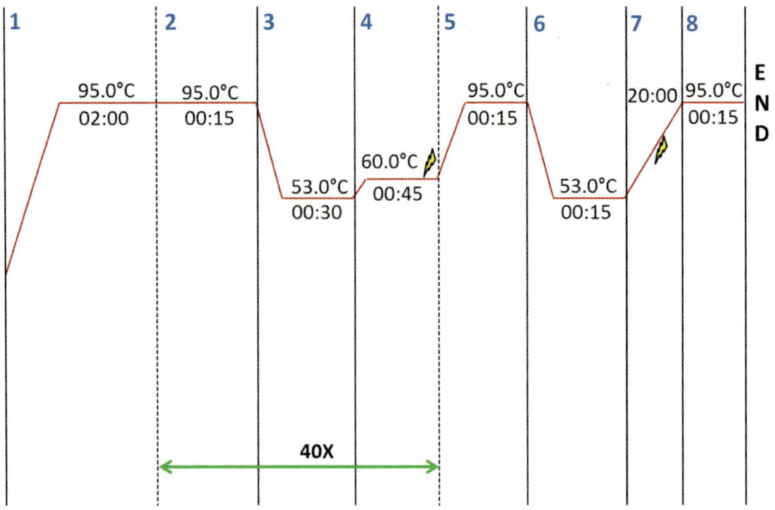

Fig. 3.7 qRT-PCR programme for analysis of expression level of *czrCBA* gene

Analysis of Result

1. Go to File → New → Assay (= relative quantification (ΔΔCt) study) → plate name Next
2. Add plates → select your file → finish
3. Bottom right corner main window → Line colour = Detector
4. Analysis → Analysis settings → Detector → All → Auto Ct (Start with default) → Auto Base line (Start with default = 3–15 cycles) → Calibrator sample → choose a sample condition, i.e. control that will compare control to treated→Endogenous Control Detector should always be set as your housekeeping gene = 16S rRNA→Apply → Ok and Reanalyse
5. Gene expression tab→your graph here will be showing the treated and untreated conditions for each gene highlighted with 16S rRNA set at zero. Export the data from the sample summary window from bottom left corner and manually graph the ddCt values of the experimental triplicates against one another to confirm that they are statistically insignificantly different (as the program is unable to do this because it treats your experimental triplicate conditions, $n=3$, as different sample sets)

Observation

The relative quantification of SYBR green is actually the relative quantification of the expression level of the gene of interest in relation to the control or house-keeping gene. The following observations should be taken care of during qRT-PCR analysis.

Positions	Name	Ct SYBR	Ct mean SYBR	Ct deviation SYBR	Expression level SYBR	Range SYBR
A1	Target gene					
A2	Target gene					
A3	Target gene					
B1	House-keeping gene					
B2	House-keeping gene					
B3	House-keeping gene					

Trouble-Shootings

Problems	Possible cause	Possible solutions
No amplification	Enzyme not activated	Check to perform full 15 min activation step before cycling
	Annealing problem	Check for optimum annealing temperature
	Extension time	Increase extension time for larger amplicons
	Poor primer design	Check for primer dimers on agarose gel, use freshly diluted primers
	Product too long	The ideal amplicon size should vary between 80 and 200 bp
	Too few cycles	Recommended PCR cycle is 40
	Impure/poor concentration template	Purify template, use upto 500 ng template
	Wrong dye layer	Check the machine setting corresponding to the experiment
Amplification in no template control	Primer dimers	It can be detected easily by using serial dilution of the template and running product on agarose gel
	Contamination	Purify template before use, repeat the experiment with all the fresh reagents
Low sensitivity	Evaporation	Do not use corner wells or use more robust seal
	Primer dimers	It can be detected easily by using serial dilution of the template and running product on agarose gel
	Annealing step	Check for the optimum annealing temperature
	Extension step	Extension time can be increased for long amplicons
	Wrong primer concentration	Recommended primer concentration is 0.4 µM
Abnormal amplification	Too many cycles performed	Reduce the number of cycles
	Wrong detection step	Check that the detection step is set in the correct stage of the cycle
	Machine needs calibrating	They may be generated by the mirror misalignment or lamp problems
	Reaction volume too low	Some instruments are set to read accurately only at volumes of at least 15 µl
High well to well variance	Poor plate choice	Never use black or frosted plates
	Low quality sealing material	Use only high quality optically clear seals
	Machine needs re-calibrating	Follow the manufacturer's guidelines
	Evaporation	Do not use corner wells or use more robust seal
	Concentration gradient formed in vial	Invert the mixture a couple of times before use

Precautions

1. Always wear gloves while performing this experiment.
2. Use sterile disposable plastic wares whenever possible.
3. Treat your glasswares and plastic wares with RNase-inactivating enzymes.
4. Designate a specific area for this work only. Treat surfaces of working area with commercially-available RNase-inactivating agents.
5. Whenever possible purchase reagents those are free of RNases.
6. Always store the synthesised cDNA at −80 °C till further use.

FLOW CHART

Prepare a master mixture containing SYBR green, water and reference dye thoroughly and dispense equal aliquots into 0.5 ml PCR tube

↓

Add equal volume of target and house-keeping primers and divide it in to two tubes

↓

Add template cDNA to each reaction tube and transfer to EpiWhite tubes in triplicates

↓

Set the PCR programme with initial denaturation at 95°C for 2 min followed by 40 cycles of 95°C for 15 s, 53°C for 30 s, 60°C for 45 s and the melting curve

↓

After completion of the program, save the assay and start analysis

↓

Normalize the obtained data with the data obtained for house-keeping gene and observe for the difference in expression level of the target gene of interest

Exp. 3.3. Gene Expression Analysis Using Reporter Gene Assay

Objective To study *czrCBA* gene expression at transcription level by luciferase reporter assay.

Introduction

Reporter genes have become an inseparable and widely used tool in study of gene expression and regulation. Transcriptional regulation coupled with gene expression of a reporter gene is widely used in study of many physiological processes like intracellular signal transduction, receptor activity, mRNA processing, protein folding and protein–protein interaction. A reporter gene is cloned along with a gene of interest in an expression vector. Then the expression vector is transferred into cells. The reporter gene either produces enzymes (e.g. luciferase) or proteins (e.g. green fluorescent protein). The cells are then assayed for the presence of reporter by directly measuring the reporter protein itself or the enzymatic activity of the reporter protein. Receptor genes are used to study whether the gene of interest is taken and stably expressed in a cell or not.

Luminescence is the light emission as a result of a chemical reaction without the production of heat or any thermal changes. Bioluminescence is the light emitted from a biological source whereas chemiluminescence is the light emitted from a non-biological source due to a chemical reaction. Bioluminescent reactions are widely used because they are derived from natural sources and can deliver 1–1000-fold higher sensitivity than any other fluorescence methods. Unlike chemiluminiscence assays, which are completely dependent on environmental conditions, bioluminescence assays are independent of the surrounding molecules.

Bioluminescence is found in many organisms including coelenterates, echinoderms, mushrooms, insects, bacteria, etc. Luciferase, found in firefly (*Photinus pyralis*) and Renilla (*Renilla reniformis*), is commonly used as reporter. Firefly luciferase assay is commonly used because the reporter activity is immediately available upon translation, as the protein does not undergo post translational processing and readily available

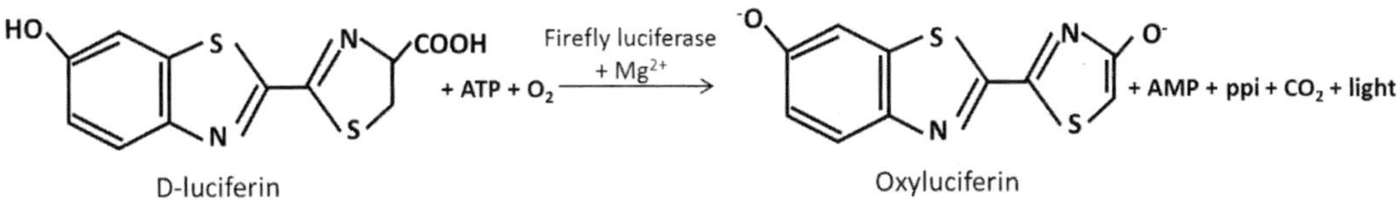

Fig. 3.8 Reaction mechanism of luciferase activity for production of CO_2 and light

immediately after translation. It produces light of very high quantum efficiency and the assay delivers very rapid detection. The gene encoding the firefly luciferase is a cDNA clone that has been incorporated in a number of reporter vectors, e.g. pGL4 expression vector.

Cells maintain inherent complex environment and the quantity of data obtained from a single reporter may be insufficient. Dual reporter assays are commonly used to obtain additional information with minimal effort. The dual luciferase assay system contains two different luciferase reporter enzymes that are expressed simultaneously in each cell. The most common dual luciferase assays measures both firefly and Renilla luciferase activity. These luciferases use different substrates and thus can be differentiated by their enzymatic specificities. Addition of the first reagent activates firefly luciferase and then adding second reagent extinguishes firefly luciferase and activates Renilla luciferase.

Principle

Two agents are necessary for a bioluminescence reaction to occur, i.e. (i) luciferase enzyme and (ii) the luciferase substrate. Firefly luciferase is a 61 kDa monomeric enzyme that catalyses the two-step oxidation of D-luciferin using ATP and Mg^{2+} as cofactors. During first step, the luciferyl carboxylate is activated by ATP to yield reactive mixed anhydride intermediate. Then the activated intermediate reacts with oxygen and creates a dioxyethane that further breaks down to form an oxidised product, oxyluciferin and CO_2 (Fig. 3.8). The reaction produces light usually in the region 550–570 nm which can be measured by luminometer. After mixing with substrate the luciferase produces light that decays rapidly over the period of time. Coenzyme A is also included to increase the sensitivity of the assay and to get a sustained and stable luminescence that decays over several minutes slowly.

Expression vectors are constructed containing promoter gene of interest viz. *luc*. When the expression vector is transferred into the cell the construct gets incorporated and ultimately gets translated by cell's own machinery. When the gene of interest undergoes transcription following translation, simultaneously the *luc* gene gets translated. The final product of the *luc* gene expression, luciferase enzyme, in the presence of its substrate luciferin produces light (Fig. 3.9). The fluorescence within the transformed cells is directly proportional to the steady state mRNA level. Due to this property, gene expression and many cellular events coupled with gene expression can be studied. Thus, luciferase assay is used in many fields of molecular biology, biochemistry, study of gene regulation, recombinant DNA technology, etc.

Strong controllable promoters are the essential requirements for both efficient expression of heterologous genes and for the optimum exploitation of homologous genes. Hence, in most of the cases, analysis of promoters depends on their fusion to reporter genes that can be analysed assayed to monitor the level of transcription.

Reagents Required and Their Role

Luciferase Substrate

In most of the experiments, Renilla luciferase substrate is used for this purpose. It is a uniquely engineered coelenterazine-based compound with protected oxidation sites. These modifications are helpful in minimizing substrate degradation and autoluminescence. It can provide a brighter output for in vivo applications.

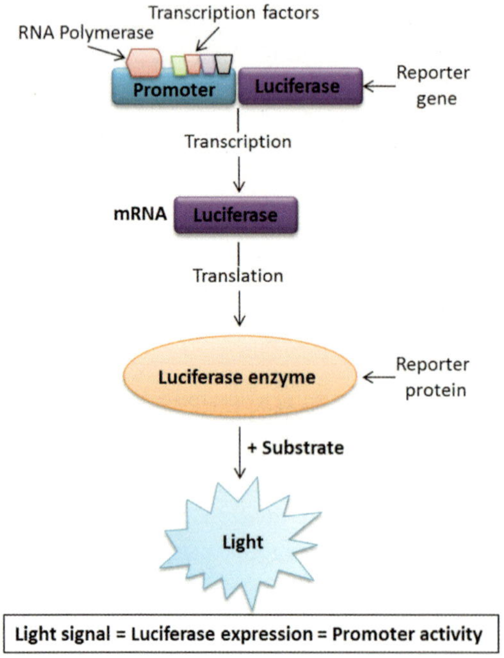

Fig. 3.9 Luciferase assay for the detection and quantification of expression level of mRNA

Phosphate Buffered Saline

All the reagents required during this experiment should be prepared using phosphate buffered saline (pH 7.4). It can be prepared using the following components in desired concentrations (Table 3.2).

Dissolve the above components in 1.8 l of deionised water. Adjust pH to 7.4 by addition of 0.1 N NaOH. Add deionised water to final volume of 2 l. Store at room temperature till further use.

Lysozyme

Lysozyme is a glycoside hydrolase that damages bacterial cell wall by catalysing hydrolysis of 1,4-beta linkages between N acetylmuramic acid and n-acetyl D glucosamine residues in a peptidoglycan. The common sources of lysozyme include a number of secretions such as tears, saliva, human milk and mucous. Thus, for lysis of the bacterial cells, lysozyme is used so that the genetic material present inside the bacterial cell can come out to the solution and perform their desired role during luciferase assay.

Procedure

Bacterial Cell Lysis

1. Prepare 1 ml of cell lysis buffer for 1–10 ml of bacterial culture by adding lysozyme to a final concentration of 1 mg/ml.
2. Centrifuge 1–10 ml aliquot of bacterial culture at 6000 rpm for 5 min at 4 °C. If required an optimal concentration may be determined after initial measurement of luciferase activity.
3. Remove the supernatant carefully without disturbing the cell pellets.
4. Resuspend the bacterial cell pellet with 1 ml of lysis buffer and mix completely by vortex.
5. Incubate at room temperature for 5–10 min.
6. Centrifuge the lysate at 14,000 rpm for 1 min at room temperature to remove insoluble debris.
7. Store the bacterial cell extract at −70 °C till further use, however storage period should not exceed 1 month.

Luciferase Assay

1. Place 20–100 µl of cell extract to an assay cuvette. Make sure to use exactly same volume for each sample.
2. Transfer the extract to a luminometer or scintillation counter device to start the chemiluminiscent reaction. It is always recommended to use the 96 well micro-titre plate for each reaction set.

Table 3.2 Components of phosphate buffered saline

Components	Final concentration (mM)	Amount to prepare 2 l of solution (g)
Na_2HPO_4	58	16.5
NaH_2PO_4	17	4.1
NaCl	68	8.0

3. Start the reaction by injecting 100 µl of luciferase assay reagent containing luciferase substrate at 15–25 °C.
4. Gently vortex the assay mixture immediately after addition of the luciferase assay reagent.
5. Start the measurement of light emission within 0.5–10 s after addition of the assay reagent for a period of 1–5 s.
6. After an almost constant light emission over a period of 20 s, the light production decreases with a half-life of about 5 min. At that time, stop the reaction.

Observation

The expression level of target gene in terms of luciferase induction can be determined by light production as measured by luminometer. Suitable internal control can be used such as the expression level of house-keeping gene to determine the expression level of target gene.

Result Table

Experimental setup	Luminiscence of target gene	Luminiscence of house-keeping gene	Expression level
Control			
Exp. 1			
Exp. 2			

Precaution

1. Luciferase assay substrate is a mild irritant. Thus, take appropriate precautions to prevent skin and eye contact.
2. Luciferase reactions are very fast. So the result should be taken as soon as possible after the addition of each reagent, preferably within 1 min.
3. For assay optimisation, optimal number of cells per well should be determined by serial dilution of cells.
4. Make sure to thaw all the reagents before starting the experiment by incubating them on ice.
5. Always wear gloves while carrying out the experimental steps.

Trouble-Shootings

Problems	Possible cause	Possible solution
Incomplete luciferase activity	Luciferase promoter might be destroyed by non-specific nuclease	Synthesise a pair of primers based on the vector sequence flanking the cloning site and do a PCR. If you don't get a band, your promoter may have been destroyed
No luciferase activity	Longer incubation time	As luciferase reaction is very fast, note the reading as soon as you add the reagents
	Expired reagents	Perform a control set of experiment, if no luciferase activity is observed then order for fresh reagents
DNA amount does not correlate with luciferase result	The expression of gene of interest may interfere the expression of TK reporters by affecting cell growth	Try to standardise the experiment by normalising the expression level of the target gene with more than one house-keeping genes

FLOW CHART

Prepare lysis buffer by adding lysozyme at a final concentration of 1 mg/ml

↓

Centrifuge 1-10 ml aliquot of bacterial culture at 6,000 rpm for 5 min at 4°C, remove supernatant carefully without disturbing the cell pellets

↓

Re-suspend the bacterial cell pellet with 1 ml of lysis buffer and mix by vortex, incubate at room temperature for 5-10 min

↓

Centrifuge the lysate at 14,000 rpm for 1 min at room temperature to remove insoluble debris, store the bacterial cell extract at -70°C till further use

↓

Place 20-100 µl of cell extract to an assay cuvette and transfer the extract to a luminometer

↓

Start the reaction by injecting 100 µl of luciferase assay reagent at 15°C to 25°C, mix the assay mixture immediately after addition

↓

Start the measurement of light emission within 0.5-10 sec after addition of assay reagent for a period of 1-5 sec

↓

After almost constant light emission over a period of 20 sec, stop the reaction

Exp. 3.4. Semi-quantitative Gene Expression Analysis

Objective To detect the level of *merA* gene expression from isolated total RNA from bacterial cell by semi-quantitative method.

Introduction

The principle of PCR gave rise to many other techniques like both quantitative as well as semi-quantitative PCR. Semi-quantitative PCR deals with the detection of level of gene expression by amplifying the gene of interest from the cDNA and subsequently determining the intensity level on agarose gel in comparison to the house-keeping gene. In this case, the intensity levels of the obtained bands are of high importance that can be determined by manually or by the software programmes. The major difference between the qRT-PCR and semi-quantitative PCR is that, semi-quantitative PCR allows the quantification of nucleic acids at the end of a PCR reaction, whereas quantification is done after each cycle in a qRT-PCR.

Though PCR was not viewed earlier to have great quantitative power, it has now been oversimplified for reproducible, precise, accurate quantification of nucleic acids. The key factors influencing the quantitative ability of a PCR program includes optimisation of the amplification step. Optimisation includes changes in nucleic acid preparation, primer designing and its use, buffer usages and cycle parameters. The recent developments in recognising the elimination of undesired hybridisation events formed during the first cycle of PCR and possessing subsequent devastating effects. This effect can be overcome by starting the reaction at high temperature, instead of room temperature, thus improving the sensitivity of the PCR by over 1000-fold. In most of the PCR conditions, the overall efficiency is less than 100 % while the typical amplification runs with a consistent efficiency of 70–80 % from 15th to 30th cycle which further depends on the amount of the starting material.

Semi-quantitative PCR is generally used for the preliminary investigation of the expression level of certain gene of interest with comparison to the house-keeping gene. This technique is widely applied in the field of environmental microbiology for quantification of degradation genes. In clinical microbiology, it is used for detection and quantification of toxic genes responsible for conferring toxicity to the pathogenic microorganisms. The preliminary result obtained by semi-quantitative PCR can be correlated with the qRT-PCR results to obtain a clear image on the expression level of the gene of interest under various environmental conditions.

Principle

PCR is the technique that relies on the amplification of a set of gene by the combinatorial action of buffer system, dNTPs, primers, polymerase and divalent cations. In this process, the copy number after the final cycle of PCR depends on the initial copy number of template taken. However, this can be taken as the advantage to analyse the expression level of target gene that indirectly determines the copy number of the template used.

When a gene is expressed, it forms RNA and subsequently translated to the functional proteins to carry out further cellular activities as per the central dogma of life. Thus, when a gene gets higher expression, it tends to form higher level of RNA and proteins. RNA can be transcribed back to cDNA by the action of reverse transcriptase and the target gene can be amplified from the synthesised cDNA to determine its copy number under experimental conditions. A normal PCR machine only amplifies the desired gene of interest and cannot show the exact number of copies of DNA present in the sample. Indirectly, DNA intercalating dyes such as EtBr can be used for this purpose that binds with DNA sample and illuminates under UV exposure. The intensity of EtBr bound with dsDNA molecule is directly proportional to the amount of DNA present on the gel. Thus, indirectly the expression level of the gene of interest can be determined by determining the intensity level of EtBr under UV light (Fig. 3.10).

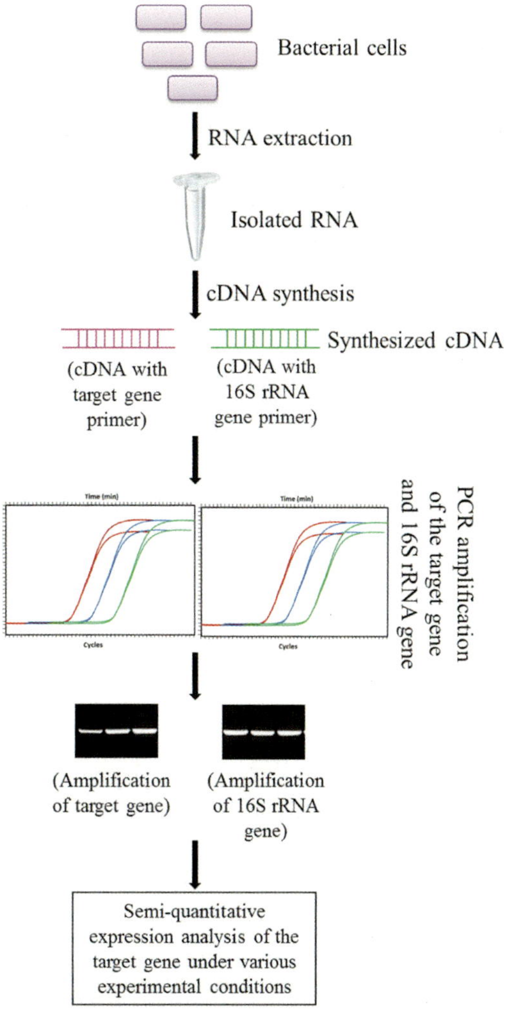

Fig. 3.10 Schematic presentation of semi-quantitative PCR analysis of target gene in normalisation with housekeeping gene i.e. 16S rRNA gene

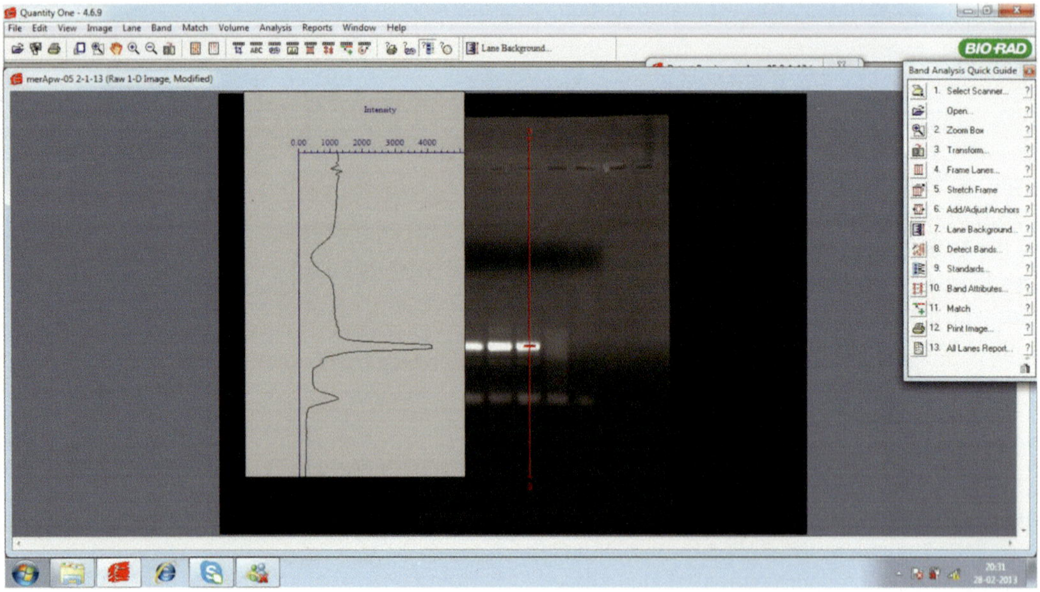

Fig. 3.11 Detection of bands in the created lanes by Quantity One software (Bio-Rad, USA)

However, based upon the intensity of gene of interest, the expression level cannot be determined. Hence, the relative level of expression is determined with respect to the expression level of any house-keeping gene. In bacterial systems, 16S rRNA gene is considered to be used as house-keeping gene, the sequence of which is present in all form of bacteria and the level of expression of this gene is considered to be unchanged under different environmental conditions. To study the expression level of a gene of interest the expression of house-keeping gene at that point of environmental condition should be normalised from the obtained data. During this experiment, the expression level of *merA* gene will be studied. *merA* is the functional gene present in the mercury resistant bacterial system that codes for mercuric ion reductase. *merA* has been reported to be at higher expression level under low salinity conditions. *MerA* is responsible for volatilising inorganic mercury to a relatively less toxic form of mercury, i.e. elemental form of mercury (Hg^0).

Quantity One® 1-D analysis software V. 4.6 supplied with thermal cycler (Bio-Rad, USA) is a useful tool for determining the intensity level of bands on gel, which can be analysed further to determine the relative expression level of the gene of interest. This procedure is called semi-quantitative PCR as it does not provide information on the exact copy number of the gene after completion of the PCR cycle as well as it is an indirect way of measurement of the expression level of the gene of interest by determining the intensity level of the EtBr upon exposure to UV light.

Reagents Required and Their Role

cDNA Template

cDNA can be prepared from the isolated pool of mRNA using reverse transcriptase in a PCR by both the target gene primer as well as 16S rRNA gene primer. The detailed procedure of synthesising cDNA has been discussed in Exp. 3.1. cDNA concentration should be uniform in all the samples to be tested and they should be of pure quality without any DNase or RNase contamination.

PCR Buffer

Each enzymatic activity is facilitated by an optimum condition of pH, ionic strength, presence of cofactors, etc. which is achieved by the addition of any buffer systems to the reaction mixture. In many cases, the enzymes shift pH in the non-buffered solutions which ultimately stops working and this process can thus be avoided by the addition of buffer to the PCR analysis. In most instances, a PCR buffer consists of 100 mM Tris-HCl, pH 8.3, 500 mM KCl, 15 mM $MgCl_2$ and 0.01 % (w/v) gelatin. In all cases, the final concentration of the PCR buffer should be 1X per reaction.

dNTPs

dNTPs are regarded as the building blocks of the synthesis of a new DNA strand. They are commercially available as a mixture of four nucleotides, i.e. dATP, dTTP, dGTP and dCTP. In each PCR around 100 μM of each dNTPs is required. They are highly sensitive to repeated freezing and thawing and hence, after 3–5 cycles they stop working properly. Long term freezing also affects the concentration of the dNTPs mixture by evaporating a small amount of water from the mixture, hence it is recommended to centrifuge the vials containing dNTPs before use. As acidic pH promotes hydrolysis of dNTPs and further interfers with PCR result, it is recommended to dilute the dNTPs sample in TE buffer rather than water.

Forward and Reverse Primer

A primer is defined to be the complimentary DNA molecule which is designed to amplify the desired DNA fragment from the whole DNA in a PCR. Primers induce the polymerase to start its action. As DNA polymerase cannot start synthesising new DNA strand directly from the template due to non-availability of free −OH group. Thus when a primer attaches to a DNA template it provides free −OH for the action of DNA polymerase to start its action. During the semi-quantitative expression analysis of a gene of interest, forward and reverse primers of both the gene of interest and house-keeping gene (in this case 16S rRNA) is used to amplify the corresponding genes.

Taq Polymerase

Taq polymerase is the thermo-stable DNA polymerase isolated from the thermophilic bacterium *Thermus aquaticus* which was originally isolated by Thomas D. Brock in 1965. The enzyme is able to withstand the protein denaturing conditions required during PCR. It has the optimum temperature for its activity between 75 and 80 °C, with a half-life of greater than 2 h at 92.5 °C, 40 min at 95 °C and 9 min at 97.5 °C and possesses the capability of replicating a 1000 bp of DNA sequence in <10 s at 72 °C. However, the major drawback of using Taq polymerase is the low replication fidelity as it lacks 3′–5′ exonuclease proof reading activity. It also produces DNA products having 'A' overhangs at their 3′ ends which is ultimately useful during TA cloning. In general, 0.5–2.0 units of Taq polymerase is used in a 50 μl of total reaction but ideally 1.25 units should be used.

Ethidium Bromide

Ethidium bromide is the intercalating agent that is used as a fluorescent tag during agarose gel electrophoresis. When exposed to UV light, it flourishes the orange colour and its intensity level increases up to 20-fold after binding to DNA molecule. The major use of EtBr in molecular biology includes in the detection of nucleic acids. The absorption maximum for EtBr in aqueous solution is 210 and 285 nm that corresponds to UV light and as a result of such excitation; it emits orange light at wave length of 605 nm. The mode of binding of EtBr in DNA molecule is to intercalate between base pairs. This binding may alter the charge, weight, conformation as well as flexibility of the DNA molecule. As the size of DNA amplified during PCR is determined by their relative movement through the gel in comparison to the molecular weight standards, mobility is highly critical in size determination.

Procedure

1. Thaw all the reagents required for PCR on ice and thaw prepared cDNA from Exp. 3.1 which was stored at −80 °C.
2. Prepare the reaction mixture separately for amplification of *merA* gene and 16S rRNA gene using the different set of primers by adding the reagents as mentioned below.

10X PCR buffer (10X)	2.5 μl
$MgCl_2$ (10 mM)	2.5 μl
dNTPs (10 mM)	0.5 μl
Forward primer (1 mM)	1.0 μl
Reverse primer (1 mM)	1.0 μl
Taq polymerase (1 U/μl)	1.0 μl
Milli-Q wate	12.5 μl
cDNA template	4.0 μl

3. As pipetting of small volume is difficult and often inaccurate, a master mixture may be prepared where constituents common to all the reactions are combined in one tube multiplying the volume for one reaction with the total number of samples. Later, the appropriate amount of master mixture is aliquoted to each tube, e.g. if 4 μl of DNA template is to be added, 21 μl of the reaction mixture should be added to prepare a total volume of 25 μl.
4. Place the tubes row wise in each well of the PCR machine.
5. Carryout the amplification with the following programme:

LID	98 °C
Initial denaturation	96 °C for 5 min
30 cycles	
Denaturation	95 °C for 15 s
Annealing	49 °C for 30 s
Extension	72 °C for 1 min
Final extension	72 °C for 10 min
Hold	4 °C for ∞

6. Run the amplified DNA product in 1 % agarose gel to obtain the clear distinguished banding pattern.
7. See and acquire the image of the gel in Gel documentation system by Quantity One software.
8. Create lanes as per the desired number of bands to be analysed.
9. Detect bands and remove the unnecessarily detected bands by remove band icon.
10. Acquire intensity of the desired bands followed by normalisation of the intensity of the bands of target gene in correspondence to the intensity of house-keeping genes (Fig. 3.11).
11. Export the band analysis report and obtain the report for peak intensity, average intensity and relative quantity (Fig. 3.12).

Observation

Analyse carefully the data obtained after the analysis of banding patterns for both gene of interest and the house-keeping gene. Plot a graph between the obtained intensity and the genes to obtain the expression level of the gene under different experimental conditions with respect to the expression pattern of the house-keeping gene.

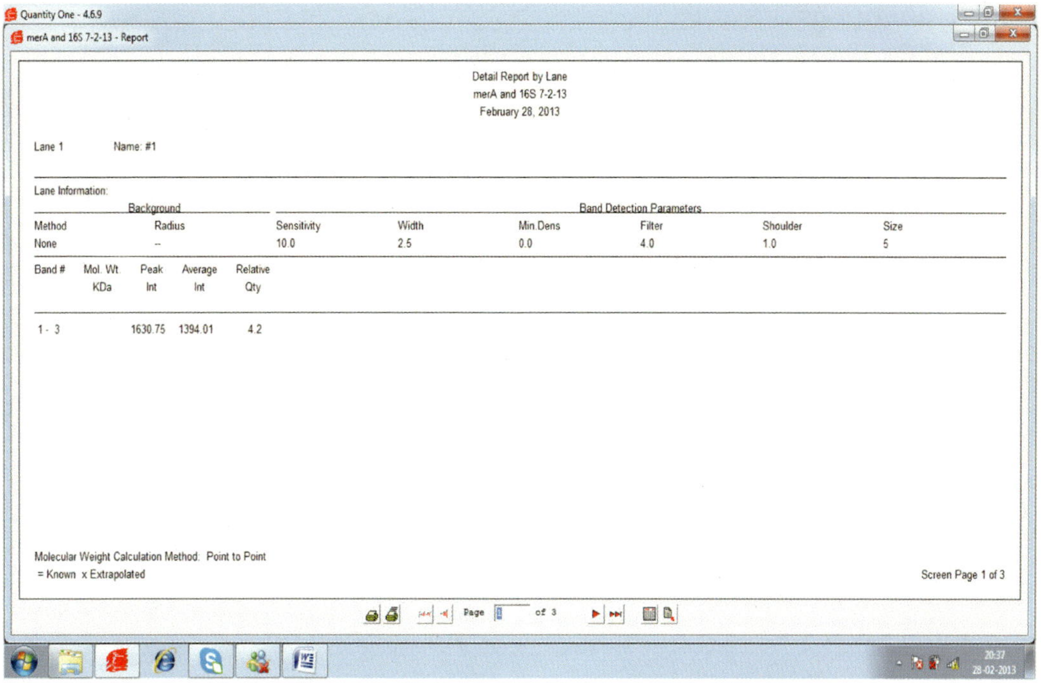

Fig. 3.12 Band analysis report generated by the software after normalisation of the intensity of target gene with house-keeping gene

Observation Table

Band no.	Band attribute	Molecular weight (KDa)	Peak intensity	Average intensity	Relative quantity
Gene of interest 1					
Gene of interest 2					
Gene of interest 3					
House-keeping gene 1					
House-keeping gene 2					
House-keeping gene 3					

Trouble-Shootings

Problems	Possible cause	Possible solutions
In correct product size	Incorrect annealing temperature	Recalculate primer Tm values by using any of the web based software
	Mispriming	Verify the primes have no additional complementary regions within the template DNA
	Improper Mg^{2+} concentration	Optimise Mg^{2+} concentration with 0.2–1 mM increments
	Nuclease contamination	Repeat the reactions using fresh solutions
No product	Incorrect annealing temperature	Recalculate the Tm values of the primers, test the correct annealing temperature by gradient, starting at 5 °C below the lower Tm of the primer pair
	Poor primer design	Check with the literature for recommended primer design, verify that the primers are non-complementary, both internally and to each other, optionally increase length of the primer
	Poor primer specificity	Verify that the oligos are complementary to the proper target sequence
	Insufficient primer concentration	The correct range of primer concentration should be 0.05–1 μM, refer to the specific product literature for ideal conditions
	Poor template quality	Analyse DNA by agarose gel electrophoresis, check A_{260}/A_{280} ratio of DNA template
	Insufficient number of cycles	Rerun the reaction with more number of cycles
Multiple/non-specific products	Premature replication	Use hot start polymerase, set up the reaction on ice using chilled components, add samples to the PCR preheated to the denaturation temperature
	Primer annealing temperature too low	Increase the annealing temperature
	Incorrect Mg^{2+} concentration	Adjust Mg^{2+} concentration with 0.2–1.0 mM increments
	Excess primer	The correct range of primer concentration should be 0.05–1 μM, refer to the specific product literature for ideal conditions
	Incorrect template concentration	For low complexity templates (i.e. plasmid, lambda, BAC DNA), use 1 pg–10 ng DNA per 50 μl reaction For high-complexity templates (i.e. genomic DNA), use 1 ng–1 μg template for 50 μl reaction

Precautions

1. Use pipette tips with filters.
2. Store materials and reagents properly under separate conditions and add them to the reaction mixture in a spatially separated facility.
3. Thaw all components thoroughly at room temperature before starting an assay.
4. After thawing, mix the components with brief centrifugation.
5. Work quickly on ice or in the cooling block.
6. Always wear safety goggles and gloves while performing the PCR reaction.

FLOW CHART

Thaw all the reagents on ice required for PCR as well as the prepared cDNA

↓

Either prepare the master mixture or reaction mixtures individually for the cDNA prepared separately

↓

Put the tubes in PCR and amplify the target gene as well as 16S rRNA house-keeping gene using the programme as described in the procedure section

↓

After PCR amplification run the product on 1% agarose gel and visualize the amplified product under UV

↓

Analyze the intensity of the bands of both target gene and house-keeping gene by Quantity One software

↓

Normalize the intensity of target gene with respect to the intensity of house-keeping gene

↓

Export the band analysis report obtained from Quantity One software for peak intensity, average intensity and relative quantity

↓

Plot the obtained data for analysing expression pattern of the target gene under various conditions

Exp. 3.5. Northern Blotting

Objective To analyse the gene expression pattern by northern blotting.

Introduction

Northern blotting is the technique used for the detection and measurement of the level of gene expression of particular genes. During the process of northern blotting, RNA samples are separated by agarose gel electrophoresis based upon their size and can be subsequently detected using a hybridisation probe complementary to the part of the target gene sequence. This technique allows the detection of expression level of a gene in terms of its RNA quantity under different environmental conditions. This technique can be used irrespective of the prokaryotic or eukaryotic system to detect up-regulation or down-regulation of a gene in comparison to the reference gene. There are many online databases for northern blot results, e.g. BlotBase. It is an online database consisting of more than 700 publications in more than 650 genes. The search result provides the information regarding blot ID, species, gene, expression level, blot image and the link to the publication that has been originated from this work.

This technique can be applied to study the gene expression in addition to other techniques like RT-PCR, RNase protection assay, microarray analysis and serial analysis of gene expression. Out of all the techniques used for gene expression analysis, microarray is regarded to be the most advanced and in most of the cases, data obtained from northern blotting is consistent to the microarray results. In addition to that, northern blot can detect small changes in gene expression which cannot be detected by microarray. However, in northern blotting one gene can be studied at a time, whereas, thousands of genes can be studied at a time in microarray. The major disadvantage of this technique is the involvement of RNAs during the experiment which is more prone to degradation through environmental contamination though it can be avoided by proper sterilisation of the glass wares to be used and by DEPC treatment of the plastic wares. In comparison to RT-PCR, this technique may be less sensitive but it has high specificity by reducing false positive results.

Though this technique possesses many advantages over other gene expression techniques, the use of certain toxic chemicals such as formaldehyde, radioactive substances, ethidium bromide, DEPC and UV light may cause health hazards upon exposure. However, it is relatively simple to perform, inexpensive and not plagued by common artefacts and the recent discoveries on hybridisation membranes and buffers have increased the sensitivity of this technique.

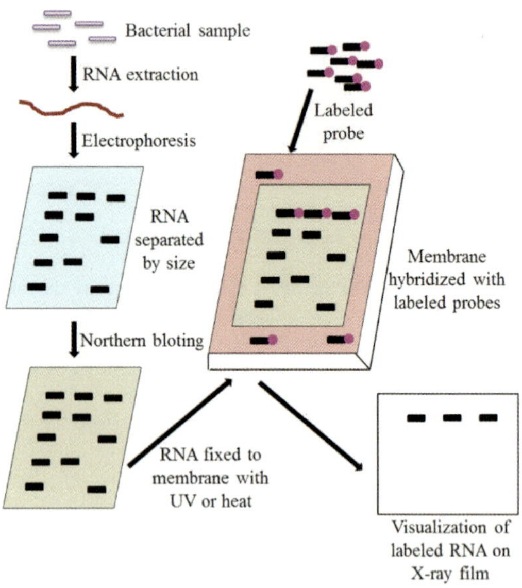

Fig. 3.13 Detailed procedure of northern blotting for detection and quantification of expression level of the desired gene of interest

Principle

RNA is a ubiquitous group of micromolecules which plays an important role in microbial metabolism which belong to the group of nucleic acids. RNA plays an important role in storing, effecting and interpreting the genetic information of an organism. Just like DNA molecules, RNA is also made up of ribose sugar and phosphate backbones along with the side chains of nitrogenous bases of purines and pyrimidines. Contrary like DNA, RNA molecules carry out their functions in single strands. It can readily hybridise with their complementary nucleotide sequences; this property of which has been explored during northern blotting and other analytical techniques of RNA.

As RNA plays a crucial role in various metabolic pathways and events that subsequently lead to the synthesis of proteins, monitoring the predominant quantity of the same may result in the useful insight of the events taking place inside the cell. In this regard, northern blotting is the most useful technique in identifying and quantifying level of RNAs and subsequently the gene expression pattern and level.

This technique involves the isolation of RNA molecules form the cell of interest followed by the separation of the isolated RNA by agarose gel electrophoresis or polyacrylamide gel electrophoresis. Ultimately, the separated RNA in the gel is transferred to a solid matrix by blotting technique. The solid matrix containing the separated RNA molecules can be treated with labelled probe that specifically hybridises with the complementary RNA sequences on the matrix. Unbound probes can be washed-off from the matrix. The labelled probe binds with the complementary RNA sequence on the matrix that can be detected by a suitable detector (Fig. 3.13).

Blotting is the process of passive transfer of the RNA molecules to the matrix by capillary action. Two different blotting techniques are used mostly, i.e. capillary transfer and electro blotting. Capillary transfer involves the passive transfer of the RNA molecules to the matrixes like filter paper, nylon membrane or nitrocellulose mem-

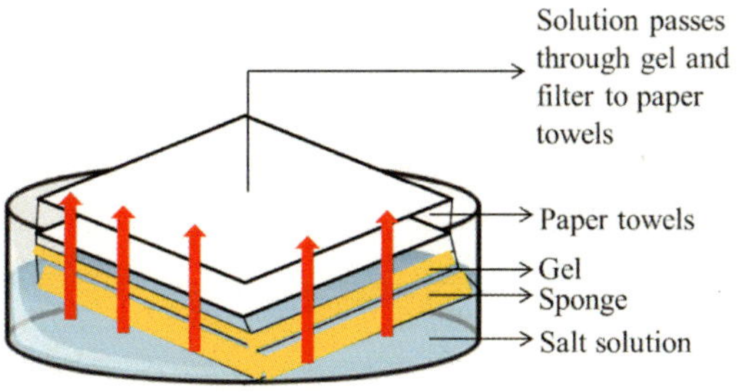

Fig. 3.14 Capillary transfer of RNA molecule to the nitrocellulose membrane

brane under capillary force. The gel can be placed on the sponge soaked in salt solution or buffer and the nitrocellulose membrane is placed over the gel on the top of which many layers of paper towels are placed. The solution passes through the gel and through the nitrocellulose membrane which takes 12–16 h. In contrary, electro blotting involves the use of electric current for the transfer of RNA molecules from the gel to the matrix (Fig. 3.14).

This step is followed by the ensuring of proper binding of the RNA molecules with the solid matrix after adsorption which can be achieved by UV treatment. UV causes extensive cross-linking to RNA molecules to ensure proper adsorption of the RNA molecules to the membrane and it can be stored for longer period by vacuum drying. Hybridisation principle is based on the fact that, two complementary single-stranded DNA molecules pair with each other which is independent to the source and nature of the nucleic acid.

Application of northern blotting involves the known RNA sequences to design a suitable detection probe. These are the labelled nucleotide sequences which are complimentary to the target RNA to be investigated. There are many ways for designing the required probes to study the expression level of RNA either by synthesising the appropriate cDNA in the laboratory or by using the appropriate plasmid; once obtained the sequence can be labelled with the appropriate radioactive probe. In most of the cases, radioactive labelling is done with ^{32}P which is inserted into the sequence in the form of ^{32}P-adCTP. Another chemical labelling of the probe can also be carried out with the enzyme which breaks down to form a chemiluminiscent substance that can be detected with the appropriate device. In another approach, the nucleotide probe can be attached with the enzyme directly for a better efficient practice of detection.

Reagents Required and Their Role

RNA Sample

A higher number of RNA copies are required for detection in northern blot analysis. The isolated RNA sample should be checked thoroughly for their quantity and purity either by spectrophotometer or Nano-drop. The detailed procedure of isolation of RNA sample has been provided in Exp. 1.4.

Agarose Gel

Agarose gel provides the suitable matrix system for the separation of RNA. A 2 % of agarose gel is suitable for the proper separation of RNA molecules. It may either be the total RNA or mRNA which may be separated by agarose gel electrophoresis. As there are so many different RNA molecules separated on the gel, generally it appears to be a smear than the discrete bands. However, addition of formamide leads to the formation of denaturing agarose gel. As RNA can form many different secondary structures that may affect the mobility in the electric field, if not maintained in the proper denatured state.

Salt Solution/Buffer

During northern blotting, saline-sodium citrate buffer (SSC) is used as hybridisation buffer. This buffer controls the stringency during washing steps of northern blotting. When SSC buffer is used at a concentration of 20X, it prevents drying of agarose gel during vacuum transfer. The 20X stock solution consists of 3 M sodium chloride and 300 mM trisodium citrate (pH 7.0).

Labelled Probe

Probe is the single-stranded nucleic acid which is radiolabelled and is used to identify the complementary nucleic acid sequence bound to the membrane. Radioactive labelling involves the enzymatic incorporation of ^{32}P, ^{33}P or ^{35}S. Radioactive labelling provides the most sensitive detection of RNA molecules with a detection limit of 0.01 pg. Non-radioactive labelling involves the attachment of probe to the antibody directed against the enzyme. The immobilised probe can be detected with the enzyme and breaks down to form chemiluminiscent or chemifluorescent substrate to produce the light signal. In addition to that, the probe can also be labelled with biotin which can be detected by streptavidin/avidin-enzyme conjugate.

Low- and High-Stringency Buffers

After hybridisation, unhybridised probes are washed-off by repeated changes of the wash buffer which can be done by low- and high-stringency buffers. Washing using low-stringency buffer removes the hybridisation solution as well as the unhybridised probes whereas; high-stringency buffers remove the partially hybridised molecules. Low-stringency buffer can be prepared by adding 2X SSC with 0.1 % SDS whereas; high-stringency buffer can be prepared by adding 0.1X SSC and 0.1 % SDS.

Maleic Acid Buffer

Maleic acid buffer is used as the blocking reagent and it should be DEPC treated. It should be dissolved properly by constant stirring on 65 °C heating block which may take several hours for proper mixing. Make sure to dissolve it properly and autoclave it and store at 4 °C till further use. It can be prepared by mixing 0.1 M maleic acid, 0.15 M NaCl and the pH should be adjusted to 7.5 by adding solid NaOH. To prepare this buffer add 10 ml of 1 M maleic acid (to prepare 1 M maleic acid, add 11.607 g of maleic acid with 100 ml of milli-Q water) and 15 ml of 1 M NaCl (to prepare 1 M NaCl, add 5.844 g of NaCl with 100 ml of milli-Q water), adjust pH to 7.5 by adding sodium hydroxide pellets and finally adjust the volume to 100 ml.

Detection Buffer

Detection buffer can be prepared by adding 0.1 M Tris Cl, 0.1 M NaCl, pH 9.5. To prepare 100 ml of detection buffer, add 10 ml of 1 M Tris Cl (mix 121.1 g of Tris base in 700 ml of milli-Q water, add HCl to adjust pH to 9.5 and finally make up the volume to 1000 ml with milli-Q water) and 10 ml of 1 N NaCl (to prepare 1 M NaCl, add 5.844 g of NaCl with 100 ml of milli-Q water), adjust pH to 9.5 and make up the volume to 100 ml.

CSPD Stock

Disodium 3-(4-methoxyspiro {1,2-dioxetane-3,2′-(5′-chloro) tricycle [3.3.1.13,7] decan}-4-yl) phenyl phosphate (CSPD) is a chemiluminiscent substrate used for alkaline phosphatase which enables fast and sensitive detection of molecules by producing visible light which can be recorded in film. CSPD stock should be used by mixing 20 µl of CSPD readily available sock with 980 µl of detection buffer

Procedure

Blotting

1. Prepare the gel and rinse it with DEPC treated water.
2. Incubate the gel in 250 ml of transfer buffer for 20 min.
3. Optionally cut-off one corner of the gel to ensure the proper orientation of the gel in subsequent steps.
4. Cut the membrane to roughly 1 mm larger than that of the gel dimension. Cut the corner of the gel as to match with the gel.

5. Float the membrane in container containing deionised water till it gets wet underneath.
6. Soak the membrane with 10X SSC for 5 min.

Transfer

1. Overlap two pieces of blotting paper so that the ends of the blotting paper drape over the edges of the support.
2. Place the blotting paper and the support in the glass baking dish.
3. The dish should be filled with transfer buffer till the top of the support. Remove all the air bubbles (if any) through a pipette when the blotting paper becomes completely wet.
4. Place the gel in inverted position in the centre of the wet blotting paper.
5. Cover the edges of the gel with parafilm which helps in preventing the liquid from flowing directly from the dish to the paper towels.
6. Put few transfer buffers on the top of the gel.
7. Place the wet-nylon membrane on the top of the gel aligning the cut corners. Remove air bubbles (if any) smoothly.
8. Cut two blotting papers to the same size of the gel, wet them in transfer buffer, place them on the top of the nylon membrane and remove any remaining air bubbles.
9. Put a glass plate on the top of the paper towel stock, additionally put 400 g weight on the top of the glass plate.
10. Incubate the set up overnight for efficient upward transfer of the RNA molecule into the membrane.
11. Dismantle the system and mark carefully the positions of the gel lanes in the membrane with a ball point pen.
12. Transfer the membrane into a buffer system containing 6X SSC at 23 °C with slow agitation for few minutes.

Pre-hybridisation

1. Place the appropriate amount of buffer (20 ml for pre-hybridisation and 5 ml for hybridisation) in a sterile tube and incubate it at 50 °C in a water bath.
2. Keep the blot membrane into a hybridisation bottle (prewarmed).
3. Add appropriate amount of prewarmed DIG Easy Hyb buffer into the tube.
4. Carefully remove any air bubbles present between the membrane and the bottle wall.
5. Incubate the blot at 50 °C for 3 h in the hybridisation chamber. Keep rotating the bottle during pre-hybridisation.

Hybridisation

1. Discard the remaining pre-hybridisation buffer and add hybridisation solution to the bottle.
2. Carefully remove the air bubbles present (if any) between the membrane and the bottle wall.
3. Incubate the blot at 50 °C for 6–16 h in a hybridisation chamber while rotating.

Stringent Wash

1. Discard the hybridisation solution.
2. Add 30 ml of low-stringency buffer to the bottle.
3. Incubate the hybridisation bottle for 5 min with rotation at room temperature.
4. High-stringency buffer should be readily available with pre-heated to 50 °C.
5. Discard the low-stringent buffer from the bottle.
6. Immediately add high-stringency buffer to the bottle and incubate the bottle at 50 °C.

Detection

1. Add 50 ml of washing buffer after discarding the high-stringency buffer.
2. Incubate the blot at room temperature for 2 min.
3. Add 30 ml of blocking solution after discarding the washing solution.

4. Incubate the solutions at room temperature for 30 min to 3 h with continuous rotating.
5. Discard the blocking solution and add 10 ml of antibody solution to the tube.
6. Incubate the membrane at room temperature for 30 min with continuous rotating and discard the antibody solution.
7. Wash the membrane twice with 30 ml of washing buffer.
8. Equilibrate the membrane with 20 ml of detection buffer for 3 min.
9. Keep the membrane in hybridisation bag or in a sealed envelope like container.
10. For each 100 cm^2 of membrane use 1 ml of ready to use CSPD (chemiluminiscent substrate for alkaline phosphatase) drop wise over the surface of the membrane until the entire surface is evenly soaked.
11. Never allow the trapped air bubbles between the membrane and the upper surface of the container.
12. Incubate at room temperature for 5 min.
13. Expose the sealed envelope to the luminescence optimised X-ray film for 15–20 min.
14. Depending upon the result obtained, adjust the exposure time to obtain a darker or lighter banding pattern.

Observation

Compare the gel picture and the bands developed on the nylon membrane after detection to find out the expressed RNA of interest. The intensity of the RNA gives the idea of the expression level of the gene of interest under the targeted environmental conditions.

Trouble-Shootings

Problems	Possible cause	Possible solution
Low sensitivity, faint bands or no bands	Incomplete transfer	After transfer of RNA to the membrane from the gel, observe the gel under UV light to see any remaining RNA in the gel
	Interest RNA not properly fixed on the membrane	Check for proper UV cross-linking or oven temperature
High background	Improper washing/contamination in buffer	Wash the membrane thoroughly
	Contaminated forceps	Be sure to clean the forceps after being used with the hybridisation solution containing labelled probe
	Probe added to the membrane	Always add the probe to the hybridisation solution not to the membrane
Non-specific bands	Membrane exposed for too long time in substrate solution	Stop the reaction after few times
Cannot strip the blot for re-probing/partially dried membrane	Inadvertently stopping the rotator during hybridisation	Store the blot at −20 °C for a prolonged time till ^{32}P decays back to the ground level
	Volume of the hybridisation solution is not sufficient to cover the blot	Add required volume of the hybridisation buffer

Precautions

1. Carry out all the experiment in the designated specified area in the laboratory.
2. Store all the reagents and chemicals in specially designated areas.
3. Wear appropriate protective laboratory clothing and disposable gloves.
4. Decontaminate all the waste materials properly by autoclaving at 121 °C.

FLOW CHART

Blotting

Prepare the gel and rinse it in DEPC treated water

↓

Incubate the gel in 250 ml of transfer buffer for 20 min, Sock the membrane with 10X SSC buffer for 5 min

Transfer

Place the blotting paper and the support in the glass baking dish

↓

Place the gel in the inverted position at the centre, cover ends with parafilm, few buffer on the top of the gel

↓

Place the wet nylon membrane on top of the gel, cut blotting papers to the same size and put them above the membrane

↓

Put a glass plate on the paper towel, put 400g weight on the top of the glass plate, and incubate overnight

Pre-hybridization

Place appropriate amount of buffer in a sterile tube and incubate at 50°C in water bath

↓

Add appropriate amount of pre-warmed DIG Easy Hyb buffer into the tube

↓

Incubate the blot for 3 h at 50°C in the hybridization chamber, keep rotating the bottle during pre-hybridization

Hybridization

Discard pre-hybridization solution and add hybridization solution to the bottle

↓

Incubate the blot at 50°C for 6-16 h in hybridization chamber with rotating

Detection

Repeated washing of the blot with low and high stringency buffer

↓

Add detection buffer and CSPD, incubate at room temperature for 5 min

↓

Expose the sealed envelope to X-ray film for 15-20 min and analyse the banding pattern

Exp. 3.6. Isolation of Metagenomic DNA

Objective To isolate metagenomic DNA from soil sample.

Introduction

Metagenomics is the study of genetic material (genomes) from a mixed community of organisms. It usually refers to the study of microbial communities. In most of the cases, microorganisms being the simpler system are studied instead of the complex systems like human to understand the biological processes of life. Microorganisms possess many such properties like complex organisms such as amino acid biosynthesis, protein synthesis and many more. They possess the unique property of degrading the waste materials in situ. Due to their genetic and biological diversity, they have become a huge asset for study in the area of scientific research. However, till now less than 1 % of the total bacteria in nature have been isolated by standard laboratory practices. This leaves the opportunity to study those 99 % of microorganisms for their genetic as well as diversity analysis. In this regard, metagenomics is relatively a new field of research that combines the molecular biology and genetic approach to study the microbial diversity and to characterise their genetic material.

Metagenomics is the comprehensive study of nucleotide sequence, structure, their regulation as well as function. It can be studied by extracting the DNA from organisms, inserting it to a model organism and study of the model organism that expresses this piece of DNA with standard laboratory conditions. Metagenomics involves the systematic investigation of classification and manipulation of the entire genetic material isolated from the environmental sample. This procedure involves isolation of genetic material, manipulation of the genetic material, library construction and analysis of the genetic material in the metagenomic library.

Many microorganisms have the potential to degrade waste materials, synthesis of bioactive compounds, production of environment-friendly plastics and synthesis of food ingredients. Isolation of DNA from these organisms can provide an insight to optimise these processes for their proper adaptation and use for social welfare. However, due to ineffective laboratory culture techniques, the potential wealth of these microorganisms has been relatively untapped, unknown as well as uncharacterised. In this regard, metagenomics represents a powerful technique to analyse abounding microbial diversity of naturally environmental samples. The advantage of this technique is to characterise effectively the genetic diversity present in the samples regardless of the availability of the laboratory culture techniques. Information obtained from the metagenomic analysis provides insight to the industrial applications, therapeutics as well as environmental sustainability. Till now, metagenomics is a new field of molecular biology and is likely to grow to become a standard technique for proper understanding of microbial diversity.

Principle

Direct cultivation or indirect molecular tools can be effectively used for exploitation of microbial diversity from soil. Isolation and cultivation of microorganisms by traditional method analyses about 0.1–1.0 % of total microbial diversity and hence, majority of the soil microbial communities remain unexplored. In most of the cases, it involves the isolation of genetic material, library construction and analysis of genetic material from the metagenomic library (Fig. 3.15).

During metagenomics, samples can be analysed from any environment by direct isolation of the genetic material. Mostly, soil or sediment microbial communities are composed of a mixture of archea, bacteria and protists displaying a diverse array of cell wall composition, which makes them susceptible for different cell wall lysis. Hence, there exists two methods of metagenomic extraction—direct extraction, where cells are lysed in the soil sample and DNA is recovered; and indirect isolation, where cells are first removed from the sample and then lysed for DNA extraction. Soil or sediments are a complex matrix of different substances such as humic acid which is a major contaminant for the DNA and is generally co-extracted during DNA isolation. Hence, humic acid removal is highly essential before further processing of DNA to get a fruitful result.

As microbial diversity of one region is different than another region, soil samples can be collected carefully from the study sites. The collected sample contains huge diversity of microorganisms and the cell of them can be broken by chemical methods such as alkaline treatment or by physical methods such as sonication. Once DNA becomes free from the cells, it can be separated from the other cellular components by taking advantage of the physical and chemical properties of DNA. There are many techniques of DNA isolation that involves density centrifugation, affinity binding, solubility or precipitation. After collection of DNA, it can be manipulated to be used in the model organism. As the size of the genomic DNA is very large it can be cut to smaller fragments by restriction digestion and the fragments are combined with the suitable vector systems. Vectors should be having self-replicating properties besides containing certain selective markers. Then the vectors containing metagenomic DNA fragments can be introduced to the suitable model organisms. This allows the growth of the model organism to express the gene of interest present in it. Analysis of DNA fragments from the metagenomic library involves the determination of physical and chemical properties of the organism. Phenotype of the organism can be studied that attributes the expression of the gene and the chemical properties and can also be analysed for the products synthesised by the model organism. It can be sequenced by Sanger method or next generation sequencing techniques

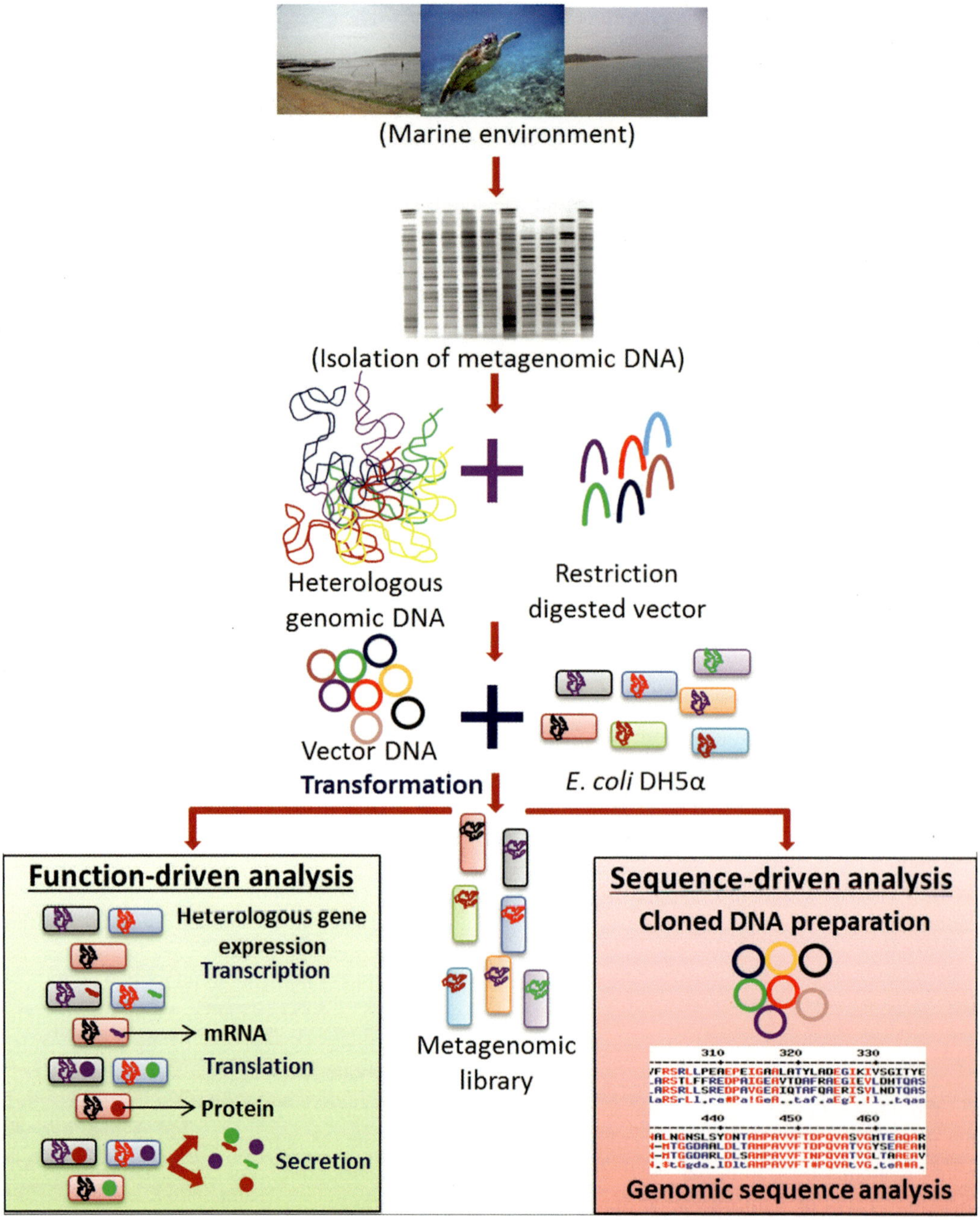

Fig. 3.15 Metagenomic analysis of bioactive compounds produced by the cultivable and uncultivable microorganisms

to get a clear idea on the metagenomes and for effective characterisation of the information coded by the DNA sequence. The information obtained from the metagenomic analysis is helpful in determining the structure, organisation, evolution and origin of the DNA that can be used further for scientific approaches of environmental benefits.

Reagents Required and Their Role

Extraction Buffer

DNA is highly sensitive to pH and hence a buffer system is required to maintain a stable pH environment throughout the process of DNA extraction. Refinements of the technique includes addition of EDTA to sequester divalent cations such as Mg^{2+} and Ca^{2+}, which prevents enzymes like DNase from degrading the DNA. The composition of extraction buffer is: 100 mM Tris-Cl (pH-8.0), 100 mM Sodium EDTA (pH-8.0), 100 mM sodium phosphate (pH-8.0) and 1.5 M NaCl. The extraction buffer can be prepared by mixing 10 ml of 1 M Tris-Cl (to prepare 1 M Tris-Cl, mix 121.1 g of Tris base in 700 ml of milli-Q water, adjust pH to 8.0 by adding HCl and make up the final volume to 1000 ml), 10 ml of 1 M Sodium EDTA (to prepare 1 m sodium EDTA, add 186.1 g of sodium EDTA to 400 ml of milliQ, adjust pH to 8.0 by adding NaOH and make up the volume to 500 ml), 10 ml of 1 M sodium phosphate (to prepare 1 M sodium phosphate, mix 6.8 ml of 1 M NaH_2PO_4 and 93.2 ml of 1 M Na_2HPO_4) and 30 ml of 5 M NaCl (to prepare 5 M NaCl, add 29.25 g of NaCl with 100 ml of milli-Q water), making up the final volume to 100 ml.

Lysis Buffer

It helps in lysis of the cell wall of both cultivable and uncultivable microorganisms, so that the genetic material can come out of the cell easily. SDS and CTAB are the detergents that facilitate the lysis of cells to release DNA. The composition of lysis buffer includes 20% SDS, 1% CTAB. Thus, to prepare 100 ml of lysis buffer, add 20 g of SDS and 1 g of CTAB with 100 ml of milli-Q water, autoclave and store at room temperature.

Chloroform: Isoamyl Alcohol

A 24:1 mixture of chloroform and isoamyl alcohol removes other membrane-bound proteins and lipids during metagenomic DNA extraction. Chloroform isoamyl alcohol is a type of detergent that binds to protein and lipids of cell membrane and dissolves them. In this way it disrupts the bonds that hold the cell membranes together and cause it to breakdown. Then it forms complexes with lipids and proteins causing them to precipitate out of solution. This is due to the fact that, lipids and proteins are non-aqueous compounds and DNA/RNA is aqueous compounds and the detergent binds to non-aqueous compounds.

Isopropanol

DNA is highly insoluble in isopropanol and hence, isopropanol dissolves in water to form a solution that causes the DNA in the solution to aggregate and precipitate out. Isopropanol has been used as a better alternative for ethanol due to its greater potential for DNA precipitation in lower concentrations. Another major advantage of using isopropanol is that it takes lesser time to evaporate.

Procedure

1. Mix 5 g of soil sample with 13.5 ml of DNA extraction buffer [100 mM tris/HCl, 100 mM sodium EDTA, 100 mM sodium phosphate, 1.5 M NaCl, pH-8.0] and 100 µl of 10 mg/ml Proteinase-K in a sterile 50 ml centrifuge tube.
2. Incubate the mixture at 37 °C for 30 min with continuous shaking at 225 rpm.
3. Add 1.5 ml of lysis buffer [20% SDS and 1% CTAB] to each sample and incubate the mixture at 65 °C for 2 h with gentle end-over-end inversions on every 20 min.
4. Centrifuge the mixture at 6000 g for 10 min at room temperature.
5. Collect the supernatant and transfer to a fresh 50 ml autoclaved centrifuge tube.
6. Mix an equal volume of chloroform/isoamyl alcohol (24:1 v/v).
7. Centrifuge the mixture at 7000 g for 30 min and collect the aqueous phase to a fresh tube.
8. Precipitate DNA by adding 0.6 volume of isopropanol with incubation at room temperature for 1 h (e.g. 60 µl of isopropanol for 100 µl of supernatant).
9. Centrifuge at 7000 g for 30 min at 4 °C and discard the supernatant.

10. Wash the pellet with 1 ml of 70 % cold ethanol.
11. Dissolve the pellet with 200 µl sterile milli-Q water or TE buffer.
12. Examine the extracted metagenomic DNA with 1 % agarose gel electrophoresis and analyse the quality of the DNA using Nanodrop.

Observation

The quantity and quality of the isolated DNA can be measured by agarose gel electrophoresis and UV-visible spectrophotometer. For a 1-cm path length, the optical density at 260 nm (OD_{260}) equals 1.0 for the following solutions:

a. 50 mg/ml solution of dsDNA
b. 33 mg/ml solution of ssDNA
c. 20–30 mg/ml solution of oligonucleotide
d. 40 mg/ml solution of RNA

Result Table

Sample	DNA content	260/280[a]	Inference
Control			
Sample I			
Sample II			

[a] For pure DNA $OD_{260/280}$ is 1.8 and for pure RNA it is 2.0. Thus the inference can be drawn from $OD_{260/280}$ values <1.8 more protein contamination and >1.8 more RNA contamination.

Trouble-Shootings

Problems	Possible cause	Possible solutions
RNA contamination	If the bacteria grow to too much higher densities than 1×10^9 cells per ml the chances of RNA contamination becomes more	Grow the bacterial cells $\leq 10^9$ cells per ml
	RNase may not have added	Add RNase (400 µg/ml) to the isolated DNA sample
Protein contamination	If the bacteria grow to too much higher densities than 1×10^9 cells per ml the chances of RNA contamination becomes more	Grow the bacterial cells $\leq 10^9$ cells per ml
		Add 2–3 volume of 100 % ethanol in DNA solution and 1/20 solution volume of 5 M NaCl. Incubate the mixture for 10 min at −20 °C. Centrifuge, discard supernatant and add 500 µl 70 % ethanol
DNA concentration is too less	Culture volume may be too less	Grow the bacterial culture up to 109 or collect more pellet by repeated centrifugation
Insoluble pellet after DNA precipitation	How long and which method has the DNA pellet been dried?	Extended drying under strong vacuum may cause an over-drying of DNA. As an acid, DNA is probably better soluble in slightly alkaline solutions as TE or 10 mM Tris buffer with a pH of 8.0 than water
Degraded DNA	Is the bacterial strain known as being 'problematic?'	Don't let the bacterial culture grow for more than 16 h

Precautions

1. Cut tips should be used to avoid mechanical disruption of DNA.
2. The incubation period with Proteinase-K may be extended depending upon the source of DNA.
3. Repetition of phenol chloroform extraction method should be performed to obtain a pure DNA.
4. DNase free plasticwares and reagents should be used during the entire experiment.
5. Phenol/chloroform is probably the most hazardous reagent used regularly in molecular biology labs. Phenol is a very strong acid that causes severe burns. Chloroform is a carcinogen. Handle these chemicals with care.
6. Wear gloves and goggles while isolating genomic DNA.

FLOW CHART

Take 5 g of sediment sample and mix with 13.5 ml of Extraction buffer in 50 ml sterile centrifuge tube

↓

Incubate at 37°C for 30 min with shaking at 225 rpm

↓

Mix 1.5 ml of 20% SDS and incubate the mixture at 65° for 2 h with gentle end over end inversion at an interval of every 20 min

↓

Centrifuge at 6,000 g for 10 min at room temperature and transfer the supernatant to a new 50 ml centrifuge tube

↓

Add equal volume of chloroform/isoamyl alcohol (24:1 v/v)

↓

Remove the aqueous phase by centrifuging at 7,000 g for 30 min at 4°C

↓

Precipitate DNA with 0.6 volume of isopropanol at room temperature for 1 h

↓

Wash with ice cold ethanol and dissolve with Milli Q water and store at 4°C till further use

Exp. 3.7. Plasmid Curing from Bacterial Cell

Objective To obtain a plasmid-cured bacterial cell.

Introduction

Plasmids are the extra chromosomal dsDNA molecules with the capability of self-replication without the help of the host chromosome. Many bacterial species from environmental sample contain plasmid DNA and the stability of the plasmid is maintained by successive partition to each daughter cell during cell division so that each cell receives a copy of plasmid DNA. In most of the cases, plasmid DNA is responsible for the functional genes mediated resistance such as antibiotic and heavy metal resistance phenomenon. In this regard, plasmid curing is the process of complete removal of plasmids from bacteria in vitro by chemical agents such as acriflavin or acridine orange.

During experiments with plasmid containing bacteria, it is often required to develop a plasmid cured derivative to obtain a clear comparison between plasmid containing and plasmid lacking bacteria. However, certain bacterial plasmids undergo spontaneous segregation and deletion; whereas, most of them are stable and hence require certain curing agents or other physical methods like elevated growth temperature and thymine starvation for spontaneous segregation of plasmids. The curing agents do not produce a fruitful result in most of the bacterial strains and there does not exist any gold standard protocol for all plasmids.

In situations where plasmid is stable and the loss of its property is difficult to determine, bacteria can be treated with the various curing agents. These agents may include the chemical or physical agents that mutate DNA, interfere with their replication, effect on structural components or enzymes of the bacterial cell. In this regard, the elimination of plasmid from bacterial culture establishes the relationship between the genetic trait and carriage of specific plasmid as the phenotypic characteristics associated with plasmid do not express with the cured strains and the reintroduction of the plasmid is responsible for the reappearance of the particular phenotypic characteristic. The efficiency of plasmid curing is dependent on the nature of the plasmid as well as the host system. In most of the instances, curing experiments are performed under conditions similar to the conditions used for routine bacterial culture. If the treatment is for longer period, then the agent should be used at a lower concentration. However, if acridine orange is used, the culture should be maintained at pH 7.6 as well as should be incubated in the dark.

Principle

The subinhibitory concentrations of chemical agents like acridine orange and SDS may lead to plasmid curing leading to the generation of cured derivatives. Acridine orange and ethidium bromide act as the intercalating agents are found to be effective against a wide variety of bacterial genera. Though most of the agents have been practised against various genera of bacteria for plasmid curing, their effect is unpredictable in many cases. However, in most of the cases, the principle of plasmid curing is not known till date, however, studies suggest their interfering ability with plasmid replication that occurs due to induction of heat or by the use of ethidium bromide or acridine orange (Fig. 3.16). In contrast, plasmid curing interfers with the growth of plasmid carrying bacteria and allowing spontaneous arise of plasmid less generations to become predominant. This occurs predominantly by the use of acridine orange, sodium dodecyl sulphate and urea. However, plasmid curing by the use of Mitomycin-C has also proved to be effective.

There is a relationship between the antibacterial and antiplasmid activities of the chemical agents with the inhibition of bacterial efflux pumps. Plasmid elimination in many compound occur only in their sublethal concentrations. As the sublethal concentration of the chemical agents allows the bacteria to grow, this is the prerequisite for plasmid elimination. This is due to the fact that, plasmid replication is dependent on bacterial replication and thus affecting the nature of subsequent generations. When bacteria carrying the *lac*-plasmid are incubated with subinhibitory concentrations of chemical agents and aliquots are streaked on drug free medium containing eosine methylene blue (EMB), cured bacterial species can be screened based upon the colour of the developed colony, i.e. deep violet colonies (*lac* + strains) and pink colonies (*lac* − strains).

The percentage of plasmid elimination can be determined by the number of pink colonies dividing with the total number of colonies. Majority of substituted phenothiazines of GroupA and GroupB possess high-plasmid-curing activities

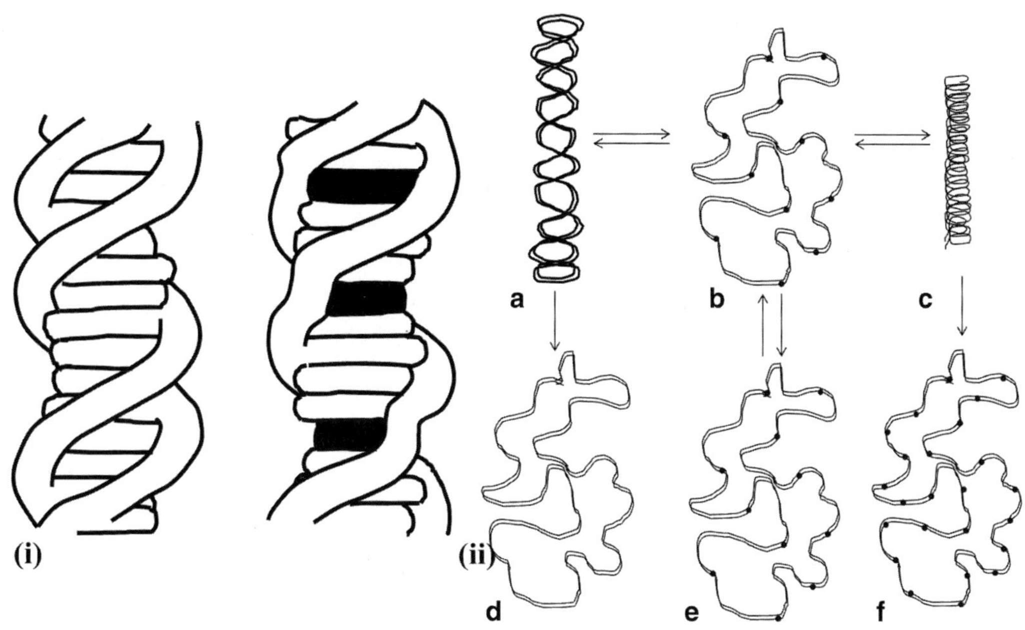

Fig. 3.16 Interaction of curing agents with the plasmids. **a′** Interaction of ethidium bromide with plasmid DNA; **b′** Effect of ethidium bromide on orientation of plasmid DNA (**a**) dye-free molecule with superhelical turns, **b** formation of positive supercoils with the addition of ethidium bromide, **c** reversible binding of the dye with the plasmid DNA generating relaxed plasmids, **d**–**f** reversible binding of dye with plasmid DNA generating plasmids that remain relaxed throughout

in their subinhibitory concentrations. 2-chloro-10-(2-dimethylaminoaethyl)-phenothiazine is the most active agent in this group with a plasmid curing activity of 90% with a MIC value of 3.1 μg/ml. In most of the cases, the inhibition of plasmid replication results from the single nick outside the replication origin of the superhelical structure, which leads to further relaxation of plasmid DNA. Intercalation of these compounds with the plasmids can be proved by an increase in melting point of the DNA and by circular dichroism. Thus, when the plasmid profile is analysed by agarose gel electrophoresis, superhelical form will be missing from the promethazine treated plasmid DNA. However, open circular and linear forms of plasmids will be present in an increased proportion.

In another approach, the replication of plasmid can be inhibited by blocking the activity of DNA gyrase. Subsequently, the superhelical turns into plasmid DNA is prevented. Promethazine and imipramine have been reported to be suitable plasmid curing agents as a specific binding site is found for promethazine interaction apart from intercalation with the dsDNA molecules. The antiplasmid compounds are also capable of inhibiting plasmid transfer as a dose-dependent inhibition of conjugational plasmid transfer have been observed in many cases leading to inhibition of both transconjugational DNA synthesis and mating pair formations.

Reagents Required and Their Role

Acridine Orange

Acridine orange is a nucleic acid binding dye which is used mostly for the determination of cell cycle information. It is permeable to the cell wall and readily interacts with nucleic acids by intercalation or electrostatic attractions. During epifluorescence microscopy the role of this dye is of worth use. DNA and RNA can also be distinguished by the use of acridine orange as it emits green colour when forms complex with DNA whereas, emits orange colour upon binding with the RNA molecule.

LB Medium

Luria Bertani (LB) broth is a rich medium that permits the fast growth and good growth yields for many species including *E. coli*. It is the most commonly used medium used in molecular biology studies for *E. coli* cell culture. Easy to make, fast growth of most *E. coli* strains, readily available and simple compositions contribute the popularity of LB broth. LB can support *E. coli* growth OD_{600} 2–3 under normal shaking incubation conditions.

Muller-Hinton Agar

It is the microbiological growth medium mostly used for antibiotic susceptibility testing. Earlier it was used to isolate *Neisseria* and *Moraxella* species. As per Clinical and Laboratory Standard Institute (CLSI) Muller-Hinton agar (MHA) medium have been recommended to be used for antibiotic susceptibility testing.

Penicillin and Ciprofloxacin Resistant *E. coli* Culture

Antibiotic resistance is the phenomenon where bacteria can survive when exposed to a concentration of antibiotics. Whereas, the bacteria that resist to a multiple number of antibiotics are called as multidrug resistant bacteria. These antibiotic resistant genotypes mostly occur in the plasmids of bacteria thus conferring resistance to them.

Procedure

1. Inoculate a loop full of culture (penicillin and ciprofloxacin resistant *E. coli*) to a 5 ml LB broth for 24 h at 37 °C.
2. Prepare 50 ml of LB broth as per the manufacturer's instructions and autoclave it.
3. After autoclaving, add acridine orange at a final concentration of 0.10 mg/ml from the stock solution and mix well.
4. Inoculate antibiotic resistant *E. coli* culture from the over-night grown LB broth to the LB with acridine orange.
5. Incubate it at 37 °C for 24 h with vigorous shaking at 180 rpm.
6. After incubation swab the grown culture in prepared MHA plates.
7. Put penicillin and ciprofloxacin discs on the swabbed plates as per the manufacturer's instructions.
8. Incubate the plates for 24 h at 37 °C and observe for the zone diameter in millimetre.
9. Follow the chart provided by the manufacturer to correlate between the zone diameter and the resistance or sensitive phenotype by the bacterial culture.

Observation

Observe the zone diameter in millimetre and compare the values for the culture before plasmid curing and after plasmid curing.

Result Table

Bacterial culture	Zone of diameter (in mm)		Inference (R/S/I)[a]
	Penicillin	Ciprofloxacin	
E. coli before curing			
E. coli after curing			

[a] Resistant/Sensitive/Intermediate

Trouble-Shootings

Problems	Possible cause	Possible solutions
Bacteria showing resistance to antibiotics	Concentration of acridine orange is not appropriate	Increase and optimise the concentration of acridine orange for the desired strain
	Antibiotic resistance phenotype not detected properly	Confirm the antibiotic resistance genotypre present in the plasmid of the strain
Bacteria showing intermediate to antibiotics	Concentration of acridine orange is not appropriate	Increase and optimise the concentration of acridine orange for the desired strain
	Acridine orange is not suitable	Use other plasmid curing agents like ethidium bromide

Precautions

1. Always wear gloves.
2. Cover your body carefully as the plasmid curing agents used are mostly carcinogenic in nature.
3. Before initiating the experiment, be sure about the antibiotic resistance genotype present in the plasmid.
4. Use properly cleaned glass wares and plastic wares for the purpose of plasmid curing.
5. Confirm the plasmid curing in bacteria by plasmid extraction and PCR amplification of the antibiotic resistant gene.

FLOW CHART

Inoculate overnight grown culture of antibiotic resistant *E. coli* to 50 ml LB supplemented with 0.10 mg/ml of acridine orange

↓

Incubate for 24 h at 37°C with 180 rpm shaking

↓

Swab the grown culture in MHA plates and put penicillin and ciprofloxacin antibiotic discs on to the plate

↓

Incubate the plates at 37°C for 24 h and measure the zone diameter in mm

↓

Observe the resistivity or sensitivity of the strain after plasmid curing

Exp. 3.8. Conjugation in Bacteria

Objective To transfer plasmid from one bacterium to another by horizontal gene transfer by conjugation.

Introduction

Bacterial conjugation is the transfer of genetic material in terms of plasmids from one bacterium to another by direct cell-to-cell contact to carry out horizontal gene transfer. The transfer of gene is carried out from a donor to a recipient cell by direct contact between them. In most of the cases, the donor cell contains 'F factor' that is responsible for the formation of the sex pilus, whereas, the recipient cell lacks the 'F factor'. Transfer of the genetic material from one cell to another is however beneficial to the recipient cell as it provides antibiotic resistance, toxic metal tolerance and potential to use new metabolites. Conjugational transfer of genetic information from donor to recipient generates transconjugants which harbour mainly the plasmids; however conjugative transposons have also been reported. Co-integration of both conjugative and non-conjugative circular plasmid results in the transfer of both the plasmids into the recipient strain, e.g. in case of Hfr, F plasmid is integrated into the bacterial chromosome and subsequently the chromosome is conjugated to the recipient strain. Thus, the conjugative transposon is transferred from one bacterium to another.

Inter-kingdom conjugation has been reported between tumour inducing plasmid of *Agrobacterium* and root tumour inducing plasmid of *A. rhizogenes*. It is a useful technique of transfer of genetic material to wide targets of organisms. In addition to that, successful transfer of genetic material has been carried out between bacteria to yeast, plants and mammalian cells. Conjugation is highly advantageous with respect to other gene transfer mechanisms as it allows minimal disruption of the target cell's membrane and it can transfer relatively large size of genetic material from one organism to another.

Conjugation has been reported in various genomes of bacteria that includes *Escherichia*, *Salmonella*, *Pseudomonas*, *Serratia* and *Vibrio*. However, the intergeneric conjugation has been reported with the following pairs of genera, i.e. *Escherichia-Shigella*, *Salmonella-Vibrio*, *Escherichia-Serratia*, *Salmonella-Serratia*, *Shigella-Salmonella*, and *Escherichia-Salmonella*. Among Gram-negative bacteria, rapid spread of antibiotic resistance is the major cause of conjugation whereas, in Gram-positive bacteria conjugation occurs via production of adhesive materials by the donor cells.

Principle

Bacteria do not have sexual mode of reproduction; however, they possess a form of sexuality called as conjugation. It is the unidirectional transfer of genetic material most preferably DNA from one organism to another. To enable the transfer of DNA from donor (also called as male) to the recipient (also called as female), the conjugation tube is formed between the two organisms. In this process, a fractional segment of the donor's chromosome passes to the recipient. An auxotroph is the bacterial mutant that requires one or more growth factors than the wild strain which is also called as prototroph. Auxotrophs can be synthesised in the laboratory by exposing the prototrophs to certain mutagenic agents such as ultraviolet irradiation or mitomycin-C.

The strains to be used in this study are one donor strain which is resistant to tetracycline (30 μg/ml) and the recipient strain is resistant to streptomycin (100 μg/ml). Hence, the donor strain cannot grow in presence of streptomycin antibiotic and the recipient strain is unable to grow in presence of tetracycline antibiotic. However, when conjugation takes place successfully, the transconjugant possesses the mechanism of resistance towards both the antibiotics and can grow on plates supplemented with both the antibiotics.

F-factor, the fertility factor is known to be the plasmids which are the extrachromosomal genetic elements. Mostly plasmids contain 23–30 genes, most of which are associated with conjugation. These genes are responsible for coding certain proteins that help in replication of DNA during conjugation and also help in synthesising certain structural proteins necessary for sex pilli synthesis on the cell surface. Sex pilli is responsible for the formation of hair like fibres and when retracts the surfaces of donor and recipient become close to one another to touch. At the area of contact the formation of a channel the conjugation bridge takes place. Once the formation of the sex pilli is over, the plasmid or F factor replicates by rolling circle mechanism and one single strand is passed through the channel to the recipient. After its arrival with the recipient organism, the enzymes synthesise the complimentary strand to form a dsDNA molecule. The double helix DNA then bends to form a loop and thus converts the F − cell to a F + cell. However, the transfer of F factor does not involve the activity of the bacterial chromosome and thus is not responsible for transfer of any new genes other than present on the F factor (Fig. 3.17).

In certain instances, F plasmids are sometimes found in association with the chromosome of the bacterial host, thus generate the high-frequency recombination (Hfr) cells. These cells are also capable of synthesising sex pillus. When the Hfr cells begin to replicate for their transfer to the recipient cell, in addition to the cellular plasmid, a fraction of the chromosomal DNA is also transferred to the recipient cell. These DNA molecules recombine with the genetic material of the host cell and are responsible for the generation of new gene variants.

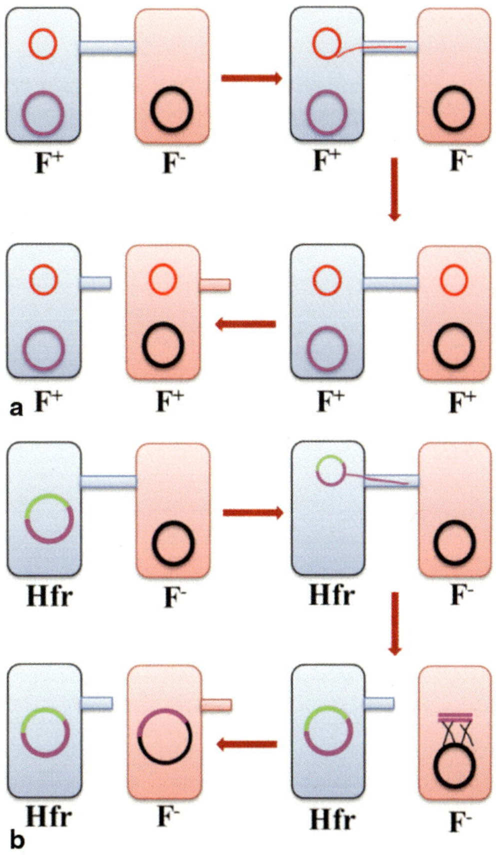

Fig. 3.17 Mechanism of conjugation in bacteria **a** mechanism of $F^+ \times F^-$ crosses, **b** mechanism of Hfr $\times$ F^- crosses

Materials Requirement and Their Role

Donor *E. coli* Resistant to Tetracycline

Many different mechanisms of tetracycline resistance in *E. coli* are found, viz. tetracycline efflux, ribosome protection and tetracycline modification. In most of the cases, tetracycline resistance is achieved by an export protein from major facilitator superfamily (MFS). This mechanism is mostly acquired by the transferable plasmids and/or the transposons.

Recipient *E. coli* Resistant to Streptomycin

Mostly a mutation in *rpsL* gene that codes for S12 polypeptide generates resistance to streptomycin. By genetically altering the number of *rspL* and *rrs* alleles present in bacterial genome, bacteria confers resistance to streptomycin. The major difference between streptomycin resistances in bacteria to tetracycline resistance is the presence of mutated genomic fragment in the bacterial genome rather than the plasmid, hence cannot be transferable by horizontal gene transfer in terms of conjugation.

Tetracycline

Tetracycline is a broad spectrum antibiotic extracted from *Streptomyces* genus of Actinobacteria. It is a protein synthesis inhibitor that binds with the 30S subunit of microbial ribosome thus blocking the attachment between charged aminoacyl tRNA with the A-site of the ribosome and subsequently preventing the introduction of new aminoacids to nascent peptide chain. The concentration required for this experiment is 30 mg/ml which can be prepared by dissolving 45 mg of tetracycline with 1.5 ml of 70% ethanol.

Streptomycin

Streptomycin is the aminoglycoside group of antibiotic derived from the actinobacterium *Streptomyces griseus*. Mode of action of streptomycin against bacterial system is by inhibiting protein synthesis that binds with the 30S subunit of bacterial ribosome, thus leading to codon misreading followed by inhibiting protein synthesis and leading to death of the bacteria.

Luria Bertani Medium

Luria Bertani (LB) agar is a rich medium that permits the fast growth and good growth yields for many species including *E. coli*. It is the most commonly used medium used in molecular biology studies for *E. coli* cell culture. Easy to make, fast growth of most *E. coli* strains, readily available and simple compositions contribute the popularity of LB broth. LB can support *E. coli* growth OD_{600} 2 to 3 under normal shaking incubation conditions.

Procedure

1. Prepare the reagents carefully and autoclave the LB agar medium, do not sterilise the antibiotic solutions.
2. After autoclaving the LB medium, cool it around 50 °C, add the respective antibiotics and then prepare the plates.
3. Using a sterile loop streak the donor cell of *E. coli* on LB plates supplemented with 30 µg/ml tetracycline and the recipient cell on LB plates supplemented with 100 µg/ml of streptomycin.
4. Incubate the plates at 37 °C for 24 h and observe for the growth of the cultures on the respective plates.
5. Carefully pick up a single colony from the over-night grown plates of donor and recipient cells and inoculate to the 5 ml of LB broths with respective antibiotics.
6. Incubate the tubes at 37 °C for 24 h with shaking at 180 rpm.
7. Inoculate 1 ml of the overnight grown culture to 25 ml of LB flasks with respective antibiotics.
8. Incubate at 37 °C with shaking at 180 rpm till 5–6 h or till the OD630 of the culture reaches 0.8–0.9.
9. Mix 0.2 ml of both donor and receipient cultures in a tube by gently pipetting and incubate at 37 °C for 1–1.5 h.
10. Add 2 ml of sterile LB broth to the tube after incubation for recovery and incubate at 37 °C for 1.5 h.
11. Plate 0.1 ml of the culture on the LB agar plates containing both the antibiotics.
12. Incubate the plates at 37 °C for 24 h and observe for the growth of the bacterial colony, i.e. the development of the transconjugants.

Observation

Observe the development of the cultures on different plates as donor cells are supposed to grow on the plates supplemented with streptomycin only whereas the recipient cells grow only in plates with tetracycline. The conjugants are supposed to develop on the plates containing both the antibiotics and the donor and recipient cells will not grow on the plates supplemented with both the antibiotics, i.e. tetracycline and streptomycin.

Result Table

Bacterial culture	No. of colonies		
	LB + Streptomycin	LB + Tetracycline	LB + Streptomycin + Tetracycline
Donor *E. coli* cell			
Recipient *E. coli* cell			
Conjugated sample			

Trouble-Shootings

Problems	Possible cause	Possible solution
Improper growth on the respective plates	Pouring of media at high temperature	Make sure to add the respective antibiotics to LB media when the temperature is in the range of 40–45 °C. Higher temperature degrades the antibiotics and hence inhibits their mode of action
	Storing of plates for a longer time	Always use plates within 1 month of preparation
	Error in pipetting	Make sure to add appropriate amount of antibiotics without any pipetting error. Always use fresh autoclaved tips while spreading the samples on the plates

Precaution

1. Always use globes during the experimental set up.
2. Never incubate the donor and recipient cells under shaking conditions.
3. Do not autoclave the antibiotic solutions.
4. Use syringe filter for antibiotic solutions.
5. Use appropriate concentrations of antibiotics for both broth and plate preparation.

FLOW CHART

Streak both donor and recipient cells on LB plates supplemented with respective antibiotics

↓

After over-night growth, inoculate single bacterial culture into 5 ml of LB broth with suitable antibiotic, incubate at 37° for 24 h at 180 rpm

↓

Inoculate 1 ml of over-night culture to 25 ml of LB broth and incubate till the growth reaches OD_{630} 0.8-0.9

↓

Mix 0.2 ml of both donor and recipient cultures in a tube and incubate at 37° for 1-1.5 h

↓

Add 2 ml of sterile LB broth to the tube after incubation for recovery and incubate at 37° for 1.5 h

↓

Inoculate 0.1 ml of the culture on the LB agar plates containing both the antibiotics

↓

Incubate the plates at 37°C for 24 h and observe for the growth of the bacterial colony for the development of transconjugants.

Exp. 3.9. Transduction in Bacteria

Objective To study the induction of generalised transduction in *Salmonella typhimurium* by lambda bacteriophage P22.

Introduction

Transduction is the process of transfer of genetic material from one bacterium to another mediated by virus. The group of viruses that infect bacteria are called as bacteriophages and they use the bacterial system machineries as hosts to multiply their number. During the course of multiplication inside the host bacterium, the virus uses the host cell's replication machineries and occasionally removes a portion of the host cell's bacterial DNA and after assembling when it infects a new host cell, the bacterial DNA may get incorporated into the new host cell's genome thus completing the transfer of genetic material. Phage particles take advantages of the bacterial cell's replicational, transcriptional and translational machinery for their replication and to synthesise now virion particles including their DNA or RNA and the protein coat.

There are many applications of transduction mediated gene transfer as the rapid spread of antibiotic resistance and correction of genetic diseases by direct modification of the genetic errors. Transduction may occur by one of the two ways, i.e. generalised transduction or specialised transduction. During generalised transduction, bacterial gene may be transferred to another bacterium mediated by bacteriophage. When the new piece of DNA gets inserted into the bacterial cell, the DNA may be adsorbed by the cell and recycled for spare parts, if the original DNA is a plasmid, it may recircularise to become a functional plasmid again or when the foreign DNA matches with the homologous region of the recipient cell's chromosome, it may exchange DNA material with it. However, during specialised transduction a restricted set of bacterial genes get transferred to another bacterium or in other terms, bacteriophages pick up only specific portions of the host's DNA for its transfer. Transduction process has been successfully employed for the introduction of gene of interest into various other cells using viruses.

The bacteriophage P22 for *Salmonella enterica* serovar Typhimurium and P1 for *E. coli* are the most preferable choice of vectors for generalised transduction purpose. Cotransduction has become a powerful tool for gene mapping and is continuing to be used widely for strain construction. Though generalised transduction is considered to be an accidental concomitant of phage infection rather than a systematic biological function, still it is the useful technique for plasmid transformation along with conjugation for gene transfer between *E. coli* and *S. typhimurium* strains. Other successful uses of this technique involves transductional mapping, strain construction, localisation of mutagenesis, moving genes location and transfer of genes between *E. coli* and *S. typhimurium*.

Principle

Bacteriophage mediated transfer of genetic element from one bacterium to another is known as transduction and has been proved to be a useful tool for use in molecular biology experiments for stable introduction of a foreign gene into a host cell's genome. However, bacteria can transfer its genetic element from one to another by different mechanisms that include conjugation, transformation or transduction. During transformation, it involves the acquisition of DNA from environment; hence it is highly susceptible to DNase. During conjugation DNA is generally acquired directly to a bacterial cell via cell to cell contact between them which is not practically possible in all the cases. In this regard, transduction is a better alternative that involves the transfer of genetic element via a bacteriophage intermediate which does not require cell to cell contact and is generally DNase resistant.

The entire process of transduction follows several steps, i.e. firstly; the phage infects a susceptible bacterium and injects its piece of DNA into the host followed by the utilisation of the host cell's machineries to synthesise phage components including phage DNA. This procedure may lead to the integration of bacterial chromosome into the phage DNA. During final stage of the phage cycle, all the phage components in cytoplasm assemble to form a complete phage particle and the cells are lysed to release the newly formed phage particles. Subsequently, when the new phage particles infect another recipient bacterium, the phage DNA along with the parts of the donor bacteria's chromosome is injected into it and hence, the transduced bacterial genes are incorporated by recombination. Though transduction is a different form of the usual gene recombination process, however the striking difference is the involvement of the phage particle. Another salient feature of transduction is that, it involves the transfer of only a small portion of the total genetic material which is transferred by the bacteriophage particle (Fig. 3.18).

Transduction practice involves two different pathways generating either generalised or specialised transduction. During the process of generalised transduction, any part of the bacterial genome is transferred to a bacterium and it does not carry the viral DNA. During the lytic cycle of the phage particle, when a phage particle infects a bacterial system it takes control of the host cell machineries and replicates its own viral DNA. By chance, if the bacterial chromosome is inserted into the viral capsid during encapsulation, it leads to generalised transduction. Similarly, during specialised transduction, a specific part of the bacterial genes that are located near the phage genome are transferred by the phage particle. Specialised transduction involves three possible outcomes, i.e. DNA may be absorbed and recycled, bacterial DNA matches with the homologous DNA of the recipient cell and exchanges it, and thus recipient cell contains DNA from both the same and other bacterial cell. Oth-

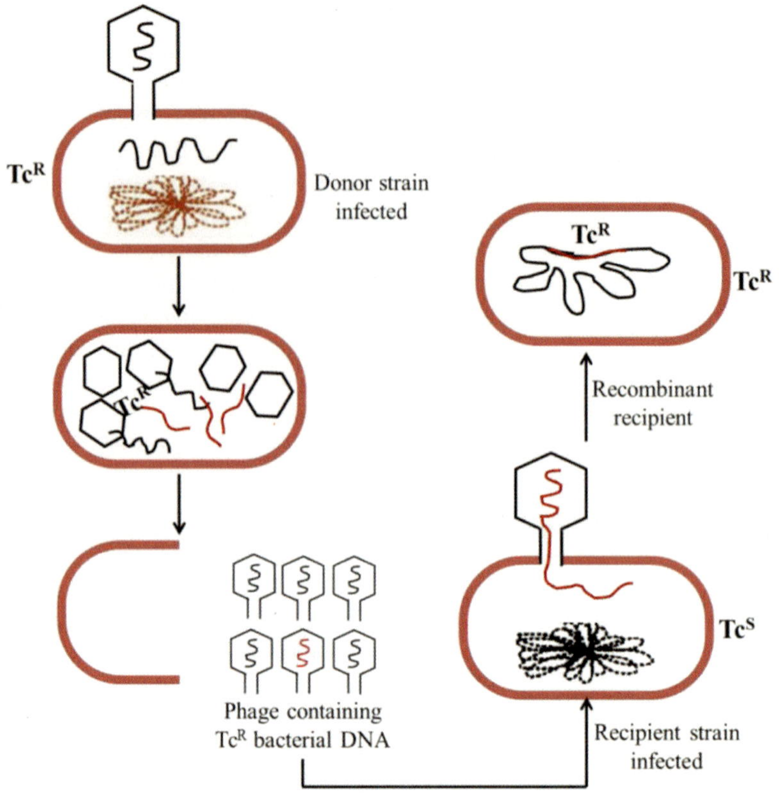

Fig. 3.18 Transfer of Tc^R region of donor bacterium to a Tc^S bacterium via P22 bacteriophage

erwise, DNA can be inserted into the genome of the recipient cell as the virus results in the generation of a double copy of bacterial genes.

During this experiment, an antibiotic resistant gene is transferred from one bacterium to another through a bacteriophage.

Reagents Required and Their Role

Donor Strain

Donor *E. coli* is the bacterial culture which is resistant to chloramphenicol. Mostly, there are three mechanisms of resistance to chloramphenicol, i.e. reduced permeability, mutation of 50S ribosomal subunit and elaboration of chloramphenicol acetyltransferase. High level of chloramphenicol resistance is mainly coded by the cat gene that codes for an enzyme chloramphenicol acetyltransferase that inactivates chloramphenicol and acetylation prevents binding of chloramphenicol to the ribosome. The resistance mechanism can also be found on the plasmid, e.g. ACCoT plasmid which is responsible for multiple drug resistance in bacteria.

Susceptible Host

Any bacterial strain that is sensitive to chloramphenicol and resistant to ampicillin can be used for this purpose as susceptible host. The ampicillin resistance gene (ampR) is mostly used as a selectable marker for routine biotechnology experiments. β-lactamase, the enzyme coded by this gene is responsible for degrading ampicillin.

P22 Phage Lysate

P22 is a bacteriophage λ which is mostly used for induction of mutation in bacteria and to introduce foreign pieces of DNA. This phage contains dsDNA as its genetic material as well as it harbours the gene expression control regions. During infection it circularizes its DNA and replicates along with the host DNA by rolling circle mechanism. In most of the transduction experiments P22 bacteriophage particle is used for studying bacterial genetics and strain construction.

Ampicillin

Ampicillin is a β-lactam group of antibiotic which is found to be efficient for both Gram-positive and Gram-negative bacterial population. It is responsible for the irreversible modification of enzyme transpeptidase that is responsible for the synthesis of bacterial cell wall. Thus it inhibits the final stage of cell division during binary fission leading to cell lysis. During this experiment, the concentration of ampicillin to be used is 100 μg/ml.

Chloramphenicol

Chloramphenicol is mostly used as a bacteriostatic agent that acts against both Gram-positive and Gram-negative bacteria. Mostly, it inhibits bacterial growth by inhibiting protein synthesis. It prevents elongation of protein chain by inhibiting peptidyl transferase activity bacterial ribosome. The concentration to be used during this experiment is 20 μg/ml.

LB Medium

LB agar is a rich medium that permits the fast growth and good growth yields for many species including *E. coli*. It is the most commonly used medium used in molecular biology studies for *E. coli* cell culture. Easy to make, fast growth of most *E. coli* strains, readily available and simple compositions contribute the popularity of LB broth. LB can support *E. coli* growth OD_{600} 2 to 3 under normal shaking incubation conditions.

1 M Calcium Chloride

For the proper attachment of phage particles and the transfer of the genetic material inside the bacterial cell, bacterial cell needs to be treated with 1M calcium chloride. It helps in easy uptake of the DNA molecule inside the bacterial cell. After addition of $CaCl_2$, it dissociates into Ca^{2+} and 2 Cl^- molecules and hence, the positive charge of Ca^{2+} cancels the negative charge of DNA thus allowing crossing the cell wall and cell membrane. Thus further steps during the experiment after addition of $CaCl_2$ should be carried out by incubating at 4 °C, otherwise the bacterial cell dies. To prepare 1M calcium chloride solution, add 85 g of $CaCl_2 \cdot 2H_2O$ in 1000 ml of milli-Q water, autoclave and store at 4 °C.

Procedure

1. Streak a loop-full of donor *E. coli* strain on LB agar plates containing ampicillin and the susceptible host strain on LB agar plates.
2. Inoculate loop-full of culture to the respective 5 ml LB broth medium with suitable antibiotics.
3. Incubate the plates at 37 °C and the tubes at 37 °C shaker for 24 h with shaking at 300 rpm.
4. Inoculate 10–15 colonies from the plate containing donor cell to 5 ml LB broth tube containing 20 μg/ml of chloramphenicol.
5. Incubate the tube at 30 °C with shaking for 2 h, prewarm 5 ml aliquot of LB broth in water bath at 60–65 °C.
6. Add 100 μl of the phage lysate to the donor tube and continue incubation for 30 min at 30 °C.
7. Add 2 ml of pre-heated sterile LB broth to the tube containing donor *E. coli* cell, mix well and incubate the tube at 42 °C for 20 min.
8. Further incubate the tube at 37° for 3 h.
9. After incubation, centrifuge the culture at 5000 rpm for 10 min. Filter the supernatant through 0.45 μm filter paper to collect the phage lysate and store at 4 °C till further use.

10. Inoculate an individual single colony from the plate containing the recipient strain in 5 ml of LB broth containing 100 μg/ml ampicillin, incubate the tube at 37 °C for 24 h with shaking at 180 rpm.
11. Inoculate 100 μl of the overnight grown culture in 5 ml of fresh LB broth containing 100 μg/ml ampicillin and incubate at 37 °C for 2 h.
12. Take 50 μl of this culture in a 2 ml micro-centrifuge tube and add 50 μl of 0.1 M $CaCl_2$ and 250 μl of phage lysate obtained from the step 9.
13. Mix well and incubate at 37 °C for 2 h, be careful not to incubate in a shaker.
14. After incubation, take 50 μl of the culture and plate on LB plates containing 20 μg/ml chloramphenicol, 100 μg/ml of ampicillin and both the antibiotics respectively.
15. Incubate all the plates at 37 °C for 24 h.
16. Observe the result on the next day for the development of transduced colonies on the plate containing both the antibiotics.

Observation

Observe the development of the cultures on different plates as donor cells are supposed to grow on the plates supplemented with ampicillin only whereas the recipient cells grow only in plates with chloramphenicol. The transduced strains are supposed to develop on the plates containing both the antibiotics and the donor and recipient cells will not grow on the plates supplemented with both the antibiotics, i.e. ampicillin and chloramphenicol.

Result Table

Style3Bacterial culture	No. of colonies		
	LB + ampicillin	LB + chloramphenicol	LB + ampicillin + chloramphenicol
Donor cell			
Recipient cell			
Transduced cell			

Trouble-Shootings

Problems	Possible cause	Possible solution
Improper growth on the respective plates	Pouring of media at high temperature	Make sure to add the respective antibiotics to LB media when the temperature is in the range of 40–45 °C. Higher temperature degrades the antibiotics and hence inhibits their mode of action
	Storing of plates for a longer time	Always use plates within 1 month of preparation
	Error in pipetting	Make sure to add appropriate amount of antibiotics without any pipetting error. Always use fresh autoclaved tips while spreading the samples on the plates

Precaution

1. Always use globes during the experimental set up.
2. Never incubate the donor and recipient cells under shaking conditions.
3. Do not autoclave the antibiotic solutions.
4. Use syringe filter for antibiotic solutions.
5. Use appropriate concentrations of antibiotics for both broth and plate preparation.

FLOW CHART

Revive both the donor strain and the susceptible host on LB plates containing suitable antibiotics

↓

Inoculate loopfull of culture to the respective 5 ml LB broth medium with suitable antibiotics and incubate the plates and tubes at 37° for 24 h

↓

Inoculate 10-15 colonies from the plate containing donor cell to 5 ml LB broth tube containing 20 μg/ml chloramphenicol

↓

Incubate at 37°C with shaking for 2 h, pre-warm 5 ml sterile LB broth in water bath to 60-65°C

↓

Add 100 μl of phage lysate to the donor cell, incubate at 37°C for 30 min

↓

Add 2 ml of pre-heated LB broth to the tube, mix well and incubate the tubes at 42°C for 20 min, continue incubating at 37°C for 3 h

↓

Centrifuge the culture at 5,000 rpm for 10 min, filter the supernatant with 0.45 μm filter paper to collect the phage lysate and store at 4°C till further use

↓

Inoculate single colony from the recipient strain in 5 ml LB broth containing 100 μg/ml ampicillin, incubate at 37° for 24 h with shaking at 180 rpm

↓

Take 50 μl of this culture in a 2 ml micro-centrifuge tube and add 50 μl of 0.1 M $CaCl_2$ and 250 μl of phage lysate. Mix well and incubate at 37°C for 2 h, do not shake

↓

After incubation take 50 μl of the culture and plate on LB plates containing both the antibiotics, incubate the plates at 37°C for 24 h and observe for the development of colonies

4 Molecular Microbial Diversity

Exp. 4.1 Plasmid Profile Analysis

Objective To study the genotyping of bacteria using plasmid profiling.

Introduction

Plasmids are small circular extra-chromosomal DNA molecules, which are capable of replicating independently in the genome. In order to understand the molecular epidemiology of the resistant plasmids, its analysis has become a major issue for their role in spreading of antibiotic and metal resistance genes. However, the major constraints in understanding plasmid epidemiology are the diversity and dynamic nature of these plasmids. Plasmid's replication system that dictates its behaviour, i.e. hostage and copy number is the major plasmid land mark from biologic stand point. Number of plasmid copies plays a critical role in imparting various characteristics to the host organism. Plasmid profile analysis is highly useful technique to serological studies of pathogenic bacteria for identification of specific serotypes with specific reference pattern to detect certain strains with possible variation in plasmid content.

Most recently, plasmid profiling has been used to determine molecular relatedness among the *Salmonella* spp. and is a highly useful technique for strain differentiation. Plasmids contain genes necessary for cell growth and survival and possess a simple mechanism for transferring DNA between serotypes and across bacterial species. In this regard, plasmid profile analysis provides a useful technique in distinguishing the bacterial species based on the presence of copy number, size and confirmation of plasmids by simply extraction of plasmid DNA, separation by agarose gel electrophoresis and visualisation under UV light with reference to the plasmids of known mass carried in a bacterial strain most preferably *E. coli.*

Plasmids are highly infectious. They are capable of being transferred between bacteria of same or different genus. All functions required for the transfer of plasmid, synthesis of pilli, are encoded by the genes present on them. Thus, after getting transferred these genes are capable of forming trans-conjugants to become donor for another round of conjugation, which is repeated for several generations. In history, most of the plasmid profile analysis was carried out to detect an outbreak by pathogenic organisms and detection of the new entry of certain plasmids that may be capable of causing the outbreak. However, plasmid profile analysis nowadays has become a useful asset to study the microbial typing in environmental strains also, keeping in mind the fact that, the strains with same plasmid profile are closely related to each other genotypically.

Principle

Plasmid profiling approach is similar to that of other typing techniques used for bacterial genotyping. In epidemiological related typing, all

S. Das, H. R. Dash, *Microbial Biotechnology- A Laboratory Manual for Bacterial Systems,*
DOI 10.1007/978-81-322-2095-4_4, © Springer India 2015

the outbreak associated strains possess the same plasmid profile and provide the similar result as other typing systems used. However, it is not necessary that all the isolates from a common origin are identical or possess the similar plasmid profile. However, it can be distinguished that if a plasmid encodes certain virulence factors, it can be concluded that the particular bacterial strain is capable of spreading the infections. In this regard, during the study of clinical microbiology, and while studying disease outbreaks, study of plasmid profiling should include control isolates from non-outbreak related patients or from the environmental isolates. Thus, in any epidemiological study, care should be taken for matching the plasmid profiling between control isolate and the infected one for an accurate conclusion.

The major advantage of using plasmid profiling over other typing techniques is that, a single set of reagents and instrument is applicable for many bacterial species. In some instances, presence or absence of plasmids also affects the plasmid profiling of the isolates. However, for any typing system to become effective it should be capable of differentiating between case isolates from non-case isolates. Hence, when plasmid profiles of a group of isolates are compared, there must be the presence of sufficient differences in the obtained patterns to deduce a strong conclusion. Plasmid profiling becomes extremely useful when it is combined with other screening or typing techniques. For clinical microbiological applications, plasmid profiling should be included as a major technique along with other techniques like replica plating, phage typing, antibiogram and others.

There exists many approaches of grouping plasmids of bacteria; one of the best approach is their ability to transfer from one bacterium to another. These plasmids are called to be conjugative plasmids that contain *tra* genes, which are capable of materializing the complex processes involved during conjugation. However, non-conjugative plasmids are not capable of starting conjugation, and hence can get transferred only with the assistance with other conjugative plasmids. There also exists an intermediate group of plasmids that can mobilise in certain instances as they contain a few genes which are required for the transfer of the genetic material. The another approach of classifying plasmids is the function-based classification that includes fertility or F-plasmids that carry *tra* genes, resistant plasmids that carry resistant genotypes for antibiotics and metals, col plasmids that possess genes capable of coding bacteriocins for killing of other bacteria, degradative plasmids, which degrade the unusual substances like toluene and salicylic acid and the virulence plasmids that can convert a non-pathogenic bacterium to a pathogenic one.

Plasmids in natural conditions are present in one of the five forms, i.e. nicked open circular in which one of the double-stranded DNA is cut. Relaxed-circular DNA is present in fully intact form, however, relaxed enzymatically by removing the supercoils. In another confirmation, linear DNA possesses free ends as both the DNA strands are cut as DNA remains linear in vivo conditions. Super coiled plasmid DNA are also called as covalently closed-circular DNA molecule, which is fully intact with both the strands uncut along with an integral twist resulting in a compact form. Another form of plasmid DNA is the supercoiled denatured DNA with unpaired regions and slightly less compact structures, and this form of plasmid DNA generally results from the use of excessive alkalinity during plasmid preparation (Fig. 4.1).

Reagents Required and Their Role

Ethylenediaminetetra-Acetic Acid

Ethylenediaminetetra-acetic acid (EDTA) binds with the divalent cations in the cell wall, thus weakening the cell envelope. After cell lysis EDTA limits DNA degradation by binding Mg^{2+} ions which are necessary cofactors for bacterial nucleases. In this way, it inhibits nucleases leading to the rupture of cell wall and cell membrane.

Sodium Hydroxide

Sodium hydroxide is used to separate bacterial chromosomal DNA from plasmid DNA. Chromosomal DNA and sheared DNA are both linear,

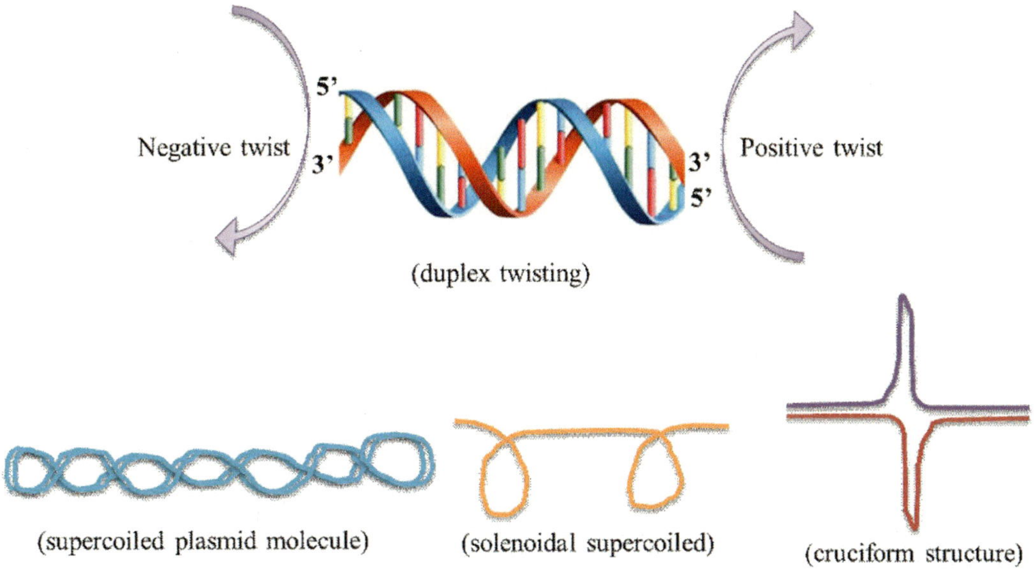

Fig. 4.1 Different confirmations of plasmid molecules giving rise to different banding patterns useful for plasmid profile analysis

whereas most of the plasmid DNA is circular. When the solution medium becomes basic due to addition of sodium hydroxide, double-stranded DNA molecules separate popularly called as denaturation and their complementary bases no longer are associated with each other. On the other hand, though plasmid DNA becomes denatured they are not separated. The circular strands can easily find their complementary strands and renature back to circular double-stranded plasmid DNA molecule once the solution is no longer alkaline. This unique property of plasmid DNA is explored by the use of NaOH to separate it from chromosomal DNA.

Potassium Acetate

Potassium acetate is used to selectively precipitate the chromosomal DNA and other cellular debris away from the desired double-stranded plasmid DNA. Potassium acetate does three things during plasmid DNA isolation: (a) it allows circular DNA to renature while sheared cellular DNA remains denatured as single stranded DNA; (b) precipitation of single-stranded DNA as large single-stranded DNA are insoluble in high salt concentration; (c) when potassium acetate is added to SDS it forms KDS, which is insoluble. This allows the easy removal of SDS contamination from the extracted plasmid DNA.

Luria Bertani Broth

It is a rich medium, which permits the fast growth as well as good growth yields of many species of bacteria. It is the most commonly used growth medium for *E. coli* cell culture during molecular biology studies. LB broth can support *E. coli* growth OD_{600} 2 to 3 under normal shaking incubation conditions.

TE Buffer

TE buffer is prepared by mixing 50 mM Tris and 50 mM EDTA in water and by maintaining the pH 8.0. The major constituent of TE buffer, Tris acts as a common pH buffer to control pH during addition of other reagents during further steps, and EDTA chelates cations like Mg^{2+}. Hence, TE buffer is helpful to solubilise DNA by protecting it from degradation.

Sodium Dodecyl Sulphate

Sodium dodecyl sulphate (SDS) is the major constituent of the alkaline lysate mixture solution used during plasmid DNA extraction. SDS is a detergent that dissolves the lipid components of the cell membrane as well as the cellular proteins.

Ethanol

Cold ethanol or isopropanol is used to precipitate the plasmid DNA. As DNA is insoluble in alcohol; upon its addition, they clump or cling together. Thus, centrifugation forms a pellet from the precipitate, which can be further separated from the undesired supernatant.

Procedure

Plasmid DNA Extraction

1. Inoculate single bacterial colony from the individual bacterial sample in 5 ml tubes containing sterile LB broth.
2. Incubate the tubes at 37 °C for 24 h with shaking of 180 rpm.
3. Centrifuge 1 ml of the overnight-grown bacterial culture in a 5 ml micro-centrifuge tube at 10,000 rpm for 2 min at room temperature to collect the cell pellets.
4. Discard the supernatant and resuspend the cell pellets with 150 µl of EDTA-Tris buffer, and vortex to mix completely.
5. Add 175 µl of 2% SDS and 175 µl of 0.4 N NaOH to the same tube. Mix vigorously.
6. Add 250 µl of cold 5M potassium acetate, mix vigorously.
7. Centrifuge at 12,000 rpm for 5 min, and transfer the supernatant to a fresh vial.
8. Add equal volume of cold ethanol to the tube. Mix the samples by inverting the tubes repeatedly for several times.
9. Immediately centrifuge the tubes at 12,000 rpm for 10 min at 4 °C.
10. Discard the supernatant and wash the pellet containing plasmid DNA with 650 µl of cold 70% ethanol by centrifuging at 12,000 rpm for 15 min.
11. Discard the supernatant and dry the cell pellet for 30 min at room temperature for complete evaporation of ethanol.
12. Resuspend the plasmid DNA pellet with 40 µl of sterile deionised water.

Agarose Gel Electrophoresis

1. Prepare 0.8% of agarose gel by mixing 0.8 g of agarose powder in 100 ml of 1X TAE buffer.
2. Dissolve agarose completely in TAE buffer by boiling in a microwave oven and allow it to cool to about 60 °C.
3. Add 2 µl of 10 mg/ml Ethidium Bromide (EtBr) to it, and mix by swirling.
4. Pour the agarose solution into the electrophoresis tank with comb in appropriate place to obtain a gel thickness of about 4–5 mm.
5. Allow it to solidify for about 20 min and remove the comb. Place the tray in the electrophoresis tank.
6. Pour 1X TAE buffer into the tank up to the mark so that the buffer covers the surface of the gel.
7. Mix 15 µl of the isolated plasmid DNA with 2 µl of gel loading dye, and carefully load the samples into the wells created by the combs.
8. Connect the electrodes to the power pack and run the electrophoresis at 60–100 V until the loading dye migrates about three quarter of the gel.
9. Disconnect the electrodes from the power pack after completion of migration of the dye font in the gel, and place it on the gel documentation system to obtain a clear image of the banding pattern obtain after completion of agarose gel electrophoresis.

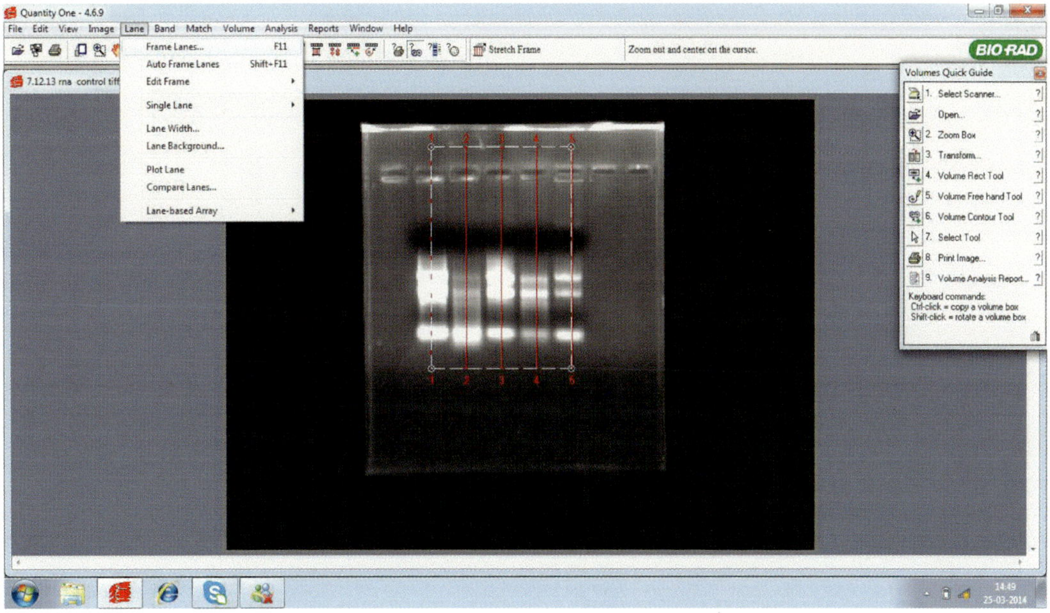

Fig. 4.2 Snap shot of the gel image showing number of lanes to be created based upon the number of bands to be analysed

Analysis of Plasmid Profile

1. Plasmids present in the respective bacterium will be separated based on their molecular weight, copy number and confirmations. Hence, a clear, distinct banding pattern is supposed to be obtained after observation under gel documentation system.
2. Analyse the banding pattern by Quantity One Software (Bio-Rad, USA). Open the gel image in Quantity One software.
3. Based on the number of organisms present, frame lanes according to the number of banding patterns present on the gel (Fig. 4.2).
4. Normalise the backgrounds of all the lanes with reference to the banding patterns.
5. Detect bands at 4% permissible limit (Fig. 4.3).
6. Designate band attributes to the number of lanes.
7. Match the banding patterns by clicking on any one of the distinct bands, which may be commonly present in all the lanes (Fig. 4.4).
8. Click on the phylogenetic tree option and choose neighbour joining method for construction of tree (Fig. 4.5).
9. Acquire the phylogenetic tree and deduce the phylogenetic relationship among them based upon their plasmid profile (Fig. 4.6).

Observation

Observe the banding pattern on agarose gel and the phylogenetic tree obtained after analysis in Quantity One software. Deduce the phylogenetic relationship among these bacteria based upon the tree obtained to know the closeness or farness among them.

Result Table

Name of organisms	Lane no.	No. of bands after agarose gel electrophoresis	Relationship
Control			
Strain 1			
Strain 2			
Strain 3			
Strain 4			
Strain 5			

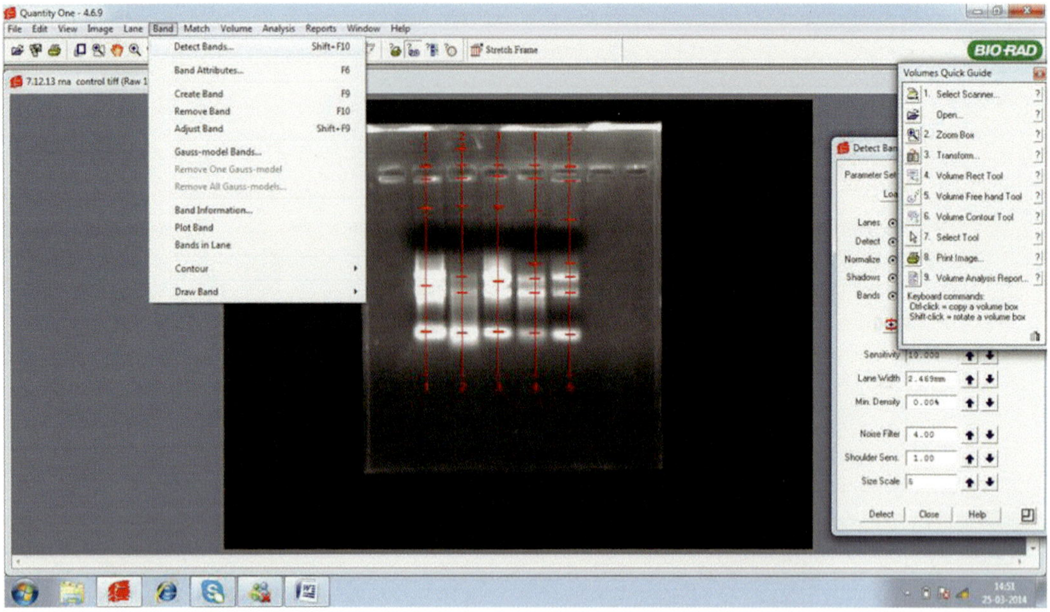

Fig. 4.3 Detection of bands with 4 % permissible limit that will detect all the bands present in the gel image

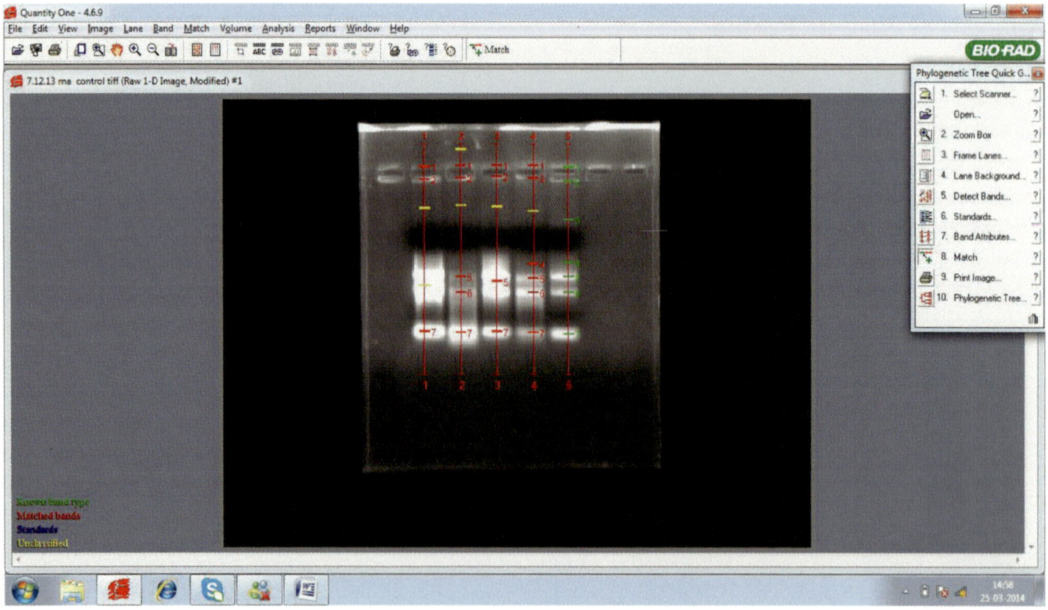

Fig. 4.4 Matching of the banding patterns present in the gel with one another

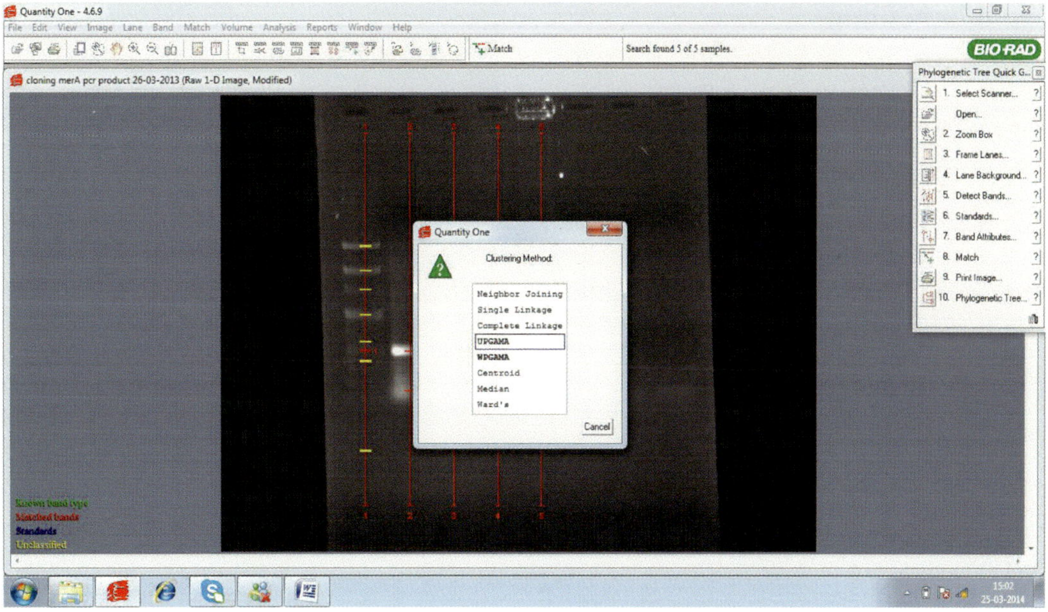

Fig. 4.5 Snap shot of the gel image showing the correct choosing of the algorithm to be used for phylogenetic tree construction

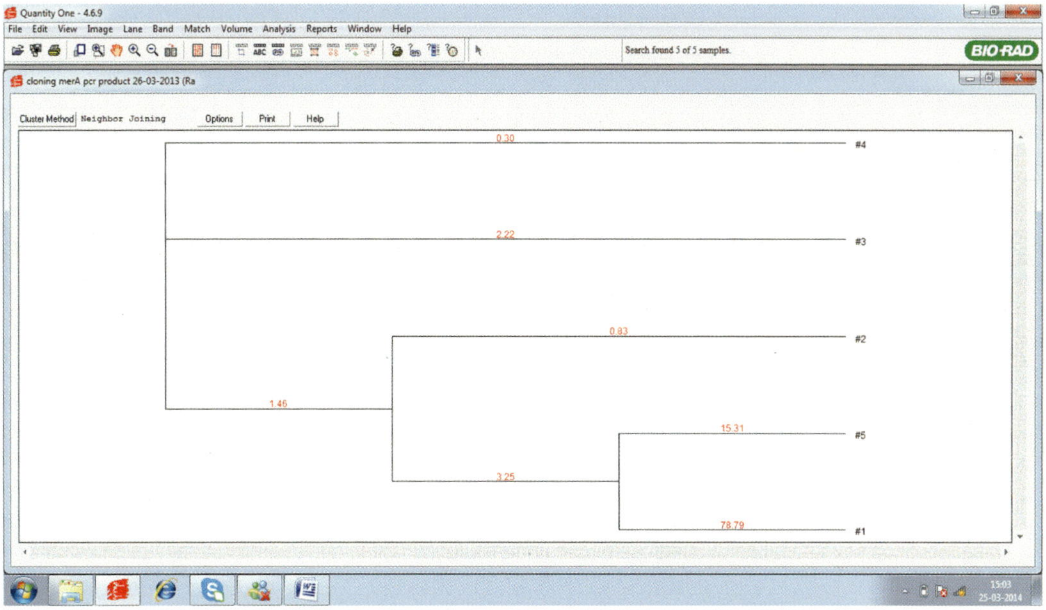

Fig. 4.6 The phylogenetic tree obtained from the analysis of banding pattern with Quantity One software to deduce the phylogenetic relationship among the bacteria tested based upon their plasmid profiling

Troubleshooting

Problems	Possible cause	Possible solution
Low yield of plasmid DNA	Growth of the culture is not proper	Grow the culture with suitable growth medium in optimum conditions with vigorous shaking
	Lysate may not have prepared completely	Incubate for 5 min before going for final centrifugation after addition of solution III
RNA contamination	Initial centrifugation has not been performed at 20–25 °C	The residual RNA may be degraded when the initial centrifugation step of lysate will be carried out at room temperature
Insoluble pellet after DNA precipitation	Pellet might have been dried excessively	As an acid, DNA is better soluble in slightly alkaline solutions such as TE or 10 mM Tris buffer with a pH of 8.0 than water Pellet may be heated for several minutes to 65 °C to enhance dissolving
Poor performance in downstream applications with the plasmid DNA	Non-complete re-suspension of pellets with Solution-II	Mixing must be done carefully (by inverting slowly), till a homogenous phase is obtained
Contamination with bacterial chromosomal DNA	Vortexing might be carried out in any step	Never vortex the solution after addition of any solutions which will result in shearing of chromosomal DNA
Smear in gel/degraded plasmid	Is the bacterial strain known as 'problematic'?	Avoid the bacterial growth for more than 16 h
	Recommended growth time of the strain exceeded	Use the recommended growth time of the bacteria

Precautions

1. Use a fresh pipette when going into different stock solutions to avoid cross-contamination.
2. Try to avoid touching the inside wall of the tube while transferring the supernatant to a fresh tube.
3. Be careful not to dislodge the pellet while transferring the supernatant to a fresh tube.
4. Always wear safety goggles and gloves.
5. Never try to mix the samples by vortexing at any step of the plasmid DNA extraction procedure.
6. Use of a cut end tip will be extremely useful to use during the extraction procedure.

FLOW CHART

Extraction of plasmid DNA

Inoculate pure culture of bacterial colony in 5 ml of LB broth and incubate at 37° for 24 h with shaking at 180 rpm

↓

Centrifuge 1 ml of the grown bacterial culture at 10,000 rpm for 2 min and collect the cell pellet

↓

Re-suspend the cell pellet with 150 μl of EDTA-Tris buffer and mix by vortexing, add 175 μl of 2% SDS and 175 μl of 0.4 N NaOH, mix vigorously

↓

Centrifuge at 12,000 rpm for 5 min and transfer the supernatant to a fresh vial, add equal volume of cold ethanol to the tube. Mix samples by inverting the tubes repeatedly several times, immediately centrifuge the samples at 12,000 rpm for 10 min at 4°C

↓

Discard the supernatant and collect the cell pellet containing plasmid DNA and wash the pellet with 650 μl of cold 70% ethanol by centrifuging at 12,000 rpm for 15 min

↓

Discard the supernatant and dry the pellets at room temperature for 30 min, re-suspend the pellet with 40 μl of sterile deionized water and store at -20° till further use

FLOW CHART

Agarose gel electrophoresis

Prepare 0.8% agarose gel by adding 0.8 g of agarose in 100 ml of 1X TAE buffer, dissolve agarose completely by boiling in microwave and allow it to cool to 60°C

↓

Add 2 μl of ethidium bromide to it and mix by swirling, pour the agarose solution to the electrophoresis tank with comb in appropriate place

↓

Allow the gel to solidify for 20 min and remove the comb, place the tray in electrophoresis chamber

↓

Pour 1X TAE till it covers the gel surface and load the samples by mixing 15 μl of the plasmid DNA sample with 2 μl of gel loading dye

↓

Run the electrophoresis at 60-100 V and observe the banding pattern under Gel documentation system

Analysis of plasmid profile

Open the gel image in Quantity One software and frame lanes according to the number of samples loaded on the gel

↓

Normalize the background of all the lanes and detect bands with 4% permissible limit

↓

Designate and attribute to the number of lanes and match the banding pattern of any one distinct band which is commonly present in all lanes

↓

Click on the phylogenetic tree construction, choose neighbour joining method and construct phylogenetic tree

↓

Acquire the phylogenetic tree and deduce the phylogenetic relationship among them based upon their plasmid profile

Exp. 4.2 Amplified Ribosomal DNA Restriction Analysis to Study Bacterial Relatedness

Objective To study the genotyping of bacteria using amplified ribosomal DNA restriction analysis (ARDRA).

Introduction

Amplified ribosomal DNA restriction analysis technique is a modification of the restriction fragment length polymorphism technique, which is applied to the 16S rRNA gene subunit of bacteria. In this technique, bacterial 16S rRNA gene is amplified by polymerase chain reaction (PCR), and subsequently, the amplified product is digested with suitable restriction enzymes to yield products of various sizes. Thus, based upon the banding pattern of the restriction products, phylogenetic relationship among the strains can be derived. Mostly tetra cutter restriction enzymes are used for this technique. However, to obtain a statistically significant result, at least three restriction enzymes are supposed to be used during this procedure.

When a suitable set of restriction enzymes are used, a unique fingerprint for each species or strain is expected to be found. When the digested DNA is run on a gel, a pattern of fragments characteristic of the community is obtained. However, the major limitation of using this technique comes during the bacterial community analysis. When a bacterial community is analysed by this technique a finger print is obtained, which does not conclude the identification of any of the bacterial strain. However, the limitation of this approach can overcome by complementing this technique with probe hybridization. The technical limitation of this technique includes the requirement of optimization prior to each set of experiment. As all restriction enzymes do not work efficiently against all organisms, the restriction map when produced from the highly conserved regions of the ribosomal genes, the generated fragments from two different groups of organisms become difficult to separate from each other.

Despite of having so many limitations, ARDRA technique is one of the most useful techniques for the study of bacterial phylogeny due to many reasons. As this technique involves the separation of 16S rRNA gene based upon their size as well as sequence, the technique becomes highly cost-effective and less time consuming, and this technique can be performed in any standard laboratory conditions with facilities for normal molecular biology facilities. The chance of getting a good result for ARDRA increases when clone of the 16S rRNA gene is used for this purpose instead of directly PCR amplified community DNA.

Principle

Exploring the 16S rRNA gene sequences coupled with restriction digestion and gel electrophoresis is known to be RDRA technique, which is also called as ribotyping. This technique was originally developed for the typing of *Mycobacterium* bacterial species. However, recent advances have development for the utilization of this technique in typing of other bacterial strains. Analysis of the pattern of bacteria is used to obtain the clusters of related bacteria, which are represented by the formation of a cardiogram or hologram. The data can be analysed further to generate phylogram or cladogram that is used further for the drawing of the phylogenetic tree. The tree drawn from this approach is the indication of the relationship of the organisms based on the restriction pattern obtained from the respective 16S rRNA genes.

Related organisms are supposed to provide the same restriction pattern after digestion with the restriction enzyme. As the size of the 16S rRNA gene is approximately 1500 bp, the chance of occurring a restriction site for a four-base-pair-cutter occurs once in 256 bp repeats. As tetra cutter restriction enzyme digests the 16S rRNA gene at particular sites to provide a unique restriction pattern, which is considered to be the signature of that organism. Thus, the restriction pattern of one organism can be compared with the restriction pattern of another to deduce the phylogenetic relationship between them. Use of at least three restriction enzymes at a time can reduce the probability of obtaining similar patterns for unrelated organisms.

As recognition site of restriction enzymes differ drastically among the bacterial species, the major principle behind the ARDRA technique involves the RFLP technique. Due to the combination of PCR and restriction digestion, a minute amount of DNA can solve the purpose by amplification of the desired 16S rRNA gene followed by restriction digestion of the DNA templates. During the first step of this technique, ribosomal DNA is amplified by PCR to avoid undesired and dominant DNA templates. Subsequently, 16S rRNA gene is digested to specific DNA fragments by the use of restriction enzymes. During the

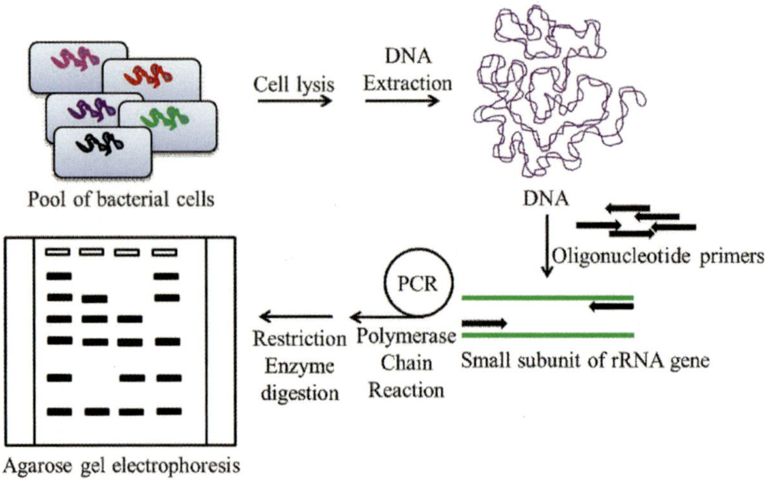

Fig. 4.7 Steps involved in amplified ribosomal DNA restriction analysis (ARDRA) technique

final step of this analysis, the fragmented DNA are subjected to high resolution agarose gel electrophoresis (Fig. 4.7). This technique has many advantages over the other techniques involved for this purpose as it provides rapid comparison of rRNA genes.

Reagents Required and Their Role

Bacterial Genomic DNA

Bacterial genomes are of small and less varying sizes ranging between 139 and 13,000 kbp. The isolation and purification of genomic DNA from bacteria is the most common prerequisite for most of the molecular biology experiments. For the analysis of ARDRA for study of molecular phylogeny, bacterial genomic DNA to be used should be of high quality and quantity. The detailed procedure for the isolation of genomic DNA from bacteria have been described during Exp. 1.1.

16S rRNA Gene Forward and Reverse Primers

Since, DNA is a double-stranded polynucleotide helical structure, one strand runs from 5′ to 3′ direction, and the other strand runs from 3′ to 5′ direction (complementary to the first strand), and the synthesis of the primer always takes place in the 5′–3′ direction, no matter where it is present. Hence, one primer is required in the forward direction and another is required in reverse direction. For proper amplification of the product, the final concentration of the 16S primers should be 0.05–1 µM. As primers have a greater influence on the success or failure of PCR protocols, it is ironic that primer designing is largely qualitative and are based on well-understood thermodynamic or structural principles. The set of primers to be used during this experiment are as follows:

16SForwardprimer(27F)05′-AGAGTTTGATCMTGGCTCAG3′

16SReverseprimer(1492R)05′-ACGGCTACCTTGTTACGA3′

dNTPs

The dNTPs are the building blocks of new DNA strand. In most of the cases they come as a mixture of four deoxynucleotides, i.e. dATP, dTTP, dGTP and dCTP. Around 100 µM of each of the dNTPs is required per PCR reaction. dNTP stocks are very sensitive to cycles of thawing and freezing and after 3–5 cycles, PCR reactions does not work well. To avoid such problems, small aliquots (2–5 µl) lasting for only a couple of reactions can be made and kept frozen at −20 °C. However, during long-term freezing, small amount of water condenses on the walls of the vial, thus changing the concentration of dNTPs solution. Hence, before using, it is essential to centrifuge the vials, and it is always recommended to dilute the dNTPs in TE buffer, as acidic pH promotes hydrolysis of dNTPs, thus interfering the PCR result.

Taq DNA Polymerase

Taq polymerase is a thermostable DNA polymerase isolated from the thermophilic bacterium *Thermus aquaticus*, which was originally isolated by Thomas D. Brock, in 1965. The enzyme is able to withstand the protein denaturing conditions required during PCR. It has the optimum temperature for its activity between 75 and 80 °C, with a half-life of greater than 2 h at 92.5 °C, 40 min at 95 °C and 9 min at 97.5 °C, and possesses the capability of replicating a 1000 bp of DNA sequence in less than 10 s at 72 °C. However, the major drawback of using Taq polymerase is the low replication fidelity as it lacks 3′–5′ exonuclease proofreading activity. It also produces DNA products having 'A' overhangs at their 3′ ends, which is ultimately useful during TA cloning. In general, 0.5–2.0 units of Taq polymerase is used in a 50 µl of total reaction, but ideally 1.25 units should be used.

PCR Reaction Buffer

Every enzyme needs certain conditions in terms of their pH, ionic strength, cofactors etc. which is achieved by the addition of buffer to the reaction mixture. In some instances, the enzyme shift pH in non-buffered solution and stops working in this process, which can be avoided by the addition of PCR buffer. In most of the PCR buffers the composition is almost same as: 100 mM Tris-HCl, pH 8.3, 500 mM KCl, 15 mM $MgCl_2$ and 0.01 % (w/v) gelatin. The final concentration of the PCR buffer should be 1X concentration per reaction.

AluI

This restriction enzyme is isolated from *Arthrobacter luteus*. It has the potential to recognise the sequences AG/CT to generate DNA fragments of blunt ends. AluI is inhibited by the presence of 6-methyladenine, 5-methylcytosine, 5-hydroxymethylcytosine and 4-methylcytosine. It has the absence of unspecific endonuclease activities as well as 5′ exonuclease or 5′ phosphatase activity, and absence of 3′-exonuclease activity.

HaeIII

HaeIII is one of the most useful restriction enzyme which has been isolated from *haemophilus aegypticus*. The enzyme has a molecular weight of 37,126. It is a tetra cutter enzyme with the recognition sequence to be GGCC. As HaeIII cuts both the strands of DNA at the same location, it yields restriction fragments with blunt ends and is capable of heat digestion at 80° after 20 min.

Agarose

It is used for the electrophoretic separation of nucleic acids. The purest form of agarose is free of DNase and RNase activities. Molecular biology grade agarose is the standard one for the resolution of DNA fragment in the range of 50 bp–50 kb, with the possibility of subsequent DNA extraction from the gel for further analysis. It possesses the following properties: gel strength (1 %)—1125 g/cm^2, gelling point (1.5 %)—36.0 °C, melting point (1.5 %)—87.7 °C, sulphate—0.098 %, moisture—2.39 % and ash—0.31 %.

Ethidium Bromide

EtBr is a fluorescent dye that intercalates between nucleic acid bases, and eases the detection of nucleic acid fragments in gel. When exposed to ultraviolet (UV) light, DNA flourishes with an orange colour, intensifying 20-fold after binding to DNA. The absorption maxima of EtBr in aqueous solution is between 210 and 285 nm that corresponds to the UV light. Hence, as a result of this excitation, EtBr emits orange light with a wavelength of 605 nm. EtBr binds with DNA, and slips in between its hydrophobic base pairs and stretches the DNA fragment, thus removing the water molecules from ethidium cation. This dehydrogenation results in the increase in fluorescence of the ethidium. However, EtBr is a potential mutagen, suspected carcinogen and it can irritate eyes, skin, mucous membranes and upper respiratory tract at higher concentrations. It is due to the fact that, EtBr intercalates into double-stranded DNA, deforms the molecule, thus, blocking the biological processes involving nucleic acids like DNA replication and transcription. Hence, there are many alternatives regarded as less dangerous and with better performance like Sybr dyes.

Procedure

Amplification of 16S rRNA Gene

1. Assemble the PCR reaction in ice by preparing a master mixture with the following compositions in the given order. For restriction digestion of 16S rRNA gene a master mixture of 100 µl volume should be prepared (Table 4.1).
2. Aliquot 18.0 µl of the master mixture in each vial. Add 2 µl of the genomic DNA from different bacterial strain in each tube.
3. Keep one negative control without template DNA and perform the PCR with the following reaction conditions.

Initial denaturation	−94 °C for 5 min	
Denaturation	−94 °C for 30 s	
Annealing	−55 °C for 30 s	30 cycles
Extension	−72 °C for 2 min	
Final extension	−72 °C for 7 min	
Hold	−4°C for ∞	

Table 4.1 Master mixture components for 16S rRNA gene amplification

Reagents	Volume (µl)
Milli-Q water	56.5
10 X buffer	10.0
10 mM dNTPs mixture	2.5
5 µM forward primer	10.0
5 µM reverse primer	10.0
Taq polymerase (5 U/µl)	1.0
Total volume	100

Table 4.2 Reaction mixture components for restriction digestion

Reagents	Volume (µl)
PCR reaction product	10.0
10 X Buffer	2.0
Restriction enzyme	1.0
Deionised water	7.0
Total volume	20.0

Restriction Digestion

1. Set the restriction digestion reaction on ice using the following conditions. A typical 20 µl reaction mixture is given below (Table 4.2):
2. Incubate the reaction mixture at 37 °C for 2 h, and inactivate by heating at 70 °C for 10 min.

Analysis of the PCR Products by Agarose Gel Electrophoresis

1. Prepare 2 % of agarose gel by mixing 2.0 g of agarose powder in 100 ml of 1 X TAE buffer.
2. Dissolve agarose completely in TAE buffer by boiling in a microwave oven, and allow it to cool to about 60 °C.
3. Add 2 µl of 10 mg/ml ETBr to it and mix by swirling.
4. Pour the agarose solution into the electrophoresis tank with comb in appropriate place to obtain a gel thickness of about 4–5 mm.
5. Allow it to solidify for about 20 min and remove the comb. Place the tray in the electrophoresis tank.
6. Pour 1 X TAE buffer into the tank up to the mark so that the buffer covers the surface of the gel.

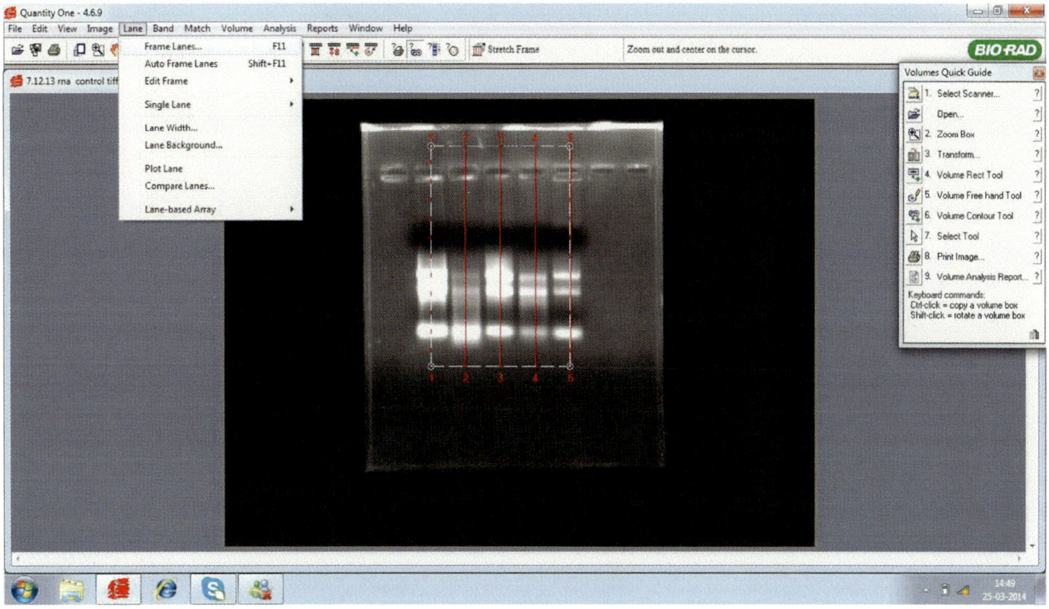

Fig. 4.8 Snapshot of the gel image showing number of lanes to be created based upon the number of bands to be analysed

7. Mix 15 μl of the isolated plasmid DNA with 2 μl of gel loading dye, and carefully load the samples into the wells created by the combs.
8. Connect the electrodes to the power pack and run the electrophoresis at 60–100 V until the loading dye migrates about three quarter of the gel.
9. Disconnect the electrodes from the power pack after completion of migration of the dye font in the gel, and place it on the gel documentation system to obtain a clear image of the banding pattern obtain after completion of agarose gel electrophoresis.

Analysis of Microbial Phylogeny Using Quantity One

1. Acquire and save the image of the gel using a gel documentation system with UV illumination (Bio-Rad, USA). Click on the band analysis button to open the band analysis quick guide.
2. Based upon the number of organisms present frame lanes according to the number of banding patterns present on the gel (Fig. 4.8).
3. Normalise the backgrounds of all the lanes with reference to the banding patterns.
4. Detect bands at 4% permissible limit (Fig. 4.9).
5. Designate band attributes to the number of lanes.
6. Match the banding patterns by clicking on any one of the distinct bands, which may be commonly present in all the lanes (Fig. 4.10).
7. Click on the phylogenetic tree option, and choose neighbour joining method for construction of tree (Fig. 4.11).
8. Acquire the phylogenetic tree and deduce the phylogenetic relationship among them based upon their plasmid profile (Fig. 4.12).

Observation

Observe the banding pattern on agarose gel and the phylogenetic tree obtained after analysis in Quantity One software. Deduce the phylogenetic relationship among these bacteria based on the tree obtained to know the closeness or farness among them.

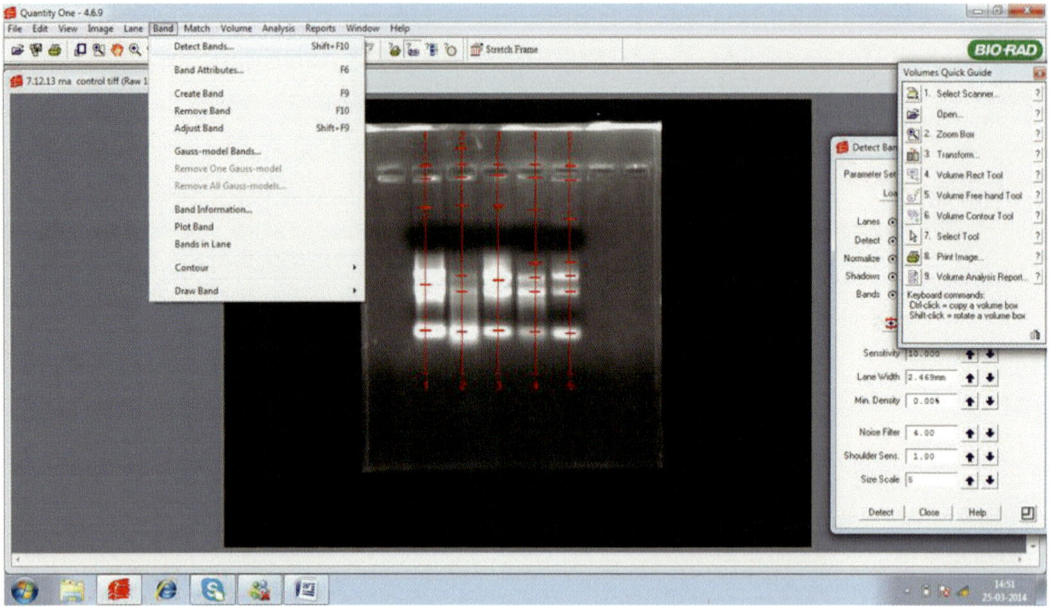

Fig. 4.9 Detection of bands with 4 % permissible limit that will detect all the bands present in the gel image

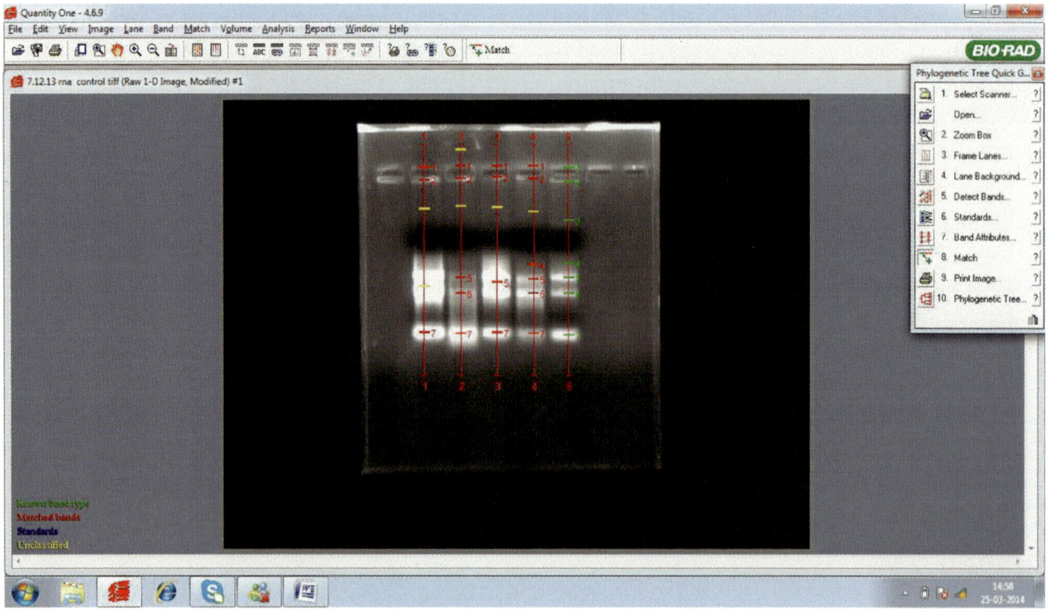

Fig. 4.10 Matching of the banding patterns present in the gel with one another

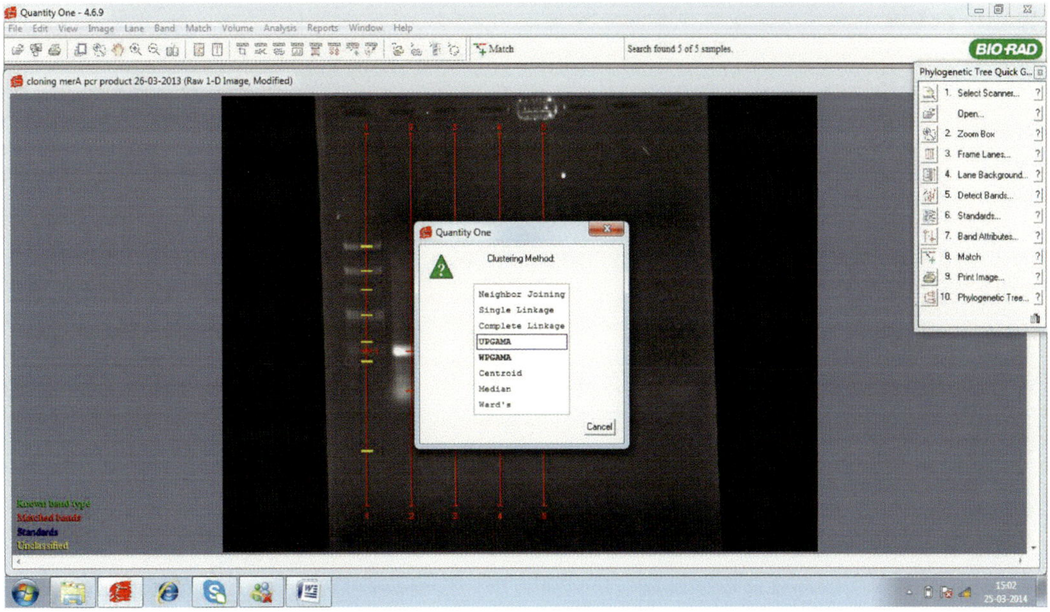

Fig. 4.11 Snapshot of the gel image showing the correct choosing of the algorithm to be used for phylogenetic tree construction

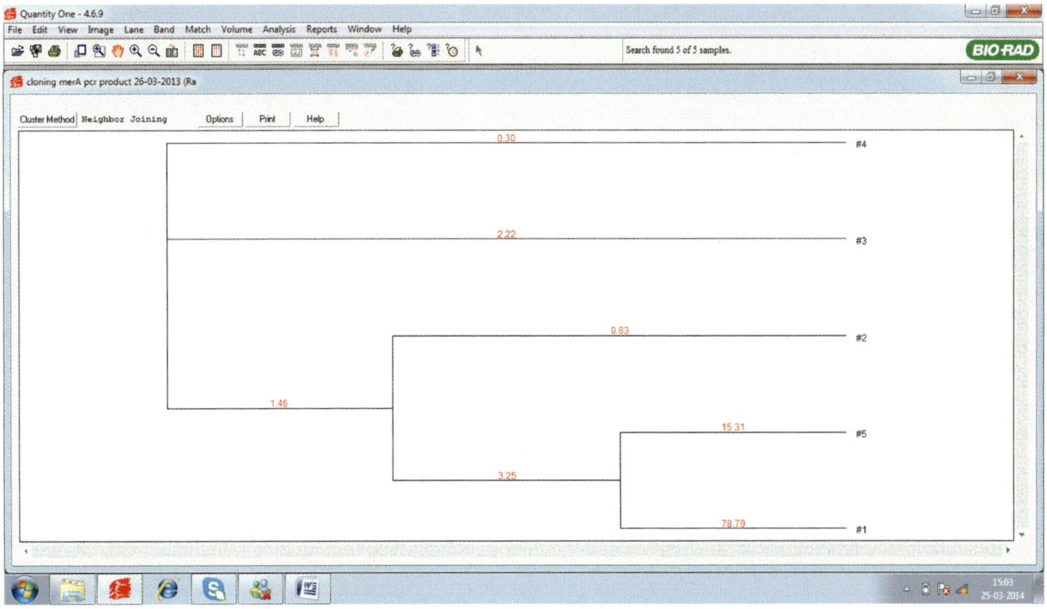

Fig. 4.12 The phylogenetic tree obtained from the analysis of banding pattern using Quantity One software to deduce the phylogenetic relationship among the bacteria tested based upon their plasmid profiling

Result Table

Name of organisms	Lane no.	No. of bands after agarose gel electrophoresis	Relationship
Control			
Strain 1			
Strain 2			
Strain 3			
Strain 4			
Strain 5			

Troubleshooting

Problems	Possible cause	Possible solution
Smear in the gel	Bad quality of template	Use a better quality of genomic DNA. Check the DNA quality and quantity by spectrophotometer and use a proper 260/280 ratio
	Degraded primers	Use fresh stock of the primers by freshly diluting the primers. Else order for fresh primers
	High concentration of restriction enzyme	Use appropriate concentration of restriction enzyme as provided by the manufacturer's instruction
	More incubation time with restriction enzyme	Inactivate the restriction digestion reaction after suitable incubation time by incubation at 70 °C for 10 min
Single band in the gel	Concentration of restriction enzyme is not appropriate	The concentration used for the restriction digestion of 16S rRNA gene is not appropriate for complete digestion of the amplified gene. Hence, appropriate concentration of the restriction enzyme should be used per reaction as per the manufacturer's instruction
	Low percentage of agarose gel cannot separate the distinct bands	Agarose gels with high concentration can distinguish the banding patterns properly due to their high resolution capability. Thus always a 2 % agarose gel should be used for a proper banding pattern during ARDRA analysis
	No restriction sites present for the used restriction enzyme	Check for the presence/absence of the restriction sites in the amplified gene for the restriction enzymes to be used. Otherwise use certain other restriction enzymes or the combination of more than one enzymes
No band	PCR amplification is not appropriate	Check all the reagents used during PCR and also check the PCR reaction conditions for proper amplification of 16S rRNA gene

Precautions

1. Always use gloves while performing this experiment.
2. Perform each step of the experiment by incubating the reagents on ice; otherwise there will be a chance for degradation of the reagents.
3. Do not store the amplified PCR product for a longer time. Always use freshly amplified samples for restriction enzyme digestion.
4. Remember to inactivate the action of the restriction enzyme after suitable time to obtain a clear banding pattern; otherwise the result on agarose gel will appear like a smear.
5. Use the acquired image immediately for analysis of phylogeny; otherwise the bands on the gel may be diffused.
6. Deal cautiously with ETBr as it has been reported to a potential carcinogenic agent.

FLOW CHART

Amplification of 16S rRNA gene

Assemble the reagents required for polymerase chain reaction by adding the components as: milli-Q water- 56.5 μl, 10X buffer- 10.0 μl, dNTPs mixture- 2.5 μl, forward primer- 10.0 μl, reverse primer- 10.0 μl, taq polymerase- 1.0 μl

↓

Aliquot 18.0 μl of the master mixture in each vial and to which add 2 μl of the genomic DNA from different bacterial strain

↓

Keep one negative control with all the PCR reagents added to it but no template DNA

↓

Perform PCR with the following reaction conditions: initial denaturation- 94° for 5 min, followed by 30 cycles of denaturation at 94° for 30 sec, annealing at 55° for 30 sec and extension at 72 °C for 2 min and final extension of 72° for 7 min followed by hold at 4°C for ∞.

Restriction digestion

Set the restriction digestion reaction on ice for a total volume of 20 μl by adding PCr reaction product 10.0 μl, 10X buffer- 2.0 μl, restriction enzyme- 1.0 μl, and deionized water- 7.0 μl.

↓

Incubate the teaction mixture at 37° C for 2 h and inactivate by heating at 70° C for 10 min

Agarose gel electrophoresis

Prepare 0.8% agarose gel by adding 0.8 g of agarose in 100 ml of 1X TAE buffer, dissolve agarose completely by boiling in microwave and allow it to cool to 60°C

↓

Add 2 µl of ethidium bromide to it and mix by swirling, pour the agarose solution to the electrophoresis tank with comb in appropriate place

↓

Allow the gel to solidify for 20 min and remove the comb, place the tray in electrophoresis chamber

↓

Pour 1X TAE till it covers the gel surface and load the samples by mixing 15 µl of the plasmid DNA sample with 2 µl of gel loading dye

↓

Run the electrophoresis at 60-100 V and observe the banding pattern under Gel documentation system

Analysis of plasmid profile

Open the gel image in Quantity One software and frame lanes according to the number of samples loaded on the gel

↓

Normalize the background of all the lanes and detect bands with 4% permissible limit

↓

Designate and attribute to the number of lanes and match the banding pattern of any one distinct band which is commonly present in all lanes

↓

Click on the phylogenetic tree construction, choose neighbour joining method and construct phylogenetic tree

↓

Acquire the phylogenetic tree and deduce the phylogenetic relationship among them based upon their ARDRA profile

Exp. 4.3 Denaturing Gradient Gel Electrophoresis (DGGE) Analysis to Study Metagenomic Bacterial Diversity

Objective To study the molecular phylogeny of the metagenomic DNA by using Denaturing Gradient Gel Electrophoresis (DGGE) analysis.

Introduction

Selective enrichment culture techniques fail to mimic the original conditions required for the growth and proliferation of bacterial cultures as in their natural habitat. Certain groups of bacteria are found to remain bound with the sediment particles, and cannot be detected by the conventional microscopic techniques. In this regard, molecular biology techniques explore new opportunities for the analysis of structure as well as species composition of the microbial communities. With the advent of the use of 16S rRNA gene sequences for analysis of bacterial phylogeny, several uncultured bacteria are in the verge of identification to add the steps further for microbial diversity studies. This technique involves the cloning of ribosomal DNA or by PCR of the 16S rRNA gene followed by analysis of the sequence clones to obtain the phylogenetic relationship.

A relatively newer approach for direct determination of the genetic diversity of complex microbial population has been reported to be denaturing gradient gel electrophoresis (DGGE). It relies on the separation of PCR amplified 16S rRNA gene fragments in polyacryl amide gel containing a linearly increasing gradient of denaturants. This technique facilitates the separation of two DNA fragments of equal size differing in their base pair sequences. DGGE is considered to be the most advanced technique for the identification of sequence variations in a number of genes from different organisms. It can be used for the direct analysis of genomic DNA from organisms with millions of base pairs genomes. Direct genome analysis is mostly carried out by the separation patterns of hybridization membranes by capillary blotting or by electroblotting followed by analysis with DNA probes. In an alternative technique, PCR is performed for the amplification of the sequence of interest and are separated by DGGE process. In certain instances, GC rich sequences are incorporated on one of the primers, which modify the melting behaviour of the DNA fragment of interest that is responsible for the detection of sequence variation of close to 100%.

DGGE has widespread applications starting from mutation detection to the analysis of microbial ecology. Nowadays, this technique has also been applied for the analysis of other functional genes like sulphur reduction, nitrogen fixation as well as ammonium oxidation.

Principle

Separation in DGGE is based on the electrophoretic mobility of the partially melted DNA molecules in polyacrylamide gels and the electrophoretic mobility of the partially melted DNA molecules are less compared to the complete helical form of the DNA molecule. The stretches of DNA fragments melting at identical temperature are known as melting domains, and they form discrete banding pattern at the same position. The melting domain with lowest melting temperature while melting at a particular position on the gel, it forms a transition of helical to partially melted molecule that halts its migration at that position. The melting temperature of these melting domains differs based upon their sequence variation. Thus, DGGE separates the DNA fragments that stop migrating at different positions on the denaturing gradient gel.

DGGE has been found to detect differences in the melting behaviour of small DNA fragments (200–700 bp) that differ by as little as a single-base substitution. When a DNA fragment is subjected to an increasingly denaturing physical condition, it melts. As the denaturing conditions increase, the partially melted fragment dissociates completely to form single strands. Hence, the discrete domains of DNA fragment become single-stranded within a narrow range of denaturing conditions. The rate of mobility of DNA fragments in acrylamide gels depends on the physical shape of the fragment. Partially melted fragments migrate much more slowly than completely double-stranded fragments. When a double-stranded fragment is electrophoreses into a gradient of increasingly denaturing conditions, it partially melts, and undergoes a sharp reduction in mobility because it changes shape.

Many fragments can be analysed simultaneously on a single denaturing gel in which the direction of electrophoresis is perpendicular to that of the denaturing gradient. When a large number of different fragments are electrophoresed, the fragments can be identified by their molecular weight in the low denaturant side of the gel. By following the S-shaped curves, the characteristic denaturant concentration at which the first domain melts can be determined. When two nearly identical set of fragments are mixed together and electrophoresed into a 'perpendicular' denaturing gradient gel. The melted domains that have sequence differences between each other will melt at slightly different positions and produce double bands.

Sequence differences are often easily detected in DNA fragments when nearly identical digests are electrophoresed in the same direction as that of the denaturing gradient. These 'parallel' gels permit the simultaneous comparison of as many sets of fragments as there are lanes on the gel, unlike the perpendicular gels. The procedures below refer almost entirely to parallel denaturing gradient gels.

Reagents Required and Their Role

40 % Acrylamide/Bis (37.5:1)

Acrylamide and Bisacrylamide are the crystalline solids with the solubility in water, ethanol, ether as well as chloroform. Most of the uses of this compound involves in the separation of charged molecules as in case of polyacryl amide gel electrophoresis. Acrylamide, also possess the characteristics of use as a carrier for the precipitation of small amounts of DNA molecules.

Formamide

Formamide is the amide derivative of formic acid, which is most commonly used for the cryopreservation of tissues and organs. During electrophoresis it is used for the stabilization of single stranded DNA molecules inhibiting its deionization. It is highly corrosive to skin and eyes, and hence it should be dealt with proper care.

Urea

Urea is the organic compound capable of denaturing the protein and DNA molecules. The denaturing ability of urea is attributed to its ability to disrupt the interpeptide or interchain hydrogen bonds. Hydrophobic interaction of urea is responsible for altering the aqueous configuration of proteins and nucleic acids, thus leading to their denaturation.

TAE Buffer

The electrophoretic mobility of DNA is dependent on the composition and ionic strength of the electrophoresis buffer. During the absence of ions, there will be a minimal electrical conductance and DNA migrates slowly. The buffer of high ionic strength and high electrical conductance is efficient, and additionally a significant amount of heat is generated. Thus, worsening the situation the gel melts and DNA is denatured. Several different buffers have been recommended for use in electrophoresis of native double-stranded DNA. These buffers contain EDTA (pH 8.0) and Tris-acetate (TAE), Tris-borate (TBE) or Tris-phosphate (TPE) at an approximate concentration of 50 mM (pH 7.5–7.8). These buffers are generally prepared as concentrated solutions and stored at room temperature, when used the working solution is prepared as 1X concentrations. TAE and TBE are the most commonly used buffers and two of them have their own advantages and disadvantages. Borate has disadvantages as it polymerizes and interacts with cis diols found in RNA. On the other hand, TAE has lowest buffering capacity but it provides the best resolution for larger DNA, which implies the need for the lower voltage and more time with a better product. Lithium borate is a relatively new buffer and is ineffective in resolving fragments larger than 5 kb.

Ammonium Persulphate

APS is the colourless crystalline salt of inorganic compound capable of forming free radicals, and is often used as an initiator for gel formation. Initiators are actually the effectors for polymerization. The rate of polymerization depends on the concentration of the initiators, and the property of the gel is also dependent on the nature of initiator used. However, the increased concentration of initiator may decrease the average polymer chain length by increasing the gel turbidity, and also in gel elasticity.

N, N, N′, N′-Tetramethylethylenediamine (TEMED)

TEMED stabilizes free radicals and improves polymerization. The rate of polymerization and the properties of the resulting gel depend on the concentrations of free radicals. Increase in the amount of free radicals results in a decrease in the average polymer chain length, increase in gel turbidity and decrease in gel elasticity. APS and TEMED are typically used at approximately equal molar concentrations in the range of 1–10 mM.

Gel Loading Dye

Loading buffer is mixed with the DNA samples to be used in agarose gel electrophoresis. The dye present in the buffer is used primarily to assess how fast the samples are running during electrophoresis, and to render a higher density to the samples than that of the running buffer. The increased density can be achieved by the addition of materials like ficoll, sucrose or glycerol. There are many colour combinations available to trace the migration rate of the DNA samples.

Ethidium Bromide

ETBr is a fluorescent dye that intercalates between nucleic acid bases and eases the detection of nucleic acid fragments in gel. When exposed to UV light, DNA flourishes with an orange colour, intensifying 20-fold after binding to DNA. The absorption maximum of EtBr in aqueous solution is between 210 and 285 nm that corresponds to the UV light. Hence, as a result of this excitation EtBr emits orange light with a wavelength of 605 nm. EtBr binds with DNA and slips in between its hydrophobic base pairs and stretches the DNA fragment, thus removing the water molecules from ethidium cation. This dehydrogenation results in the increase in fluorescence of the ethidium. However, EtBr is a potential mutagen, suspected carcinogen, and it can irritate eyes, skin, mucous membranes and upper respiratory tract at higher concentrations. It is due to the fact that, EtBr intercalates into double-stranded DNA, deform the molecule, thus blocking the biological processes involving nucleic acids like DNA replication and transcription. Hence, there are many alternatives regarded as less dangerous and with better performance like Sybr dyes.

Table 4.3 Preparation of 40 % Acrylamaide/Bis (37.5:1)

Reagents	Amount
Acrylamaide	38.93 g
Bis-acrylamaide	1.07 g
dH2O	Up to 100 ml

Table 4.4 Preparation of 50X TAE buffer

Reagents	Amount	Final concentration
Tris base	242.0 g	2 M
Acetic acid, glacial	57.1 ml	1 M
0.5 m EDTA, pH 8.0	100 ml	50 mM
dH_2O	Up to 1000 ml	

Procedure

Preparation of Reagents

1. Prepare the following reagents required for DGGE analysis as mentioned in Tables 4.3, 4.4 and 4.5.
2. Filter the 40 % Acrylamide/Bis solution through a 0.45 μM filter paper and store at 4 ° till further use. Similarly, autoclave 50X TAE buffer, and store at room temperature till further use.
3. Degas denaturing solution for 10–15 min, filter through 0.45 μM filter paper, and store at 4 °C in an amber bottle till further use. Do not store this solution for more than a month.

Table 4.5 Preparation of 100 % denaturing solution

Reagents	6 %	8 %	10 %	12 %
40 % Acrylamide/Bis	15 ml	20 ml	25 ml	30 ml
50X TAE buffer	2 ml	2 ml	2 ml	2 ml
Formamide (deionised)	40 ml	40 ml	40 ml	40 ml
Urea	42 g	42 g	42 g	42 g
dH_2O	Up to 100 ml	Up to 100 ml	Up to 100 ml	Up to 100 ml

Table 4.6 Preparation of denaturing solutions

Reagents	10 %	20 %	30 %	40 %	50 %	60 %	70 %	80 %	90 %
40 % formamide	4	8	12	16	20	24	28	32	36
Urea (g)	4.2	8.4	12.6	16.8	21	25.2	29.4	33.6	37.8

4. For denaturing solutions less than 100 %, use the volumes for acrylamide, TAE and water described above in the 100 % denaturing solution. Use the amounts indicated below for urea and formamide (Table 4.6).
5. Prepare other reagents required for the running of denaturing gradient gel electrophoresis as indicated in Tables 4.7, 4.8, 4.9 and 4.10.

Preheating the Running Buffer

1. Fill the electrophoresis tank with 7 l of 1X TAE running buffer.
2. Place the temperature control module on the top of the electrophoresis tank. Attach the power cord to the temperature control module and turn the power, pump and heater on. The clear loading lid should be on the temperature control module during preheating.
3. Set the temperature controller to the desired temperature. Set the temperature ramp rate 2000 C/h to allow the buffer to reach the desired temperature the quickest.
4. Preheat the buffer to the set temperature. It can take 1–1.5 h for the system to heat the buffer up to the set temperature. Heating the buffer in a microwave helps reduce the preheating time.

Table 4.7 Preparation of 10 % ammonium persulphate

Reagents	Amount
Ammonium persulphate	0.1 g
dH_2O	1.0 ml

Table 4.8 Preparation of dye solution

Reagents	Amount	Final
Bromophenol blue	0.05 g	0.5 %
Xylene cyanol	0.05 g	0.5 %
1X TAE buffer	10.0 ml	1X

Table 4.9 Preparation of 2X gel loading dye

Reagents	Amount	Final
2 % Bromophenol blue	0.25 ml	0.05 %
2 % Xylene cyanol	0.25 ml	
100 % Glycerol	7 ml	
dH_2O	2.5 ml	
Total volume	10 ml	

Table 4.10 Preparation of 1X TAE running buffer

Reagents	Amount (ml)
50X TAE buffer	140
dH_2O	6860
Total volume	7000

Assembling the Parallel Gradient Gel Sandwich

1. Assemble the gel sandwich on a clean surface. Lay the large rectangular plates down first, and then place the left and right spacers of equal thickness along the short edges of the larger rectangular plate. To assemble parallel gradients gels, place the spacers so that the grooved opening of the spacers face the sandwich clamps. When properly placed, the grooved side of the spacers and the notches will face the sandwich clamps, and the hole is located near the top of the plates.
2. Place the short glass plate on top of the spacers so that it is flush with the bottom edge of the long plates.
3. Loosen the single screw of each sandwich clamp by turning each screw counter clockwise. Place each clamp by the appropriate side of the gel sandwich with locating arrows facing up and towards the glass plates.
4. Grasp the gel sandwich firmly. Guide the left and right clamps onto the sandwich so that the long and short plates fit the appropriate notches in the clamp. Tighten the screws enough to hold the plates in place.
5. Place the sandwich assembly in the alignment slot of the casting stand with the short glass plate forward. Loosen the sandwich clamps and insert an alignment card to keep the spacers parallel to the clamps.
6. Align the plates and spacers by simultaneously pushing inward on both clamps at the locating arrows while at the same time pushing down on the spacers with your thumbs; tighten both clamps just enough to hold the sandwich in place. Pushing inward both the clamps at the locating arrows will insure that the spacers and glass plates are flush against the sides of the clamps.
7. Remove the alignment card. Remove the sandwich assembly from the casting stand, and check that the plates and spacers are flush at the bottom. If the spacers and glass plates are not flush, realign the sandwich and spacers to obtain a good seal.
8. When a good alignment and seal are obtained, tighten the clamps screw until his finger tight.

Casting Parallel Denaturing Gradient Gels

1. Place the grey sponge on to the front-casting slot. The cam shafts on the casting strand should have the handles pointing up and pulled out. Place the sandwich assembly on the sponge with shorter plate. When the sandwich is placed correctly, press down on the sandwich and turn the handle of the camshafts down so that the cams lock the sandwich in place. Position the gel sandwich assembly by standing it upright.
2. Use the longer piece of Tygon tubing for keeping gel solutions from the syringe in the Y fitting. The short piece of Tygon tubing will conduct the gel solution from Y fitting to the gel sandwich. Connect one end of the 9 cm Tygon tubing to the Y fitting and connect luer coupling of the other end of the 9 cm tubing connect luer fitting on to the two long pieces tubing. Connect the luer fitting to 30 ml syringes.
3. Label one of the syringes as LO (for the low density solution) and other as HI (for the high density solution). Attach a plunger cap onto each syringe plunger head. Position the plunger "head" in the middle of the plunger cap and tighten enough to hold the plunger in place. Position the cap in the middle for proper alignment with the lever to the middle of the syringe, keeping the volume gradations visible. Make sure that the lever attachment screw is in the same plane as the flat or back side of the sleeve.

4. Rotate the cam wheel counter clockwise to the vertical of the start position. To set the desired delivery volume, loosen the volume adjustment screw. Place the volume setting indicator located on the syringe holder to the desired volume setting. Tighten the volume adjustment screw. For 16 × 16 cm gels (1 mM thick), set the volume setting indicator to 14.5.
5. From the stock solutions, pipette out the desired amount of the high and low density gel solutions into two disposable test tubes.
6. Add the final concentration of 0.09 % (v/v) each of ammonium persulphate and TEMED solutions. The 0.095 (v/v) concentrations allow about 5–7 min finishing casting the gel before polymerization. Cap and mix by inverting several times with the syringe connected to the tubing, withdraw all of the high density solution in to the HI syringe. Do the same for the low density solution in to the LO syringe.
7. Carefully remove the air bubbles from the LO syringe by turning the syringe upside down (plunger cap towards the bench), and gently tapping the syringe. Push the gel solution to the end of the tubing. Do not push it out of the tubing as loss of solution will disturb the volume required to cast the desired gel.
8. Place the LO syringe in to the gradient delivery system syringe holder (LO density side) by holding the syringe by the plunger and inserting the lever attachment screw in to the lever groove. Do not handle the syringe. It will dispense the gel solution out of the syringe. Casting a parallel gel is referred to as a top filling method, so place the LO syringe on the correct side of the gradient system.
9. Carefully remove the air bubbles from the HI syringe by turning the syringe upside down (plunger cap towards the bench) and gently tapping the syringe. Push the solution to the end of the tubing. Do not push it out of the tubing as loss of solution will disturb the volume required to cast the desired gel.
10. Place the HI syringe in to the gradient delivery system by holding the syringe in the plunger and inserting the lever attachment screw in to the lever. Do not handle the syringe, it will dispense the gel solution out of the syringe.
11. Slide the tubing from the low density syringe over one end of the Y fitting. The same is used for the high density syringe.
12. Attach 19 gauge needle to the coupling. Hold the bevelled side of the needle at the top-centre of the gel sandwich and cast. For convenience, the needle can be tapped in place.
13. Rotate the cam wheel slowly and steadily to deliver the gel solution. It is important to cast the gel solution at a steady pace to avoid any disturbances between the gel solutions within the gel sandwich.
14. Carefully insert the comb to the desired well depth and straighten. The gel may take around 60 min to get solidified.

Running and Visualisation of the Gel

1. After solidification of the gel, load 20 µl of the DNA sample by mixing with gel loading buffer.
2. Connect the electrodes to the power supply and run the electrophoresis program for 16 h at 100 V.
3. After completion of the gel run, stain the gel with 10 mg/ml ETBr solution and observe the gel under UV light in a gel documentation system.

Observation

Observe the banding pattern on the gel. Each band represents one operational taxonomic unit (OTUs). The gel image can also be analysed by Quantity One software (Bio-Rad, USA) to obtain the relationship among the bacterial community present in the different samples based upon the sequences of 16S rRNA gene.

Result Table

Sample no.	Lane no.	No. of OTU	Relationship predicted

Troubleshooting

Problems	Possible causes	Possible solutions
Acrylamide clots during gel casting	Polymerization occurs rapidly if temperature of the gel stock solution is high	Chill the stock solution before use
	Gels do not reach the upper edge of the plate	Make sure to reach the gel solution up to the upper edge of the gel
	Leaking of gel during injection	Always keep a flat surface at the bottom while casting the gel
Bands appear as smear	Structural characteristic of BioRad DCode system	Insert a magnetic bead to the electrophoresis tank and continuously stir the buffer during electrophoresis
	Sample not stored under proper conditions	Store the samples at −20 °C to avoid degradation
	Diffused sample during loading	Take maximum care not to diffuse the samples while loading
Poor pattern reproducibility	Use of old gel stock	Always use freshly prepared reagents and buffers
	Degraded formamide	Check the colour of formamide, if it is found to be coloured use fresh stock of deionised formamide
Upper buffer falls off	More space left between the core and the plate	Ensure that the gasket is sandwiched between the core and the plate with no space in between

FLOW CHART

Assemble the gel sandwich for parallel denaturing gradient gel electrophoresis

↓

Prepare the reagents for running gel (40% acrylamide, 50X TAE, formamide)

↓

Prepare one 30% denaturing solution and one 50% denaturing solution for 6% gel

↓

Add APS and TEMED (0.09% v/v) respectively in both the solutions

↓

Mix the 30% and 50% gel by inverting the tubes upside down twice very quickly

↓

Pipette out the desired amount of 30% gel in the marked low density syringe and 50% gel in the marked high density syringe respectively and set the syringes in the loading cascade

↓

Rotate the cam wheel of the loading cascade very slowly to deliver the gel solution by placing the tygon tubing into the gel sandwich

↓

Carefully insert the comb to the well depth and straighten. Let the gel polymerize for about 60 min

↓

After polymerization, mix your samples with the loading dye and load into the wells

↓

Place the gel assembled gel sandwich in the preheated running buffer solution and run for 5 h at 60V

↓

Transfer the gel into EtBr solution (1 μg/ml) for 30 min and visualize under Gel Documentation System

Exp. 4.4 Pulsed Field Gel Electrophoresis (PFGE) Analysis

Objective To study the molecular phylogeny amongst bacteria by pulse field gel electrophoresis (PFGE) analysis.

Introduction

Pulsed field gel electrophoresis is the technique used for the separation of DNA fragments of large molecular size by applying the electric field of periodically changing direction to the gel matrix. Conventional electrophoresis uses a single electric field for migration of biological molecules, though, the gel matrix is based upon the mass-to-charge ratio. In this case, the distance of migration is directly proportional to its size or mass. However, this conventional electrophoresis can separate DNA fragments effectively up to ~20 kb. The DNA molecules of larger fragments co migrate independent of their size and appear as a large band at the top of the gel. In this regard, PFGE overcomes this problem by alternating the electric field between spatially distinct pairs of electrodes, which can separate DNA fragments up to ~10 Mb. It can be achieved by reorientation and movement of DNA fragments at different speeds through the pores of agarose gel.

This technique involves the similar procedure to that of the normal agarose gel electrophoresis except the constantly running voltage in one direction. In this case, the voltage is switched periodically among three directions, one electric field runs through the central axis of the gel and the other two run at an angle of 60° from either side. The pulse times are maintained equally from each direction, which results in a net forward migration of DNA. For large DNA fragments, pulse time is increased for each direction ranging from 10 s at 0 h to 60 s at 18 h. This technique takes a longer time to complete compared to the normal gel electrophoresis as the size of the DNA fragments are quite large and the DNA molecule does not move in a straight line in the gel.

PFGE is mostly used for the study of genotyping or genetic fingerprinting of the organisms, and is considered to be the gold standard protocol for epidemiological studies of pathogenic microorganisms. Thus, this is quiet useful for the discrimination between pathogenic strains, environmental or food borne isolates with clinical infections. There are different types of pulse field gel electrophoresis units available in market that involves CHEF (clamped homogenous electric field), PACE (programmable autonomously controlled electrodes), DR (dynamic regulation), FIGE (field inversion gel electrophoresis) and AFIGE (asymmetric field inversion gel electrophoresis) (Fig. 4.13).

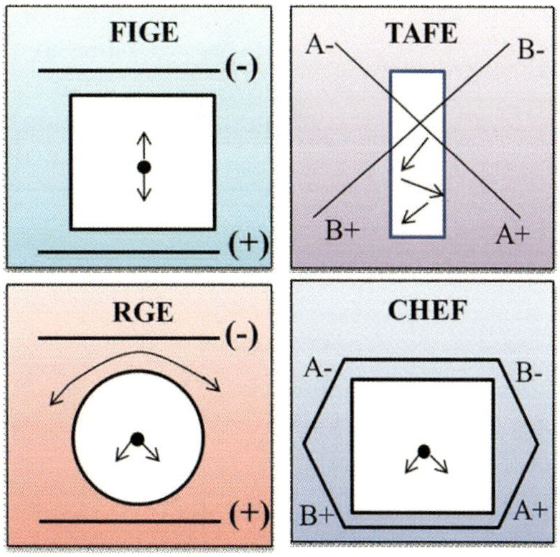

Fig. 4.13 Different types of pulse field gel electrophoresis systems available for separation large-sized genomic DNA fragments

Principle

CHEF coupled with programmable autonomously controlled electrode gel electrophoresis is the most common pulse field technique used for DNA fingerprinting analysis. The systems contain three major components, i.e. a power module to generate the electrode voltage and stores switching function parameters, the cooling module to maintain temperature at 14 °C and the electrophoresis chamber. The chamber mostly contains 24 horizontal electrodes, which are clamped to eliminate DNA lane distortion. The electrodes are arranged in a hexagon pattern, which offers an orientation angle of 60 and 120° in comparison to the gel systems between the two perpendicular electrodes. The resolution of PFGE depends on the number and configuration of the electrodes used, which are responsible for the shape of the applied electric field. It has been reported that, for a high resolution separation the most effective electrode angles should be more than 110°. In PACE each of the electrode's voltage can be controlled independently, and hence can generate unlimited number of electric fields of different voltage gradients, orientations and sequentially interval time. Thus, both the CHEF

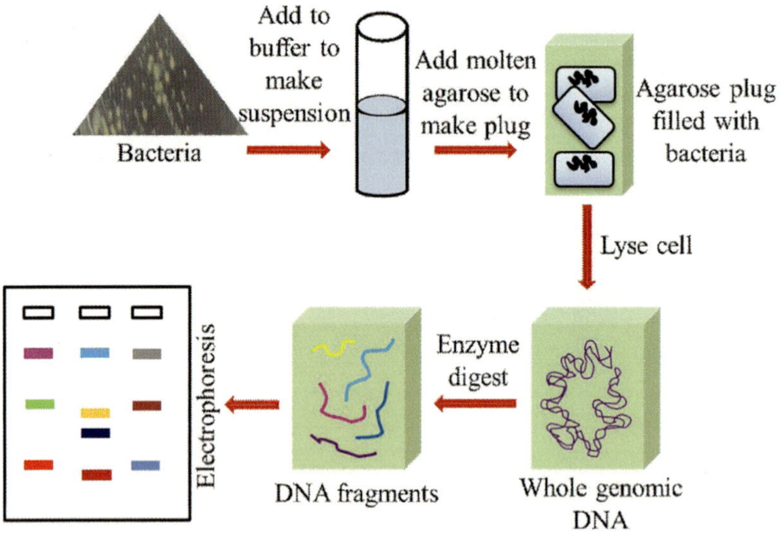

Fig. 4.14 Common steps involved in pulse field gel electrophoresis (PFGE) for separation of large fragments of DNA sequences

and PACE technologies can be configured simultaneously to distinguish between large molecular weight DNA molecules (Fig. 4.14).

The quality of the PFGE results is dependent on the performance and expertise in each step of the procedure, which involves cell lysis and release of intact chromosomal DNA, restriction digestion of chromosomal DNA, separation of DNA fragments and analysis of DNA fragment length polymorphism. In spite of the bigger size of genomic DNA molecules, they can be fractionalised and analysed by performing PFGE. This procedure has a wide variety of applications for all organisms from bacteria to viruses to mammals. This technique has the potential to separate small, natural, linear chromosomal DNA molecules ranging from size 50 kb to multimillion base pair chromosome molecules. However, in routine practice, PFGE involves the separation of DNA molecules between few kilo bases and 10 Mega base pairs.

PFGE has wide applications including the resolution of large DNA molecules for the study of bacterial genomes, application of restriction digestion produce discrete banding patterns useful for fingerprinting and physical mapping of the chromosome, to establish the degree of relatedness among different strain of the same species, estimation of genome size of a bacterium and construction of chromosomal maps useful for characterization of bacterial species, application of this technique to chromosomal DNAs from fungi, parasitic protozoa, bacterial genomes as well as mammals. PFGE has been widely applied for construction of yeast artificial chromosome libraries and construction of transgenic mice. Nowadays, it has been used for the study of radiation induced DNA damage and repair, size organization and variation of mammalian centromeres.

The analysis of the entire genome of bacteria and other organisms has represented a revolutionary approach to genetics, and its availability to produce vast amount of information has conceptualised new approaches for the study of biological research. However, at present scenario it is clear that a single technique cannot be considered to be as a gold standard for construction of bacterial genome maps. Hence, a combinatorial approach of many powerful new approaches together may provide a useful technique for establishing the degree of relatedness among different strains of the same species.

Reagents Required and Their Role

Brain Heart Infusion Agar

Brain Heart Infusion Agar is an enriched nonselective medium used for the enrichment and isolation of most of the anaerobic and fastidious microorganisms. Hemin and vitamin K1 have been added as growth factors for most of the anaerobic microorganisms. The composition of BHI medium is calf brain infusion—200.0 g/l, beef heart infusion—250.0 g/l, proteose peptone—10.0 g/l, dextrose—2.0 g/l, sodium chloride—5.0 g/l, disodium phosphate—2.5 g/l and final pH—7.4. BHI is mostly used for analysis of food safety, water safety and for antibiotic sensitivity testing.

TE Buffer

It can be prepared by mixing 50 mM Tris and 50 mM EDTA in water, and by maintaining the pH at 8.0. As a major constituent of TE buffer, Tris acts as a common pH buffer to control pH during addition of other reagents, while EDTA chelates cations like Mg^{2+}. Thus, TE buffer is helpful to solubilise DNA while protecting it from degradation.

Sodium Dodecyl Sulphate (SDS)

In this case, 10% SDS will solve the purpose. SDS is a strong anionic detergent that can solubilise the proteins and lipids that form the membranes. This will help the cell membranes to break down and expose the chromosomes that contain DNA.

Proteinase K

To get a quality DNA product, 20 mg/ml of proteinase-K is a very good enzyme that degrades most types of protein impurities. It is also responsible for the inactivation of nucleases, thus preventing damage of isolated DNA.

Lysozyme

Lysozyme is also known as muramidase, which damages bacterial cell wall by catalysing hydrolysis if 1,4 beta linkage between NAM and NAG residues of peptidoglycan. During PFGE, lysozyme is used for the lysis of bacterial cells for the release of genetic material, which will be explored further for restriction enzyme digestion.

Agarose

It is used for the electrophoretic separation of nucleic acids. The purest form of agarose is free of DNase and RNase activities. Molecular biology grade agarose is the standard one for the resolution of DNA fragment in the range of 50 bp–50 kb, with the possibility of subsequent DNA extraction from the gel for further analysis. It possesses the following properties: gel strength (1%)—1125 g/cm^2, gelling point (1.5%)—36.0°C, melting point (1.5%)—87.7°C, sulphate—0.098%, moisture—2.39% and ash—0.31%.

ASCI Restriction Enzyme

This restriction enzyme is obtained from an *E. coli* strain harbouring *ascI* gene from *Arthrobacter* sp. This enzyme recognises GGCGCGCC sites and exhibits optimum activity at 37°C in the specific buffer solution.

Ethidium Bromide

EtBr is a fluorescent dye that intercalates between nucleic acid bases and eases the detection of nucleic acid fragments in gel. When exposed to UV light, DNA flourishes with an orange colour, intensifying 20-fold after binding to DNA. The absorption maxima of EtBr in aqueous solution is between 210 and 285 nm that corresponds to the UV light. Hence, as a result of this excitation, EtBr emits orange light with a wavelength of 605 nm. EtBr binds with DNA and slips in between its hydrophobic base pairs and stretches the DNA fragment, thus removing the water molecules from ethidium cation. This dehydrogenation results in the increase in fluorescence of the ethidium. However, EtBr is a potential mutagen, suspected carcinogen and it can irritate eyes, skin, mucous membranes and upper respiratory tract at higher concentrations. It is due to the fact that, EtBr intercalates into double-stranded DNA, deform the molecule, thus blocking the biological processes involving nucleic acids like DNA replication and transcription. Hence, there are many alternatives regarded as less dangerous and with better performance like Sybr dyes.

Procedure

Revival of Bacterial Culture

1. Streak the pure culture of *Vibrio parahaemolyticus* strains on Brain Heart Infusion Agar plates.
2. Incubate the plates at 37 °C for 14–18 h and observe for the confluent growth pattern.

Preparation of Plugs

1. Place 10 % SDS solution into a 55–60 °C water bath for warming.
2. Weigh 0.50 g of agarose into a 250 ml screw cap bottle and add 50.0 ml of TE buffer to it.
3. Loosen the cap and boil in microwave for 30 s. Mix gently and repeat at 10 s interval till agarose is dissolved completely.
4. Add 2.5 ml of warm 10 % SDS stock solution and swirl to mix. Recap flask and keep it back at water bath for 15 min to equilibrate the agarose.
5. Suspend grown bacterial culture in TE buffer by spinning sterile swab containing bacterial culture from the petriplate.
6. Adjust concentration of cell suspensions to McFarland 0.5 by diluting the culture or adding additional cells.

Casting of Plugs

1. Transfer 400 µl of adjusted cell suspensions to the labelled 1.5 ml micro-centrifuge tube. Add 20 µl of thawed lysozyme stock solution to each tube and mix gently.
2. Add 20 µl of proteinase-K to each tube and mix gently by pipetting up and down. Add 400 µl of melted agarose with 0.5 % SDS to 400 µl of cell suspension, mix gently by pipetting up and down.
3. Immediately dispense few amount of the mixture to appropriate plug moulds by avoiding the formation of bubbles.
4. It is always wise to prepare two plugs from the same cell suspension for possible repeat testing. Allow the plugs to solidify for 10–15 min at room temperature. Optionally, incubate the plugs at 4 °C for 5 min.

Lysis of Cells in Plugs

1. Prepare cell lysis buffer with the following composition: 50 ml of 1 M Tris, 100 ml of 0.5 M EDTA, 100 ml of 10 % Sarcosyl, and dilute the mixture to 1000 ml with sterile water.
2. Trim excess agarose from top of the plug and transfer the plugs with the help of a spatula to the appropriately labelled tubes.
3. Add appropriate amount of lysis buffer and proteinase-K and make sure the plugs are inside the buffer completely.
4. Incubate the tubes at 54–55 °C in a shaker water bath with constant vigorous agitation.
5. Preheat milli-Q water to 54–55 °C for the washing of plugs twice with 10–15 ml of water.

Washing of Plugs After Lysis

1. Remove the tubes from the water bath and carefully pour-off the lysis buffer without removing the plugs.

Table 4.11 Reagents for preparation of ASCI restriction enzyme master mixture

Reagents	µl/Plug slice
Milli-Q water	175.5
10 X restriction buffer	20
BSA (10 mg/ml)	2
ASCI	2.5
Total volume	200

2. Add 10–15 ml of preheated sterile milli-Q water and incubate the tubes at 54–55 °C with shaking for 10–15 min.
3. Pour-off water from the tubes and repeat washing for one more time. Repeat washing with TE buffer for three times.
4. Add 5–10 ml of TE buffer and store the plugs at 4 °C till further use.

Restriction Digestion of DNA in Agarose Plug

1. Prepare a mister mixture by diluting the restriction buffer with milli-Q water to prepare 1 X buffer.
2. Add 200 µl of diluted 1X restriction buffer to a 1.5 ml micro-centrifuge tube.
3. Carefully remove the plugs and cut 2.0–2.5 mm wide slices from each test sample, replace rest of the plug into the original tube with TE buffer and store at 4 °C.
4. Prepare ASCI restriction enzyme master mixture according to the following table (Table 4.11).
5. Add 200 µl of restriction enzyme master mixture to each tube. Close the tubes and mix by tapping gently. Make sure the plug slices are under enzyme mixture.
6. Incubate samples and control plug slices in water bath for 2 h.

Preparation of Gel and Loading Digested Plug Slices into the Wells

1. Prepare 1 % agarose gel following protocol as described in Exp. 1.6.
2. Remove restriction digested plugs from tubes and load into appropriate wells. Gently push plugs to the bottom and front of the wells. Manipulate position of the plug slice and make sure there are no air bubbles.
3. Fill wells of the gel with melted 1 % agarose gel and allow hardening for 3–5 min. Remove excess agarose from sides, carefully placing the gel inside the electrophoresis chamber, then closing the cover of the chamber.

Electrophoresis Parameters and Analysis of Results

1. Select the following conditions on CHEF mapper, i.e. Auto Algorithm, 49 kb low MW, 450 kb high MW, and maintain run time to 18–19 h.
2. When electrophoresis run is over, turn-off the instrument, remove and stain the gel with ETBr.
3. Acquire the gel image on a gel documentation system under UV light and analyse the fingerprinting using Quantity One software.

Observation

Observe the banding pattern under UV light in a gel documentation, and analyse the banding pattern for the similarity among the test organisms. The similar genotypes are supposed to produce a clear similar banding pattern, thus concluding their similar genotypic origin.

Result Table

Sample no.	Lane no.	No. of bands obtained	Predicted relationship

Troubleshooting

Problems	Possible causes	Possible solutions
Ghost or Shadow bands	Poor plug quality	Wash proteinase-K and enzyme inhibitors properly from the plug. Check for the cell concentration and use less amount of bacterial culture
	Poor enzyme quality	Use fresh lot of enzyme, do not use expired enzyme or the vials that are opened frequently
	Enzyme digestion not optimal	Check for the inclusion of BSA in master mixture, do not use enough enzyme (star activity), use optimum incubation time and temperature, choose correct buffer
Dark bands in wells	Cell concentration too high	More cells lead to more DNA and subsequently incomplete lysis, use less starting culture
	Inadequate washing	Perform the washing step carefully, repeat washing with milli-Q water and TE buffer, three times each
High background	Degraded DNA	Check for the quality of DNA sample, never use degraded DNA samples
	High cell concentration	Do not use high cell concentrations, use less culture so that restriction digestion can be performed optimally
	Incomplete lysis	Check for the exact quantity of lysozyme, SDS, proteinase-K and incubation time for proper lysis of cells
	Inadequate washing	Washing of the plugs is the most important step during this experiment and hence perform this step carefully
	Incomplete restriction	Check for the quality and quantity of restriction enzyme to be used. Check for the suitable incubation time for proper digestion of the DNA fragments

Precautions

1. Follow the protocol without any modifications.
2. Always start with a pure culture.
3. Use good quality reagents.
4. Take a less amount of culture; more amount may cause hindrance in the banding pattern.
5. Use good PFGE equipment like CHEF from the reputed manufacturers.
6. Always wear gloves during the experiment.

FLOW CHART

Revival of bacterial culture

Streak the pure culture of bacterial strain in Brain Heart infusion Agar plates

↓

Incubate the plates at 37°C for 14-18 h and observe the confluent growth pattern

Preparation of plugs

Weigh 0.50 g of agarose to a 250 ml of screw cap bottle and mix with 50 ml of TE buffer, dissolve by heating in a microwave, add 2.5 ml of 10% SDS solution and swirl to mix

↓

Suspend bacterial culture to TE buffer from the grown plates

↓

Adjust the concentration of bacterial culture to 0.5 McFarland by diluting the culture or addition of additional culture

Casting of plugs

Transfer 400 µl of cell suspension to a 1.5 ml tube; add 20 µl of lysozyme and proteinase-K and 400 µl of agarose with SDS

↓

Immediately dispense the mixture to the plug mold by avoiding air bubbles

↓

Allow the plugs to solidify for 10-15 min at room temperature, optionally incubate the plugs at 4°C for 5 min

FLOW CHART

Lysis of cells in plugs

Prepare cell lysis buffer with the following composition: 50 ml of 1 M tris, 100 ml of 0.5 M EDTA, 100 ml of 10% Sarcosyl and dilute the mixture to 1000 ml with sterile water

↓

Trim excess agarose from top of the plug and add appropriate amount of lysis buffer and proteinase-K and make sue the plugs to dip inside the buffer completely

↓

Incubate the tubes at 54-55°C in a shaker water bath with constant vigorous shaking

↓

Pre-heat Milli-Q water to 54-55° for the washing of the plugs

Washing of plugs

Remove the tubes from water bath and carefully pour off the lysis buffer without removing the plugs

↓

Add 10-15 ml of pre-heated sterile milli-Q water and incubate the tubes at 54-55°C for 10-15 min

↓

Pour off water from the tubes and repeat washing for one more time, repeat washing with TE buffer for three times

↓

Add 5-10 ml of TE buffer and store the plugs at 4°C till further use

Restriction digestion of DNA

Prepare slices of plugs (2.0-2.5 mm size) and add 200 μl of restriction enzyme master mixture to each tube, close tubes and mix by gentle tapping

↓

Make sure the plug slice dip inside the enzyme mixture

↓

Incubate samples and the control plug slices in water bath for 2h

FLOW CHART

Preparation of gel and loading of the digested plug slices

Prepare 1% agarose gel following protocol as described in Experiment 1.6

↓

Remove restriction digested plugs from tubes, load into appropriate wells, gently push plugs to the bottom and front of the well, make sure no air bubbles

↓

Fill well of the gel with melted 1% agarose gel, allow to harden for 3-5 min, remove excess agarose from sides and carefully place the gel inside the electrophoresis chamber, close the cover of the chamber

Electrophoresis parameters and analysis of results

Select the following conditions on CHEF mapper i.e. auto algorithm, 49 bp low MW, 450 kb high MW and maintain run time to 18-19 h

↓

When electrophoresis run is over, turn off the instrument, remove and stain the gel with ethidium bromide

↓

Acquire the gel image on a gel documentation system under UV light and analyse the fingerprinting by Quantity One software

Exp. 4.5 Multiplex PCR for Rapid Characterization of Bacteria

Objective Rapid characterization of a bacterial species by multiplex PCR approach using different set of primers.

Introduction

Multiplex PCR is a modification of normal PCR protocol for rapid detection of deletions, duplications, presence or absence of gene segments. This can be achieved by amplifying the genomic DNA sample in a PCR with the help of multiple primers and *Taq* polymerase. While targeting more than one gene at a time, much more information on an organism can be obtained from a single run of PCR, which may require several assays and more reagents to obtain the result. In this regard, the annealing temperature of all the primers to be used should be optimised so that they can be able to amplify the correct set of genes at a single PCR reaction.

The common applications of multiplex PCR include pathogen identification, SNP genotyping, mutation analysis, gene deletion analysis, template quantitation, linkage analysis, RNA detection, forensic studies as well as diet analysis. By employing multiplex PCR approach, considerable amount of time and effort can be saved due to simultaneous amplification of multiple sequences of genes in a single PCR run. Nowadays, multiplex PCR analysis has become a rapid and convenient tool for assay in clinical and laboratory practices. For the development of an efficient multiplex, PCR assay strategic planning and multiple attempts are required to optimise the reaction conditions. An optimal combination of annealing temperature and buffer concentration is highly essential during multiplex PCR for a highly specific amplification product. However, before going for clinical use the sensitivity and specificity of the procedure should be evaluated thoroughly using standard purified nucleic acids with suitable internal and external controls. As the number of microbial agents can be detected by PCR, it becomes highly desirable for practical purposes for simultaneous detection of multiple agents that cause similar or identical clinical syndromes and share similar epidemiological features.

However, there are many constraints during multiplex PCR analysis for a set of genes. During this case, the primer sets should be chosen having similar annealing temperature, similar length of amplified products, less chance of mispriming and self-priming. Nowadays, many commercially available kits are of use for the rapid analysis of degraded DNA samples for forensic application.

Principle

During this experiment a septaplex PCR assay will be performed for rapid identification of *Vibrio cholerae* by simultaneous detection of the virulence and *intsxt* genes. As cholera is a life-threatening disease, its rapid detection and characterization is the prime objective for all the researchers. Conventional techniques used to detect and classify *Vibrios* isolated both from clinical and environmental samples mostly take several days to complete. It may involve the enrichment of the culture in alkaline peptone water followed by growth on TCBS agar, slide agglutination test with suitable antisera, assay for cholera toxin and other confirmatory tests. This traditional procedure is highly laborious, time consuming and expensive. In addition to that, biochemical properties of many strains are highly similar as in case of *V. cholerae*, *V. mimiscus* and other *Vibrio* spp. that renders huge ambiguity in identification procedure. In this regard, the septaplex PCR to be demonstrated here will be of great use for rapid identification and characterization of *Vibrio* spp.

Development of a multiplex PCR should follow a rational approach for inclusion or exclusion of different gene segments and organisms. Wherever possible, multiplex PCR should avoid the use of nested primers used for the second round of amplification as it has a major contribution for the false positive results due to carryover contaminations. Most of the difficulties encountered during multiplex PCR analysis can be avoided by the use of a hot stat PCR and nested PCR. During this case, the use of hot stat PCR eliminates most of the non-specific reactions, whereas nested PCR increases the sensitivity and specificity of the test reaction through two independent round of amplification using two discrete set of primers. However, the second round of amplification may alter the reaction by cross-contamination, and further complicate the automation.

There are various applications of multiplex PCR analysis for the detection and differentiation of human retroviruses; other transfused transmitted viruses and others. This technology has become a valuable tool for differentiation, subgrouping, subtyping and genotyping of many bacterial and viral species. Commercial development of PCR has facilitated the widespread introduction of this procedure, and has improved dramatically by the ease of use of the technology, still multiplex PCR are yet in their infancy stage.

Reagents Required and Their Role

Bacterial Genomic DNA

Bacterial genomes are of small and less varying sizes ranging between 139 and 13,000 kbp. The isolation and purification of genomic DNA from bacteria is the most common prerequisite for most of the molecular biology experiments. For the analysis of ARDRA for study of molecular phylogeny, bacterial genomic DNA to be used should be of high quality and quantity. The detailed procedure for the isolation of genomic DNA from bacteria have been described during Exp. 1.1.

Table 4.12 Target gene and the primer sequences. (Mantri et al. 2006)

Target gene	Primer sequence (5′–3′)	Amplicon size (bp)	Gene accession no.	Primer site
O139 rfb-F	AGCCTCTTTATTACGGGTGG	449	Y07786	12,288–12,307
O139 rfb-R	GTCAAACCCGATCGTAAAGG		Y07786	12,717–12,736
O1 rfb-F	GTTTCACTGAACAGATGGG	192	X59554	13,195–13,213
O1 rfb-R	GGTCATCTGTAAGTACAAC		X59554	13,368–13,386
ISRrRNA VC-F	TTAAGCSTTTTCRCTGAGAATG	295–310	AF114723	227–248
ISRrRNA VM-R	AGTCACTTAACCATACAACCCG		AF114723	501–522
ctxA F	CGGGCAGATTCTAGACCTCCTG	564	X00171	588–609
ctxA R	CGATGATCTTGGAGCATTCCCAC		X00171	1129–1151
toxR F	CCTTCGATCCCCTAAGCAATAC	779	M21249	277–298
toxR R	AGGGTTAGCAACGATGCGTAAG		M21249	1034–1055
tcpA-F Clas/El Tor	CACGATAAGAAAACCGGTCAAGAG	620	X64098	3379–3402
tcpA-R Clas	TTACCAAATGCAACGCCGAATG		X64098	3977–3998
tcpA-R El Tor	AATCATGAGTTCAGCTTCCCGC	823	X74730	3235–3256
sxt-F	TCGGGTATCGCCCAAGGGCA	946	AF099172	90–109
sxt-R	GCGAAGATCATGCATAGACC		AF099172	1016–1035

Primers

Primers are the small nucleotide sequences for the amplification of targeted DNA sequences. In this experiment a set of primers are to be used, which has been listed in Table 4.12.

dNTPs

The dNTPs are the building blocks of new DNA strands. In most of the cases, they come as a mixture of four deoxynucleotides, i.e. dATP, dTTP, dGTP and dCTP. Around 100 µM of each of the dNTPs is required per PCR reaction. dNTP stocks are very sensitive to cycles of thawing and freezing, and after 3–5 cycles PCR reactions does not work well. To avoid such problems, small aliquots (2–5 µl) lasting for only a couple of reactions can be made and kept frozen at −20 °C. However, during long-term freezing, small amount of water condenses on the walls of the vial, thus changing the concentration of dNTPs solution. Hence, before using it is essential to centrifuge the vials and is always recommended to dilute the dNTPs in TE buffer, as acidic pH promotes hydrolysis of dNTPs and interferes with the PCR result.

Taq Polymerase

Taq polymerase is a thermostable DNA polymerase isolated from the thermophilic bacterium *Thermus aquaticus* which was originally isolated by Thomas D. Brock in 1965. The enzyme is able to withstand the protein denaturing conditions required during PCR. It has the optimum temperature for its activity between 75 and 80 °C, with a half-life of greater than 2 h at 92.5 °C, 40 min at 95 °C and 9 min at 97.5 °C, and possesses the capability of replicating a 1000 bp of DNA sequence in less than 10 s at 72 °C. However, the major drawback of using Taq polymerase is the low replication fidelity as it lacks 3′–5′ exonuclease proofreading activity. It also produces DNA products having 'A' overhangs at their 3′ ends, which is ultimately useful during TA cloning. In general, 0.5–2.0 units of Taq polymerase is used in a 50 µl of total reaction, but ideally 1.25 units should be used.

PCR Reaction Buffer

Every enzyme needs certain conditions in terms of their pH, ionic strength, cofactors etc. which is achieved by the addition of buffer to the reaction mixture. In some instances, the enzyme shift pH in non-buffered solution and stops working in this process, which can be avoided by the addition of PCR buffer. In most of the PCR buffers the composition is almost same as: 100 mM Tris-HCl, pH 8.3, 500 mM KCl, 15 mM $MgCl_2$ and 0.01 % (w/v) gelatin. The final concentration of the PCR buffer should be 1X concentration per reaction.

Procedure

1. Grow *Vibrio cholerae* strains in LB broth for 24 h at 37 °C with shaking at 180 rpm.
2. Isolate genomic DNA from the bacterial cell following protocol as described in Exp. 1.1. Store the isolated genomic DNA at −20° till further use.
3. Thaw all the reagents required for the set up of PCR reaction in ice.
4. Prepare the reaction mixture for PCR reaction by adding the reaction components as follows: 5.0 µl of 10X PCR buffer, 2 mM each dNTPs from a stock of 10 mM dNTPs, 2.5 µl each of the forward and reverse primer with a stock concentration of 10 µM, 1 µl of the 2.5 U/µl of Taq polymerase, 29.5 µl of milli-Q water and 2 µl of template DNA.
5. Mix the reaction components well, and if possible, centrifuge briefly.
6. Run PCR with the cyclic conditions: initial denaturation at 94 °C for 5 min followed by 30 cycles of denaturation at 94 °C for 30 s, annealing at 55 °C for 30 s, extension at 72 °C for 2 min and final extension at 72 °C for 7 min followed by holding at 4 °C for ∞.
7. Run the amplified PCR product by agarose gel electrophoresis with 2 % agarose gel.
8. Observe the banding pattern in gel documentation system under UV light.
9. Analyse the banding pattern by Quantity One software and draw the phylogenetic relationship.

Observation

Observe the banding pattern for the presence and absence of the targeted genes in the test organism. Compare the banding pattern with the control set of organism and draw the phylogenetic relationship between the test organisms.

Result Table

Organism	O139	O1	ISRrRNA	ctxA	toxR	tcpA	sxt
Positive control	+	+	+	+	+	+	+
Test organism							

Troubleshooting

Problems	Possible cause	Possible solution
Incorrect product size	Incorrect annealing temperature	Recalculate primer Tm values by using any of the web-based software
	Mispriming	Verify the primes have no additional complementary regions within the template DNA
	Improper Mg^{2+} concentration	Optimise Mg^{2+} concentration with 0.2–1 mM increments
	Nuclease contamination	Repeat the reactions using fresh solutions
No product	Incorrect annealing temperature	Recalculate the Tm values of the primers, test the correct annealing temperature by gradient, starting at 5 °C below the lower Tm of the primer pair
	Poor primer design	Check with the literature for recommended primer design, verify that the primers are non-complementary, both internally and to each other, optionally increase length of the primer
	Poor primer specificity	Verify that the oligos are complementary to the proper target sequence
	Insufficient primer concentration	The correct range of primer concentration should be 0.05–1 μM, refer to the specific product literature for ideal conditions
	Poor template quality	Analyze DNA by agarose gel electrophoresis, check A_{260}/A_{280} ratio of DNA template
	Insufficient number of cycles	Rerun the reaction with more number of cycles
Multiple/ non-specific products	Premature replication	Use hot start polymerase, set up the reaction on ice using chilled components, add samples to the PCR preheated to the denaturation temperature
	Primer annealing temperature too low	Increase the annealing temperature
	Incorrect Mg^{2+} concentration	Adjust Mg^{2+} concentration with 0.2 to 1.0 mM increments
	Excess primer	The correct range of primer concentration should be 0.05–1 μM, refer to the specific product literature for ideal conditions
	Incorrect template concentration	For low complexity templates (i.e. plasmid, lambda, BAC DNA), use 1 pg–10 ng DNA per 50 μl reaction For high complexity templates (i.e. genomic DNA), use 1 ng–1 μg template for 50 μl reaction

Precautions

1. Use pipette tips with filters.
2. Store materials and reagents properly under separate conditions, and add them to the reaction mixture in a spatially separated facility.
3. Thaw all components thoroughly at room temperature before starting an assay.
4. After thawing, mix the components with brief centrifugation.
5. Work quickly on ice or in the cooling block.
6. Always wear safety goggles and gloves while performing the PCR reaction.

FLOW CHART

Take overnight grown culture of *Vibrio cholerae* and isolate genomic DNA by using protocol described as in Experiment 1.1., Store at -20°C till further use

↓

Thaw all the reagents on ice

↓

Prepare the PCR mixture by adding the components i.e. 5.0 µl of 10X PCR buffer, 2 mM each dNTPs from a stock of 10 mM dNTPs, 2.5 µl each of the forward and reverse primer with a stock concentration of 10 µM, 1 µl of the 2.5 U/µl of Taq polymerase, 29.5 µl of milli-Q water and 2 µl of template DNA

↓

Run PCR with the cyclic conditions initial denaturation at 94°C for 5 min followed by 30 cycles of denaturation at 94°C for 30 sec, annealing at 55°C for 30 sec, extension at 72°C for 2 min and final extension at 72°C for 7 min followed by hold at 4°C for ∞

↓

Resolve the PCR products by agarose gel electrophoresis with 2% agarose gel

↓

Observe the banding pattern under UV light in a Gel Documentation System∞

↓

Analyse the banding pattern and draw phylogeny by using Quantity One software∞

Exp. 4.6 ERIC and REP-PCR Fingerprinting Techniques

Objective To reveal genetic heterogeneity among *Vibrio cholerae* by using Enterobacterial Repetitive Intergenic Consensus (ERIC) and Repetitive Extragenic Palindromic (REP)-PCR.

Introduction

Bacterial genomes are considered to be highly streamlined with a number of short interspaced repetitive sequences found in them. However, very little things are known so far regarding their origin, evolution, generation and their functions. These sequences are found in some species of bacteria while others are devoid of it suggesting they have a certain function, and the gene has been acquired by them. These repetitive sequences are reported to act as binding sites for many proteins like DNA polymerase and gyrase to carry out the desired function, which may be incidental. Most of the short repetitive sequences form imperfect palindromes having the potential to form secondary structures and help in mRNA stability; otherwise most of them form the non-functional junks.

There are various PCR fingerprinting techniques available for the amplification of polymorphic DNA through specific selection of primer annealing sites. These molecular typing methods of bacteria are extremely reliable, rapid and highly discriminative as well as reproducible. The fundamentals behind this technique is the use of PCR with primer sets that are complimentary to the highly conserved, naturally occurring, repetitive DNA sequences present in multiple copies in distinct intergenic positions throughout the genome of most Gram-negative and Gram-positive bacteria. REP-PCR based molecular typing method was first used by Versalovic et al. (1994), and the technique has advanced dramatically in the years to become completely automated (Healy et al. 2005). In comparison to REP-PCR, ERIC-PCR is highly sensitive and useful

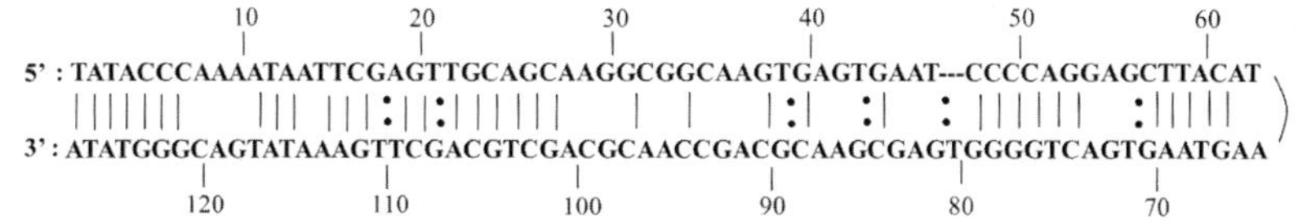

Fig. 4.15 The ERIC sequences. The 127 bp sequence showing hair pin structure and the complementary sequences have been shown here

in detecting microorganisms from any environment. BOX-PCR is the superior to all the techniques creating distinct fingerprinting patterns; however, ERIC and REP-PCR are the methods, which are primarily used for genotyping (Frye and Healy 2006). REP-PCR is having a consensus sequence of 38 bp in addition to the 5 bp in the stem loop of the palindrome structure. In the same line, ERIC-PCR is 126 bp and also found in the extragenic regions. However, BOX-PCR has three subunits, BOX-A, BOX-B and BOX-C having 59, 45 and 50 nucleotide lengths, respectively. Most importantly, BOX-PCR does not share any sequence relations with either ERIC or REP-PCR (Olive and Bean 1999). The arbitrarily primed polymerase chain reaction (AP-PCR) can amplify fragments of DNA from any genome varying the size distribution of amplified fragments among species. Thus, the closely related taxa possess similar fragment distributions, while that of the distantly related taxa are more divergent, hence, providing considerable phylogenetic information (Espinasa and Borowsky 1998).

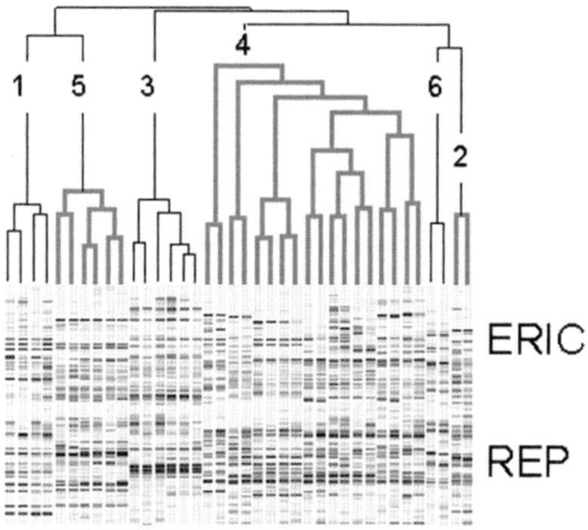

Fig. 4.16 Deducing phylogenetic relationship among the bacterial species using PCR fingerprinting techniques, i.e. ERIC and REP-PCR

Principle

Enterobacterial repetitive intergenic consensus (ERIC) sequences are the intergenic repetitive units, which are different from most of the other bacterial repeats as they are distributed across a wider range of bacterial species. ERIC sequences were first found in *E. coli*, *Salmonella typhimurium* and other members of Enterobacteriaceae family including *Vibrio cholerae*. ERIC sequences are the imperfect palindromes of 127 bp repeats (Fig. 4.15). In addition to that, there is a huge variation in size due to the deletion of internal sequences and insertion of about 70 bp at specific internal sites. The distinguished feature of ERIC sequences is that, they are found in the intergenic regions within the transcribed regions. The technique involves the exploitation of the number of copies of ERIC sequences present in the test organism. However, till now nothing is known so far regarding the mobility and nature of these genetic elements. It is not clear also about the functional role of the gene copies.

Repetitive sequence based polymerase chain reaction (REP-PCR) yields DNA fingerprints comprised of multiple-different sized DNA molecules that contain unique sequences of chromosomal segments occurring between repetitive sequences. In contrast to hybridization technique, where the DNA fingerprints reflect the presence or absence of gene fragment in differentially sized chromosomal restriction fragments, REP-PCR reflects varying distances between oligonucleotide primer binding sites at repetitive sequence targets (Fig. 4.16). The amplicons of different sizes are fractioned further

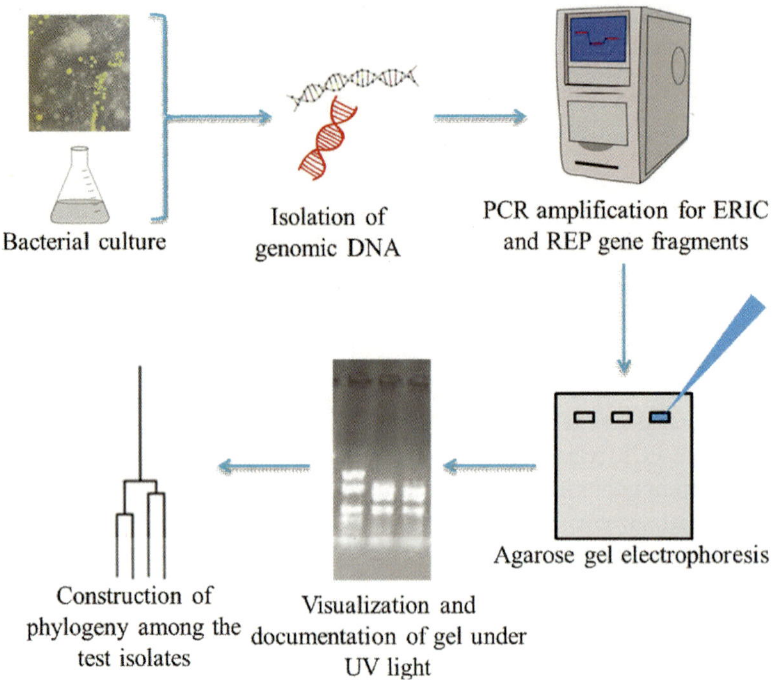

Fig. 4.17 Steps involved in ERIC and REP-PCR mediated fingerprinting of bacteria

by electrophoresis to constitute DNA finger printing patterns specific for individual bacterial clones or strains. Thus, the unique bar codes or DNA finger prints define each bacterial chromosome without the measurement of gene expression level or enzyme function. Genotypic or molecular approaches differ with relation to the level of resolution of individual bacterial species into distinct categories.

Both the techniques involve a similar procedure except the use of a suitable primer set for the amplification of the conserved genes. These techniques can be applied to both the clinical as well as environmental isolates; and genomic DNA can be isolated using any standard protocol followed by amplification of ERIC and REP gene fragments that can be characterised further for the generation of fingerprints of the specific bacterial species (Fig. 4.17). This technique is highly time shaving and economical, rendering to its advantage over other fingerprinting techniques. The fingerprint of many bacterial strains can be derived by the single run of PCR using two sets of primers. However, the genomic DNA to be used as template for this reaction should be of high quality, otherwise it may interfere with the involved steps of PCR.

Reagents Required and Their Role

Bacterial Genomic DNA

Bacterial genomes are of small and less varying sizes ranging between 139 and 13,000 kbp. The isolation and purification of genomic DNA from bacteria is the most common prerequisite for most of the molecular biology experiments. For the analysis of ARDRA for study of molecular phylogeny, bacterial genomic DNA to be used should be of high quality and quantity. The detailed procedure for the isolation of genomic DNA from bacteria have been described during Exp. 1.1.

PCR Reaction Buffer

Every enzyme needs certain conditions in terms of their pH, ionic strength, cofactors etc. which is achieved by the addition of buffer to the reaction mixture. In some instances, the enzyme shift pH in non-buffered solution and stops working in this process, which can be avoided by the addition of PCR buffer. In most of the PCR buffers the composition is almost same as: 100 mM Tris-HCl, pH 8.3, 500 mM KCl, 15 mM $MgCl_2$ and 0.01 % (w/v) gelatin. The final concentration of the PCR buffer should be 1X concentration per reaction.

Divalent Cations

The mechanism of DNA polymerase requires the presence of divalent cations. Most essentially, they shield the negative charge of the triphosphate and allow the hydroxyl oxygen of the 3′ carbon to attack the phosphorus of the alpha phosphate group attached to the 5′ carbon of the incoming nucleotide. All enzymes that break the phosphoanhydride bonds of nucleoside di- and tri-phosphates require the presence of divalent cations. A concentration of 1.5–2.0 mM of $MgCl_2$ is optimal for the activity of Taq DNA polymerase. If Mg^{2+} will be too low, no PCR product will be visible; whereas, if Mg^{2+} is too high, undesired PCR product may be obtained.

dNTPs

The dNTPs are the building blocks of new DNA strand. In most of the cases they come as a mixture of four deoxynucleotides, i.e. dATP, dTTP, dGTP and dCTP. Around 100 μM of each of the dNTPs is required per PCR reaction. dNTP stocks are very sensitive to cycles of thawing and freezing, and after 3–5 cycles, PCR reactions does not work well. To avoid such problems, small aliquots (2–5 μl) lasting for only a couple of reactions can be made and kept frozen at −20 °C. However, during long-term, freezing, small amount of water condenses on the walls of the vial, thus changing the concentration of dNTPs solution. Hence, before using it is essential to centrifuge the vials, and it is always recommended to dilute the dNTPs in TE buffer, as acidic pH promotes hydrolysis of dNTPs and interferes with the PCR result.

ERIC Primers

Two sets of forward and reverse primers are used for the amplification ERIC fragment from the bacterial genome. In this regard, two oligonucleotide primers are used, i.e. ERIC1, 5′-ATGTAAGCTCCTGGGGATTCAC-3′ and ERIC2, 5′-AAGTAAGTGACTGGGGTGAGCG-3′.

REP Primers

Two sets of forward and reverse primers are used for the amplification ERIC fragment from the bacterial genome. In this regard, two oligonucleotide primers are used, i.e. REPIR1: 5′-IIIICGICGICATCIGGC-3′ and REP2-I: 5′-ICGICTTATCIGGCCTAC-3′.

Taq Polymerase

Taq polymerase is a thermostable DNA polymerase isolated from the thermophilic bacterium *Thermus aquaticus*, which was originally isolated by Thomas D. Brock in 1965. The enzyme is able to withstand the protein denaturing conditions required during PCR. It has the optimum temperature for its activity between 75 and 80 °C, with a half-life of greater than 2 h at 92.5 °C, 40 min at 95 °C and 9 min at 97.5 °C, and possesses the capability of replicating a 1000 bp of DNA sequence in less than 10 s at 72 °C. However, the major drawback of using Taq polymerase is the low replication fidelity as it lacks 3′–5′ exonuclease proofreading activity. It also produces DNA products having ‘A’ overhangs at their 3′ ends, which is ultimately useful during TA cloning. In general, 0.5–2.0 units of Taq polymerase is used in a 50 μl of total reaction, but ideally 1.25 units should be used.

Thiosulphate Citrate Bile Salts Sucrose Agar

Thiosulphate citrate bile salts sucrose (TCBS) agar is the selective medium for the isolation of *V. cholera*, *V. parahaemolyticus* as well as other *Vibrios*. This medium contains high concentrations of sodium thiosulphate and sodium citrate to inhibit the growth of *Enterobacteriaceae*. Sucrose acts as the fermentable carbohydrate for the metabolism. In addition, alkaline pH of the medium inhibits the growth of other bacterial and facilitates *Vibrio* spp. Indicator substances, such as thymol blue and bromothymol blue, are included in the medium to sense the pH of the medium. Thus, sucrose fermenter *Vibrio* sp. can be distinguished from the non-sucrose fermenters by the development of yellow colour colonies.

Procedure

1. Streak pure culture of the *Vibrio cholerae* test strains on TCBS agar plates. Incubate at 37 °C for 24 h.
2. Inoculate two to three bacterial colonies in LB broth tubes supplemented with 1 % sodium chloride solution. Incubate the tubes at 37 °C for 24 h with shaking at 180 rpm.
3. Extract genomic DNA from the overnight grown culture using procedure as described in Exp. 1.1. Check for the quality and quantity of isolated genomic DNA and store at −20 °C till further use.

For ERIC PCR

1. Thaw all the reagents required for PCR analysis on ice.
2. Prepare the reaction mixture on ice as follows (Table 4.13):
3. Briefly centrifuge the samples for proper mixing of all the components.

Table 4.13 Reagents and the volume to prepare reaction mixture for ERIC PCR

Reagents	Volume (µl)
10X PCR buffer	2.5
$MgCl_2$ (25 mM)	1.5
dNTPs (2.5 mM)	2.5
ERIC1 (10 mM)	2.5
ERIC2 (10 mM)	2.5
Taq polymerase (5 U/µl)	0.5
DMSO	0.5
Milli-Q water	10.5
Template DNA	2.0
Total volume	25.0

4. Set the PCR reaction with the following parameters

Initial denaturation	−94 °C for 2 min	
Denaturation	−94 °C for 45 s	
Annealing	−52 °C for 1 min	30 cycles
Extension	−70 °C for 7 min	
Final extension	−70 °C for 10 min	
Hold	−4 °C for ∞	

5. Prepare 1.8 % agarose gel and load 10 µl of PCR amplified products, and run for electrophoresis for 4–5 h.
6. Observe for banding pattern under UV light in a Gel documentation system, and using Quantity one software (Bio-Rad, USA) deduce phylogeny among the test organisms.

For REP-PCR

1. Thaw all the reagents required for PCR analysis on ice.
2. Prepare the reaction mixture on ice as follows (Table 4.14):
3. Briefly centrifuge the samples for proper mixing of all the components.
4. Set the PCR reaction with the following parameters

Initial denaturation	−94 °C for 2 min	
Denaturation	−94 °C for 45 s	
Annealing	−46 °C for 3 min	30 cycles
Extension	−70 °C for 7 min	
Final extension	−70 °C for 10 min	
Hold	−4 °C for ∞	

Table 4.14 Reagents and the volume to prepare reaction mixture for REP-PCR

Reagents	Volume (μl)
10X PCR buffer	2.5
$MgCl_2$ (25 mM)	1.5
dNTPs (2.5 mM)	2.5
REP1 (10 mM)	2.5
REP2 (10 mM)	2.5
Taq polymerase (5 U/μl)	0.5
DMSO	0.5
Milli-Q water	10.5
Template DNA	2.0
Total volume	25.0

5. Prepare 1.8% agarose gel and load 10 μl of PCR amplified products and run for electrophoresis for 4–5 h.
6. Observe for banding pattern under UV light in a Gel documentation system, and using Quantity one software (Bio-Rad, USA) deduce phylogeny among the test organisms.

Observation

Observe for the banding pattern under UV light using a Gel documentation system, and deduce the phylogeny among the test organisms based on the banding pattern and presence or absence of the particular bands by using Quantity One software.

Result Table

Sample no.	Lane no.	No. of bands obtained	Predicted relationship

Troubleshooting

Problems	Possible cause	Possible solution
Incorrect product size	Incorrect annealing temperature	Recalculate primer Tm values by using any of the web-based software
	Mispriming	Verify the primes have no additional complementary regions within the template DNA
	Improper Mg^{2+} concentration	Optimise Mg^{2+} concentration with 0.2–1 mM increments
	Nuclease contamination	Repeat the reactions using fresh solutions
No product	Incorrect annealing temperature	Recalculate the Tm values of the primers, test the correct annealing temperature by gradient, starting at 5 °C below the lower Tm of the primer pair
	Poor primer design	Check with the literature for recommended primer design, verify that the primers are non-complementary, both internally and to each other, optionally increase length of the primer
	Poor primer specificity	Verify that thè oligos are complementary to the proper target sequence
	Insufficient primer concentration	The correct range of primer concentration should be 0.05–1 μM, refer to the specific product literature for ideal conditions
	Poor template quality	Analyze DNA by agarose gel electrophoresis, check A_{260}/A_{280} ratio of DNA template
	Insufficient number of cycles	Rerun the reaction with more number of cycles
Multiple/ non-specific products	Premature replication	Use hot start polymerase, set up the reaction on ice using chilled components, add samples to the PCR preheated to the denaturation temperature
	Primer annealing temperature too low	Increase the annealing temperature
	Incorrect Mg^{2+} concentration	Adjust Mg^{2+} concentration with 0.2–1.0 mM increments
	Excess primer	The correct range of primer concentration should be 0.05–1 μM, refer to the specific product literature for ideal conditions
	Incorrect template concentration	For low complexity templates (i.e. plasmid, lambda, BAC DNA), use 1 pg–10 ng DNA per 50 μl reaction. For high complexity templates (i.e. genomic DNA), use 1 ng–1 μg template for 50 μl reaction

Precautions

1. Use pipette tips with filters.
2. Store materials and reagents properly under separate conditions, and add them to the reaction mixture in a spatially separated facility.
3. Thaw all components thoroughly at room temperature before starting an assay.
4. After thawing, mix the components with brief centrifugation.
5. Work quickly on ice or in the cooling block.
6. Always wear safety goggles and gloves while performing the PCR reaction.

FLOW CHART

For ERIC PCR analysis

Thaw all the reagents required for PCR analysis, assemble the following reaction mixtures i.e. 2.5 µl of 10X PCR buffer, 1.5 µl of 25 mM mgCl2, 2.5 µl of 2.5 mM dNTPs, 2.5 µl of 10 mM ERIC1, 2.5 µl of 10 mM ERIC2, 0.5 µl of 5U/µl Taq polymerase, 0.5 µl of DMSO, 10.5 µl of milli-Q water, and 2.0 µl of template DNA, mix thoroughly by gently vortexing

↓

Set the PCR reaction with the following conditions i.e. initial denaturation at 94°C for 2 min followed by 30 cycles of denaturation at 94°C for 45 sec, annealing at 52°C for 1 min, extension at 70°C for 7 min and final extension of 70°C for 10 min followed by hold at 4°C for ∞

↓

Run the amplified product by 1.8 % of agarose gel electrophoresis, acquire image under UV in a gel documentation system and deduce phylogeny among them by using Quantity One software.

For REP PCR analysis

Thaw all the reagents required for PCR analysis, assemble the following reaction mixtures i.e. 2.5 µl of 10X PCR buffer, 1.5 µl of 25 mM $mgCl_2$, 2.5 µl of 2.5 mM dNTPs, 2.5 µl of 10 mM REP1, 2.5 µl of 10 mM REP2, 0.5 µl of 5U/µl Taq polymerase, 0.5 µl of DMSO, 10.5 µl of milli-Q water, and 2.0 µl of template DNA, mix thoroughly by gently vortexing

↓

Set the PCR reaction with the following conditions i.e. initial denaturation at 94°C for 2 min followed by 30 cycles of denaturation at 94°C for 45 sec, annealing at 46°C for 1 min, extension at 70°C for 7 min and final extension of 70°C for 10 min followed by hold at 4°C for ∞

↓

Run the amplified product by 1.8 % of agarose gel electrophoresis, acquire image under UV in a gel documentation system and deduce phylogeny among them by using Quantity One software.

Computer-Aided Study of Molecular Microbiology

5

Exp. 5.1 Analysis of Gene Sequences

Objective To analyse the 16S rRNA gene sequences obtained after sequencing reaction.

Introduction

Ribosomes are the essential components of the protein synthesis machinery and therefore, are ubiquitously distributed and functionally conserved in all organisms. Ribosomes consist of two major subunits—the small ribosomal subunit reads the mRNA, while the large subunit joins amino acids to form a polypeptide chain. Each subunit is composed of one or more ribosomal RNA (rRNA) molecules and a variety of proteins. In prokaryotes the complete 70S monosome comprises larger 50S unit that includes 5S and 23S rRNA, plus a smaller 30S unit that includes 16S rRNA. In all the gene pools, the 16S rRNA gene is the most conserved and least variable DNA sequence in all cells. 16S rDNA gene sequences are highly conserved within living organisms of the same genus and species, but that they differ between organisms of other genera and species. This RNA is not translated to protein; the ribosomal RNA is the active component. Thus, it refers to the "rRNA gene" or "rDNA" to designate the DNA in the genome that produces the ribosomal RNA. 16S rRNA genes lack the inter-species horizontal gene transfer found with many prokaryotic genes. They contain diagnostic variable regions interspersed among highly conserved regions of primary and secondary structures, permitting phylogenetic comparisons to be inferred over a broad range of evolutionary distance. The sequence of 16S rRNA gene and its analysis is a molecular figure print in the identification of bacteria. The sequences data need to be processed prior to submission. The sequence data can be processed by a number of available tools. Nearly a dozen formats are available for sequences. Formats were designed so as to be able to hold the sequence data and other information about the sequence. Each sequencer and analysis package stores data in its own format. An efficient sequencer can give a maximum of 1000 nucleotides per run. The aim of the current study is to edit the 16S rRNA gene sequence using BioEdit v7.2.5.

Example of Tools for Sequence Analysis

- SEQTOOL: http://www.seqtools.dk/downloads_a.htm
- BioEdit: http://www.mbio.ncsu.edu/bioedit/bioedit.html
- MEGA: http://www.megasoftware.net/

S. Das, H. R. Dash, *Microbial Biotechnology- A Laboratory Manual for Bacterial Systems*,
DOI 10.1007/978-81-322-2095-4_5, © Springer India 2015

Principle

BioEdit is the biological sequence editor, which intends for providing basic functions related to protein and nucleic acid sequence editing, alignment, manipulation and analysis. By using this software many other functions can also be performed including hydrophobicity or hydrophilicity analysis, interactive 2D matrix data plotting, as well as sequence manipulating. The advanced versions of the software added many advanced features for data analysis for fat saving and opening of large sequences, the sequence capability of which has been expanded to 20,000. BioEdit is a C++ program written in Borland's C++ builder. Hence, this software is much simpler and efficient, providing an easy mode of sequence alignment, output and other analysis.

Procedure

1. Open a raw sequence file of ABI format in BioEdit software.
2. Two windows will appear—one for dendogram and another for the nucleotide sequences automatically decoded by the software programme.
3. Open both the forward and reverse nucleotide sequences in a single protocol for analysis (Fig. 5.1).
4. Copy reverse sequence and paste it with forward sequence task window. Select any one of the sequences and do reverse complement (sequence > nucleic acid > reverse complement) (Fig. 5.2).
5. Generate a contig sequence (sequence > accessory application > CAP contig assembly). A contig sequence will appear; copy the sequence and perform BLAST search (Fig. 5.3).
6. Select and download the sequence (in FASTA format) that has maximum similarity with the query sequence (Fig. 5.4).
7. Open the downloaded sequence and paste it with contig sequence in same window (File > select sequence > open > edit > copy sequence(s) > paste sequence(s)).
8. Select contig sequence and subject sequence, perform ClustalW multiple alignment (accessory application > ClustalW multiple alignment) (Fig. 5.5).
9. Look for the gaps, matched and unmatched sequences. Edit if required, then copy sequence to the clipboard in FASTA format.
10. Save the sequence file in compatible format, i.e. txt, FASTA.
11. Further the edited sequence can be submitted to GenBank and the phylogeny of the sequence can be deduced.
12. When there are ambiguities during editing of the sequences, use ambiguous genetic codes as described in Tables 5.1, 5.2 and 5.3.

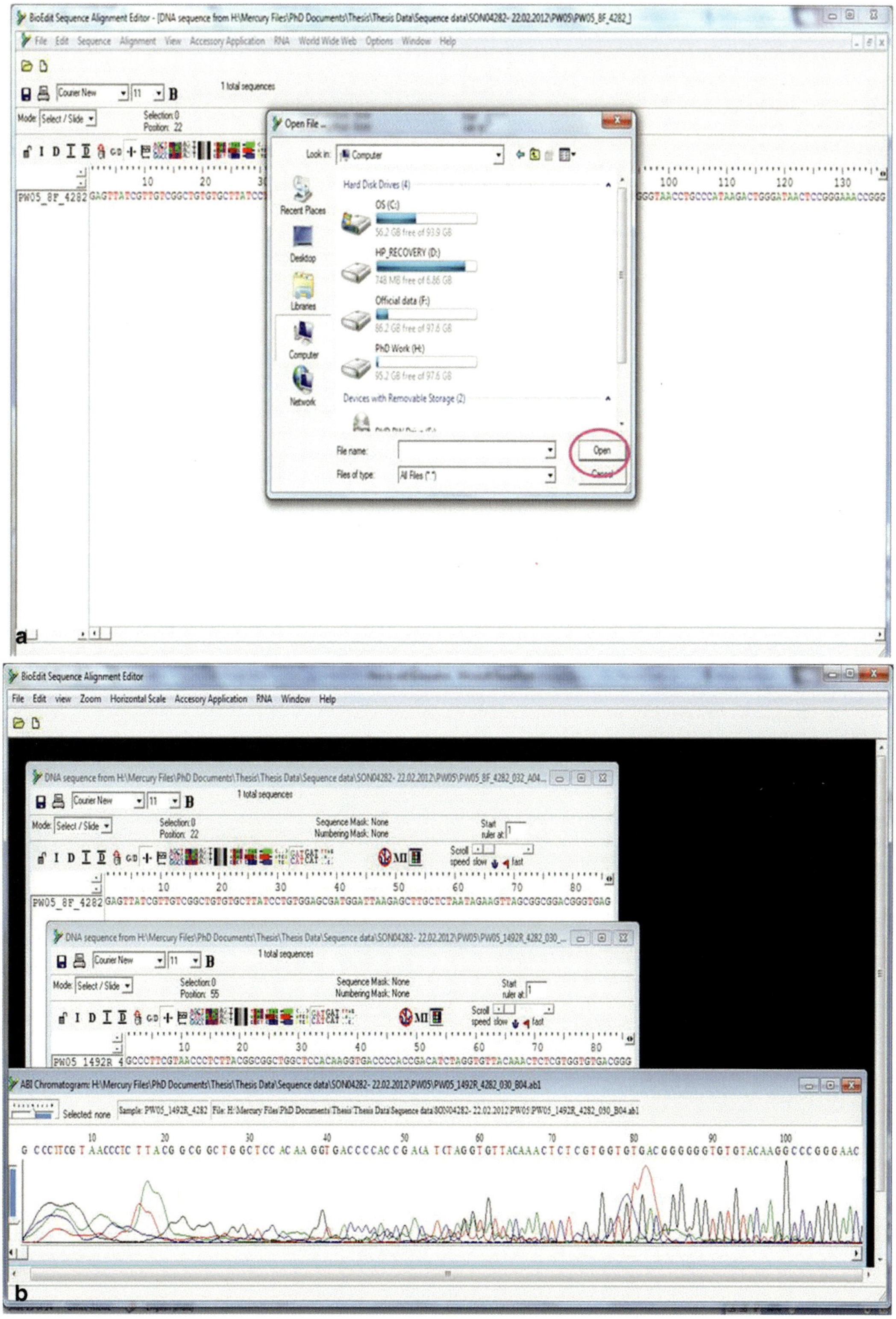

Fig. 5.1 Dendrograms and their corresponding nucleotides in BioEdit

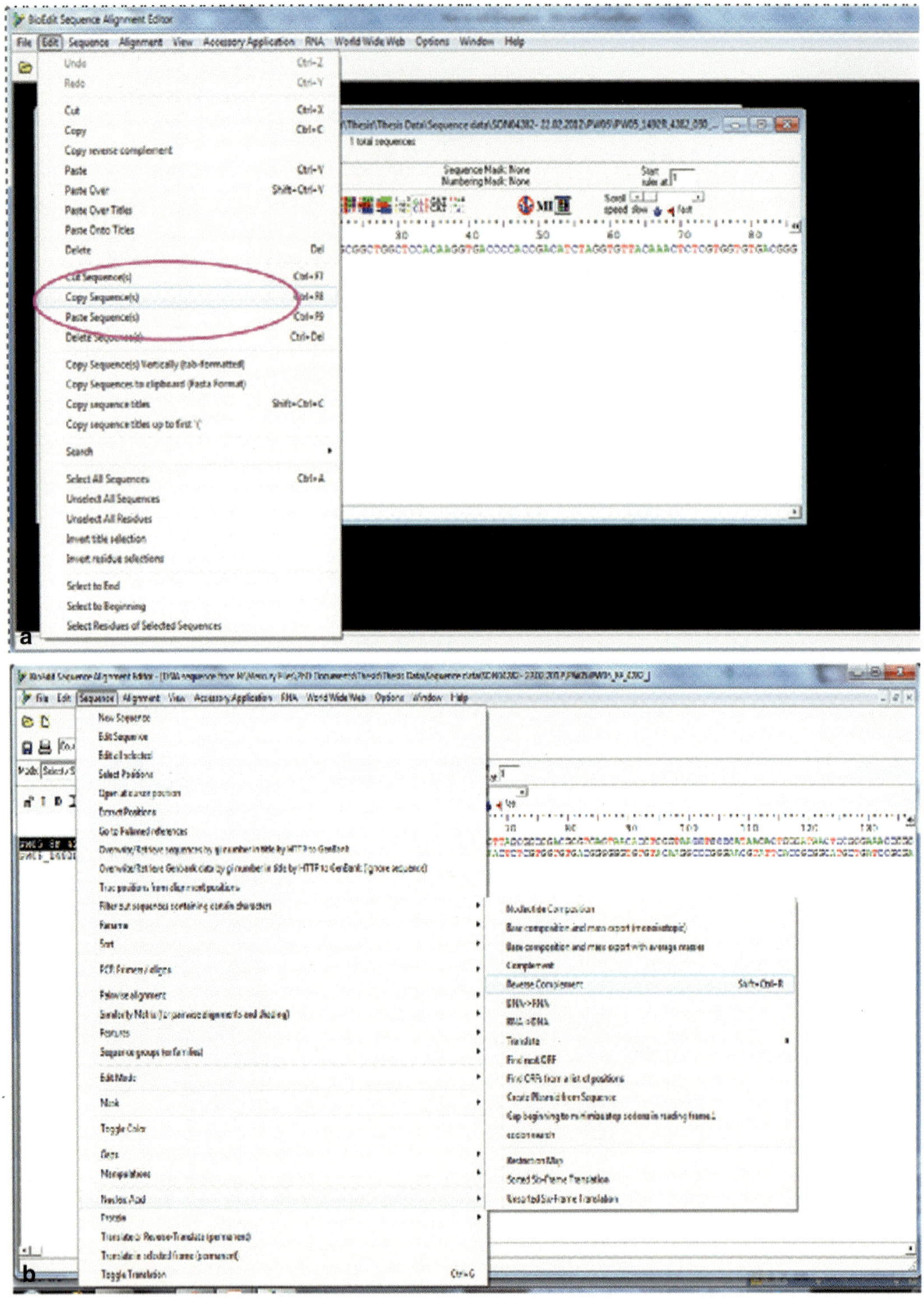

Fig. 5.2 Reverse complement of reverse sequence file in BioEdit

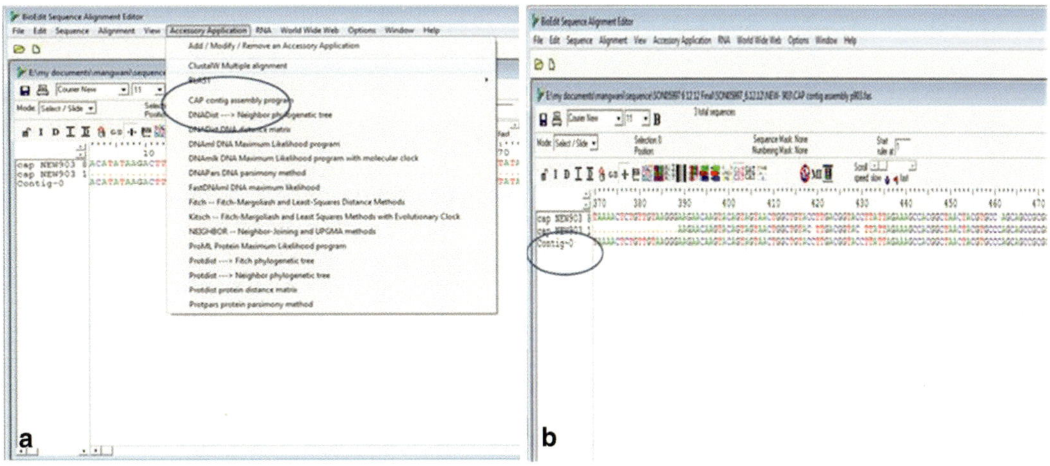

Fig. 5.3 Contig assembly of both forward and reverse sequences

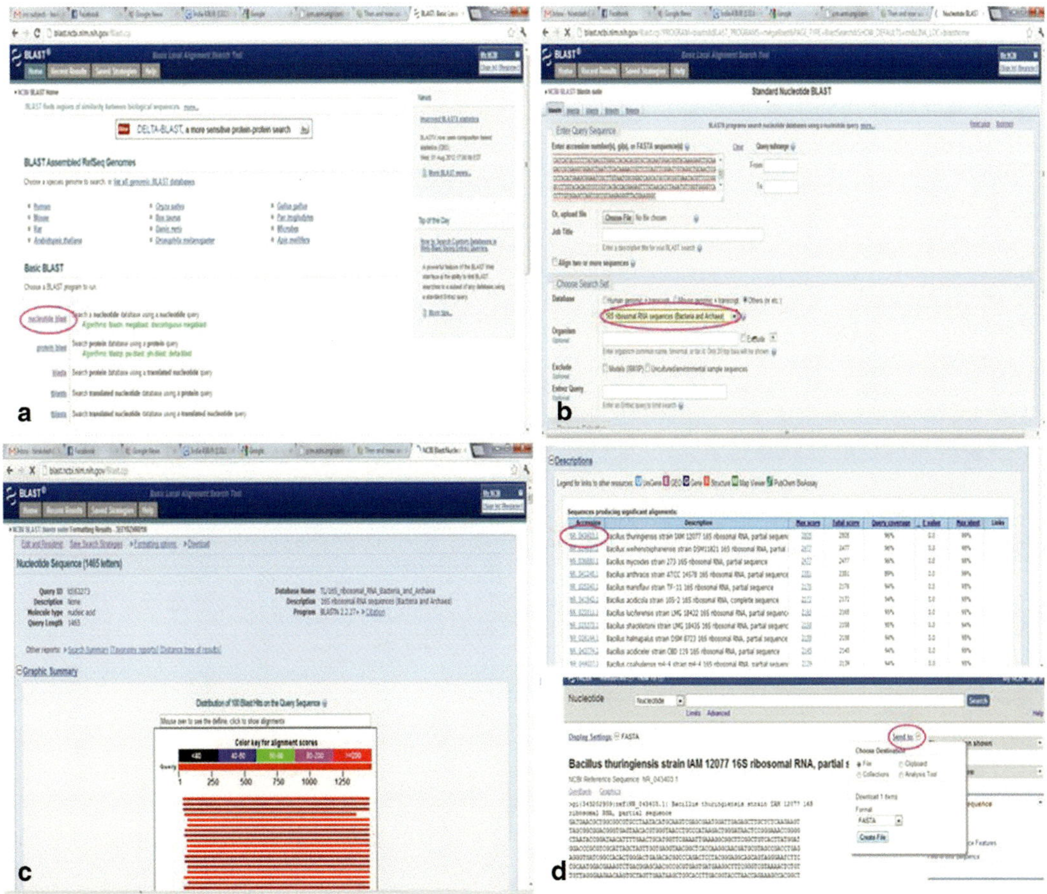

Fig. 5.4 Selecting and downloading the sequences in FASTA format

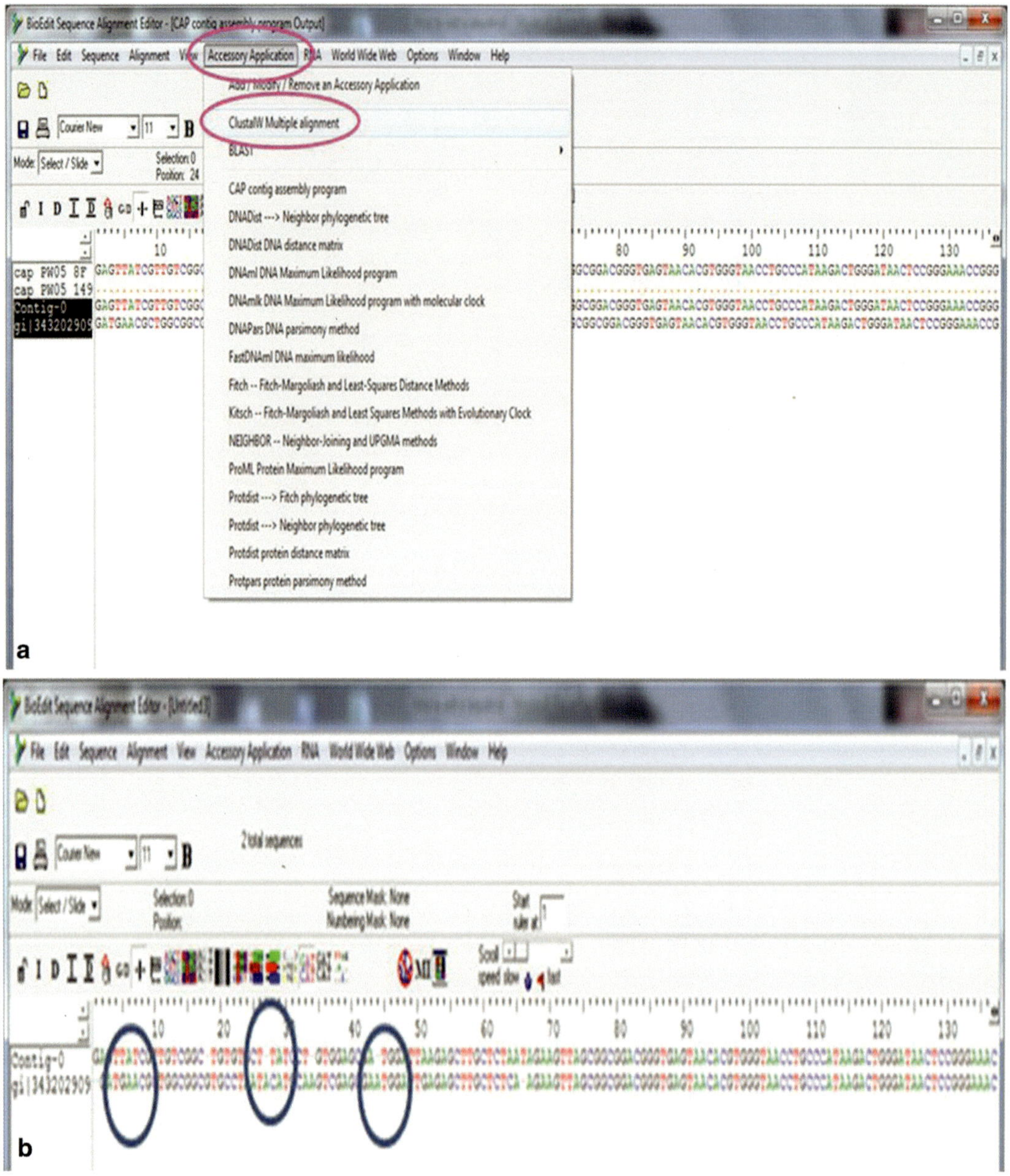

Fig. 5.5 Multiple alignment of the sequences using ClustalW

Table 5.1 Useful nucleotide abbreviations during nucleic acid editing

Base	Meaning	Complement
A	A	T
C	C	G
G	G	C
T	T	A
U	T	A
M	A or C (aMino)	K
R	A or G (puRine)	Y
W	A or T (weak, 2H-bonds)	W
S	C or G (strong, 3H-bonds)	S
Y	C or T (pYrimidine)	R
K	G or T (Keto)	M
V	A or C or G (not T; V > T)	B
H	A or C or T (not G; H > G)	D
D	A or G or T (not C; D > C)	H
B	C or G or T (not A; B > A)	V
X	A or C or G or T	X
N	A or C or G or T (aNy)	N
.	Not A, C, G, T	.

Table 5.2 Genetic code and the corresponding abbreviations

UUU	Phe	F	UCU	Ser	S	UAU	Tyr	Y	UGU	Cys	C
UUC	Phe	F	UCC	Ser	S	UAC	Tyr	Y	UGC	Cys	C
UUA	Leu	L	UCA	Ser	S	UAA	STOP	*	UGA	STOP	*
UUG	Leu	L	UCG	Ser	S	UAG	STOP	*	UGG	Trp	W
CUU	Leu	L	CCU	Pro	P	CAU	His	H	CGU	Arg	R
CUC	Leu	L	CCC	Pro	P	CAC	His	H	CGC	Arg	R
CUA	Leu	L	CCA	Pro	P	CAA	Gln	Q	CGA	Arg	R
CUG	Leu	L	CCG	Pro	P	CAG	Gln	Q	CGC	Arg	R
AUU	Ile	I	ACU	Thr	T	AAU	Asn	N	AGU	Ser	S
AUC	Ile	I	ACC	Thr	T	AAC	Asn	N	AGC	Ser	S
AUA	Ile	I	ACA	Thr	T	AAA	Lys	K	AGA	Arg	R
AUG	Met	M	ACG	Thr	T	AAG	Lys	K	AGG	Arg	R
GUU	Val	V	GCU	Ala	A	GAU	Asp	D	GGU	Gly	G
GUC	Val	V	GCC	Ala	A	GAC	Asp	D	GGC	Gly	G
GUA	Val	V	GCA	Ala	A	GAA	Glu	E	GGA	Gly	G
GUG	Val	V	GCG	Ala	A	GAG	Glu	E	GGG	Gly	G

* signifies no aminoacid coded by the nucleotide sequences, rendering to the end during protein synthesis.

Table 5.3 Single-letter amino acid identifiers and their three-letter abbreviations

A	Ala	G	Gly	M	Met	R	Arg	W	Trp
C	Cys	H	His	N	Asn	S	Ser	X	Tyr
D	Asp	I	Ile	P	Pro	T	Thr	Y	(any)
E	Glu	K	Lys	Q	Gln	V	Val	*	(end)
F	Phe	L	Leu						

* signifies no aminoacid coded by the nucleotide sequences, rendering to the end during protein synthesis.

FLOW CHART

Launch BioEdit

↓

Open Forward 16S rDNA sequence file and backward sequence file in same task window

↓

Open backward sequence file
Go to edit and select copy sequence(s) and paste the sequence with forward sequence

↓

Select both sequence and do reverse compliment (select sequence >nucleic acid > reverse compliment)

↓

Make contig sequence (accessory application > CAP contig assembly program)

↓

Copy the contig sequence and perform BLSAT search

↓

Download the sequence showed maximum identity in FASTA format

↓

Perform ClustalW

↓

Look for the gaps, match and unmatched sequences; edit the contig sequence if required

↓

Save the contig sequence in .txt or .fas format.

Exp. 5.2 Submission of Sequences to GenBank

Objective To submit sequences to NCBI GenBank to obtain accession number.

Introduction

The GenBank is a database that has an annotated collection of all publically available nucleotide sequence and their protein translations. Only original sequence can be submitted to GenBank. It is an open access database, created and maintained by the National Centre for Biotechnology Information (NCBI) as part of the International Nucleotide Sequence Database Collaboration (INSDC). The National Centre for Biotechnology Information is a part of the National Institutes of Health in the United States. There are several options for submitting sequence to GenBank. BankIt, a web-based form, is used for direct submissions of sequence, whereas, Sequin is the stand-alone submission program. After submission of the sequence to GenBank, it gets examined for originality, quality assurance check is performed and an accession number is assigned to the sequence data. After approval, the submitted sequences are released to the public database, where the entries are retrievable. In general, submissions like the bulk submissions of expressed sequence tag (EST), sequence-tagged site (STS), genome survey sequence (GSS) and high-throughput genome sequence (HTGS) data are most often submitted by large-scale sequencing centres.

DNA sequence records in the public databases (DDBJ/EMBL/GenBank) are essential components for computer analysis in molecular biology studies. Accurate and informative biological annotation of sequence data is critical in determining the function of that gene sequence and for efficient similarity search. No journals print the full sequence data anymore; rather a database accession number is mentioned for the publication process. However, the submission process is governed by the international collaborative agreement by which any sequence submitted to any of the databases appears in other databases within few days of their public release. Furthermore, the sequence records are distributed worldwide to various user groups and centres, including those that reformat the records for use within their own suites of programs and database.

Recently, there has been an exponential growth of the sequence databases; however, in the early days, sequences were submitted by individuals who were studying the genes of interest. Thus, the program suitable for this type of submission allows the manual annotation of arbitrary biological information. Recently, a significant contribution to the sequence databases came from phylogenetic and population studies, also the complete genome sequences are increasing at an exponential phase.

There are two approaches of sequence submission to the databases, i.e. the web-based approach by using BankIt and the multi-platform program that can use the direct network connection, i.e. Sequin. Sequin is an ASN.1 editing tool that takes the full advantage of NCBI data model and has become a platform for many sequence analysis tools that have been developed in NCBI over the years.

Principle

A resolution has been passed to streamline the sequence submission process during which, the person concerned with the sequence submission should furnish the full sequence of the insert, base sequence of the host flanking sequences either in EMBL/GenBank format and most preferably NCBI's Sequin (ASN.1) format. According to the documented DNA sequences it should be annotated with the INSDC feature table definition document with the following descriptions and features including their location on the sequence, i.e. definition (title describing the sequence record), source and organism (according to the NCBI taxonomy database), size (in base pairs), molecular type (DNA), topology (linear/circular), reference (references with authors, titles, journals, etc.), source (regions/sources of GMO insert and host organism), STS (PCR amplicon of the detection method) and primer bind (primer name, sequence of forward and reverse primer and probe).

Most journals require DNA and amino acid sequences cited in the articles to be submitted in public sequence repository (DDBJ/EMBL/GenBank-INSDC) as part of the publication process. Data exchange between DDBJ, EMBL and GenBank occurs daily so whichever one is most convenient, without any regards for where the sequence may be, is published. Sequence data submitted in advance of publication can be kept confidential, if requested. GenBank provides accession numbers for submitted sequences, usually within 2 working days. This accession number serves as an identifier for the submitted data, and allows the community to retrieve the sequence upon requirement. The accession number should be included in the manuscript, preferably in a footnote on the first page of the article, or as required by the journal procedures.

Procedure

1. Go to the link http://www.ncbi.nlm.nih.gov/genbank/submit/.
2. Select submission tool. There are several options for sequence submission, for step-by-step guided submission, select BankIt (Fig. 5.6).
3. Sign in to use BankIt (Fig. 5.7). You need a user account. If you do not have an account

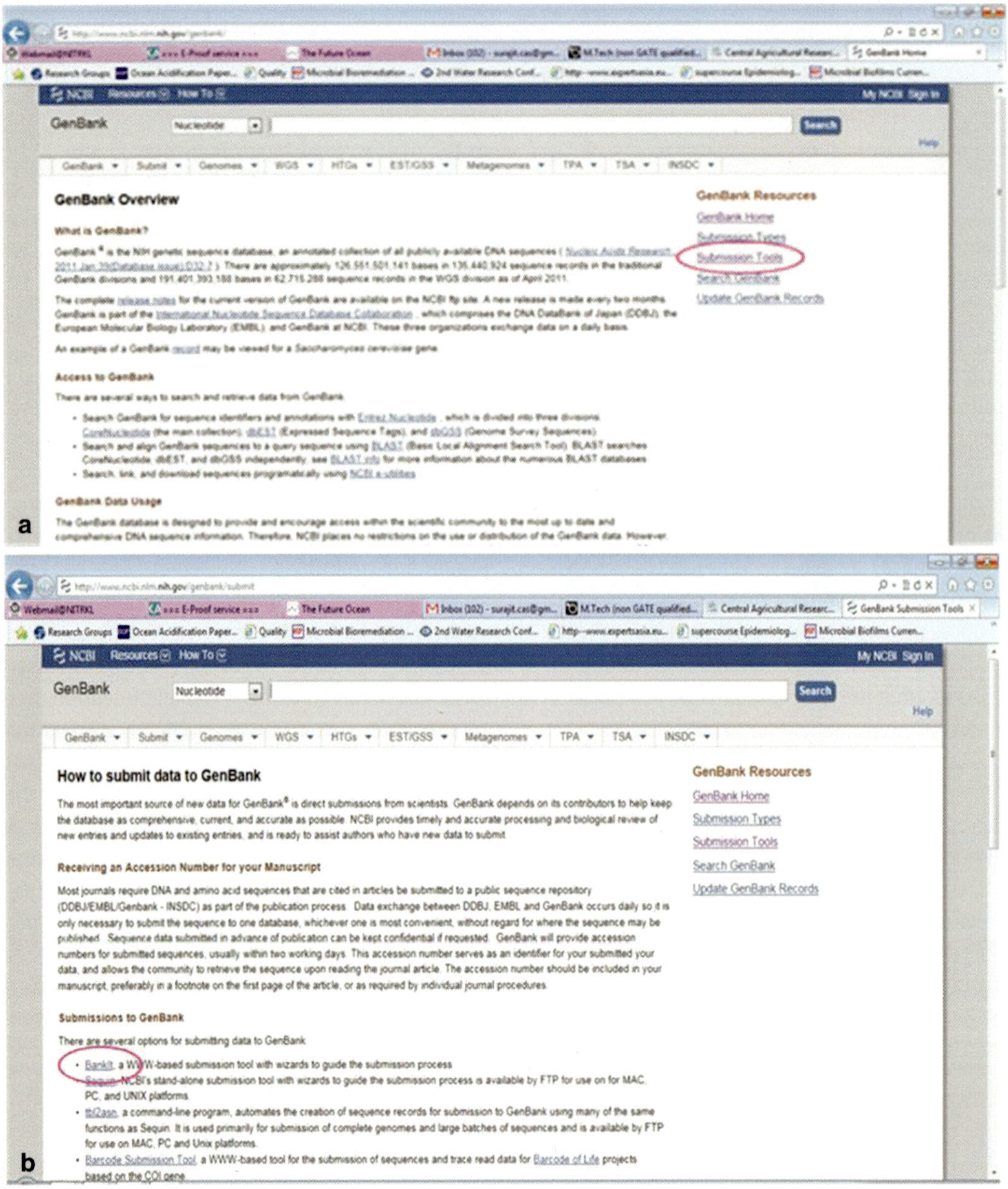

Fig. 5.6 GenBank submission tool BankIt

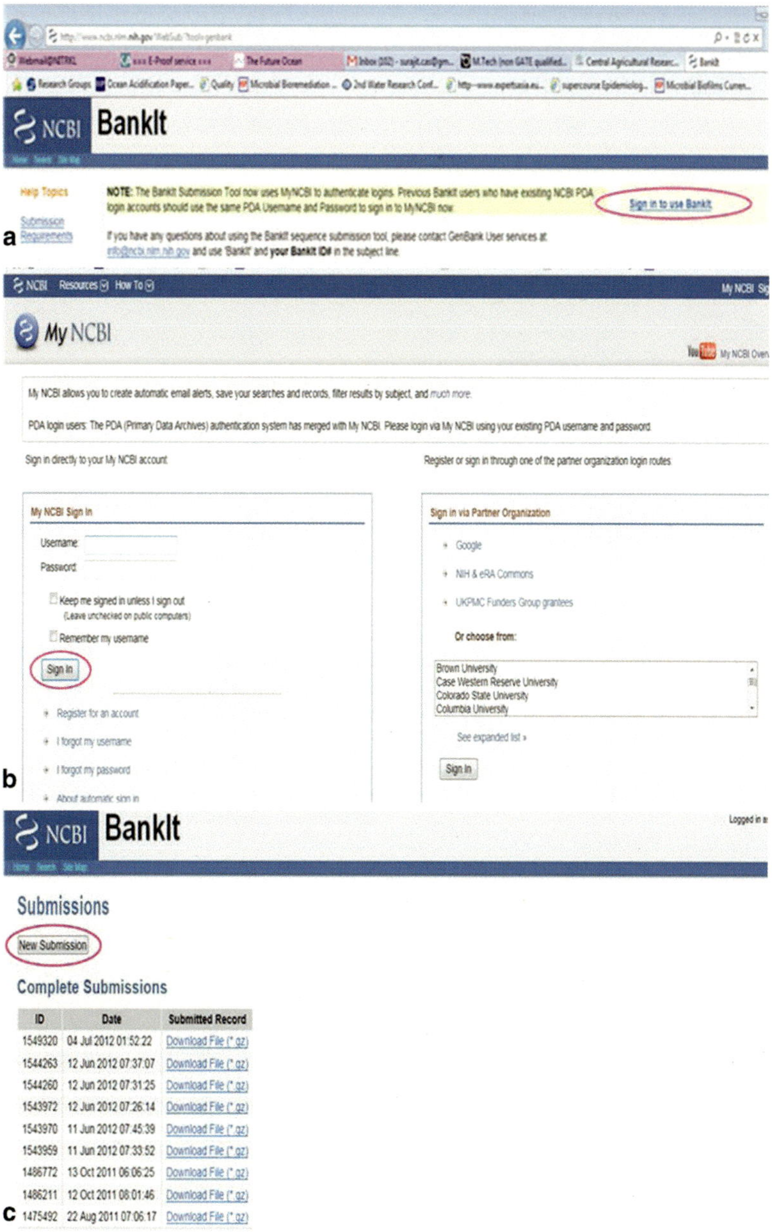

Fig. 5.7 Signing in to the user account to submit a new sequence

Fig. 5.8 Filling up of sequences and other information

you have to create one of your own, which takes a few minutes, with your personal data.

4. After you sign in, a submission option will appear. Click New Submission. A reference section will appear. Add the required information like Author's name and status of sequence.
5. Go to the next step, i.e. nucleotide, and put the required information, such as release date, type of sequence, sequence and then click on continue (Fig. 5.8).
6. Add organism name, submission category and sequence technology (Fig. 5.9).
7. Fill up the information regarding source modifier and primers (Fig. 5.10).
8. Add features and finish submission (Fig. 5.11).

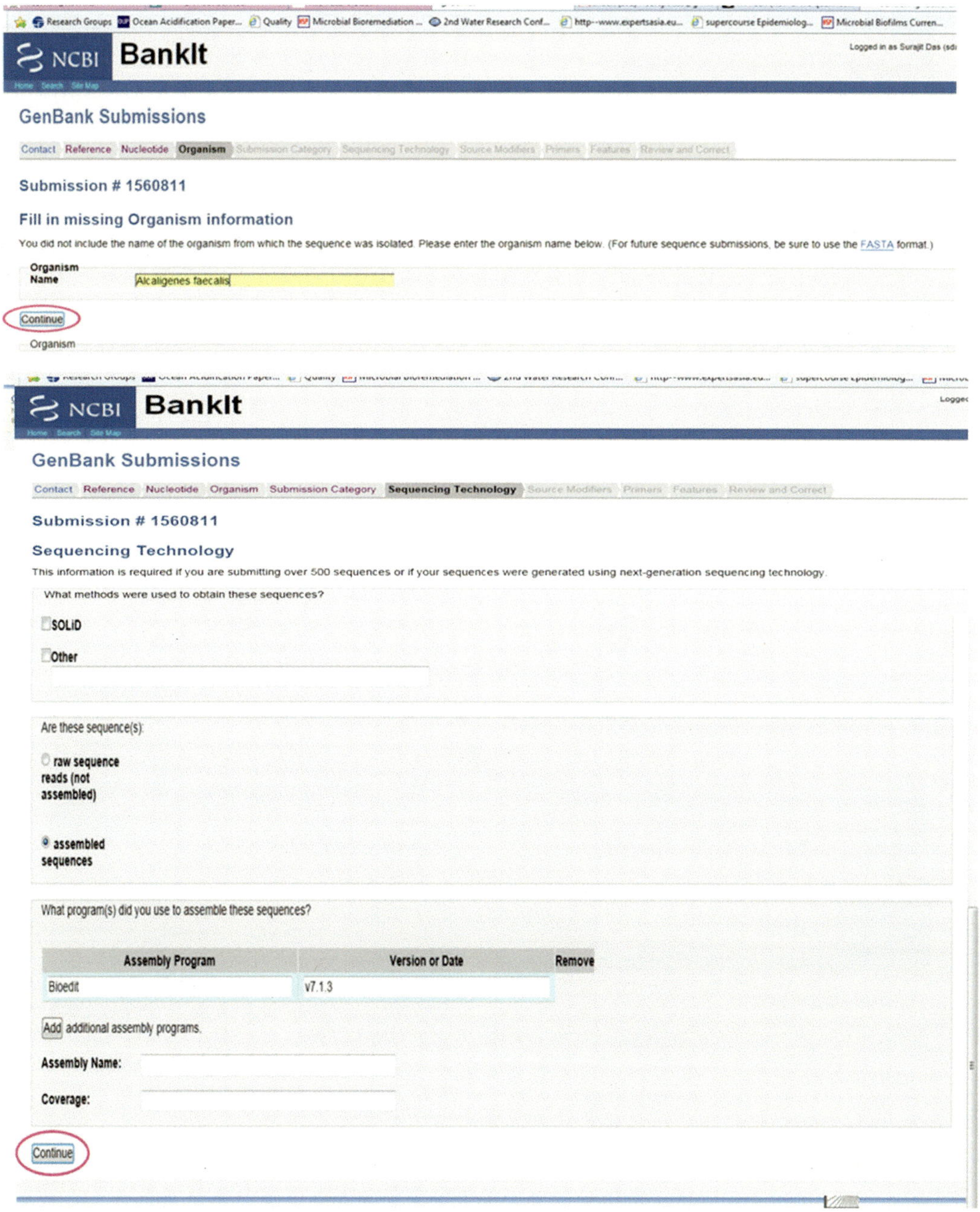

Fig. 5.9 Filling up of organism name, submission category and sequence technology

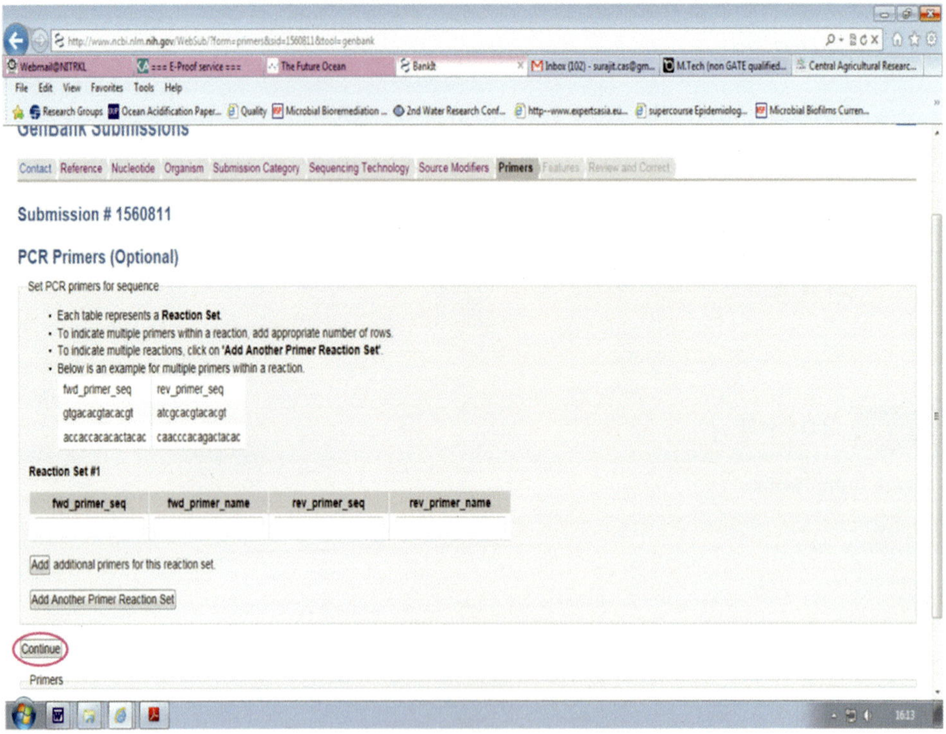

Fig. 5.10 Information regarding source modifier and primers

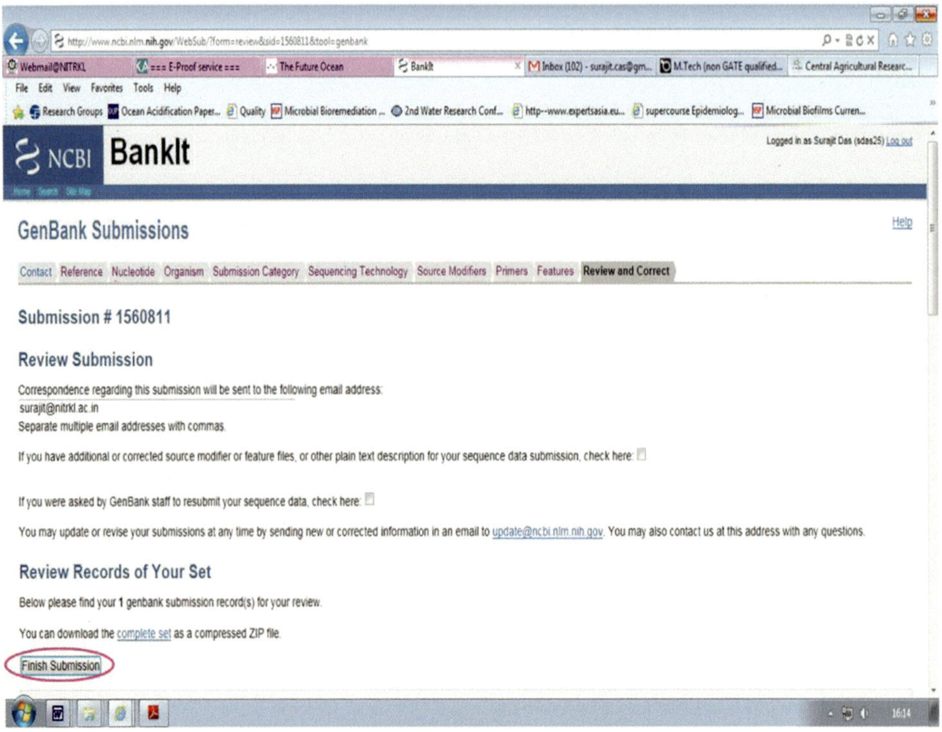

Fig. 5.11 Last step of sequence submission at GenBank

FLOW CHART

Go to the link http://www.ncbi.nlm.nih.gov/genbank/submit/

↓

Select submission tool, there are several options for sequence submission, for step to step guidance submission select BankIt

↓

Sign in to use BankIt, for this purpose you need a user account, if you already have an account sign in or else open a new account that takes few minutes with some basic information

↓

After sign in a submission option appears and click on to New Submission, a reference section appears, add required information like author name and status of sequence

↓

Go to next step i.e. nucleotide and fill required information like release date, type of sequence, sequence and continue to the next step

↓

Add organism name, submission category and sequencing technology

↓

Fill up the information regarding source modifiers and primers

↓

Add features of the sequence and finish submission

Exp. 5.3 Phylogenetic Trees

Objective To draw a phylogenetic tree from the 16S rRNA gene sequence data to deduce the phylogenetic relationship between bacteria by using MEGA5.

Introduction

Phylogeny or phylogenetic tree is a diagram that illustrates the lines of evolutionary descent of different species, organisms or genes from a common ancestor. Phylogenies are helpful for organizing facts of biological diversity, for structuring classifications and for providing insight into events that occurred during evolution. Furthermore, because these trees show descent from a common ancestor, and because much of the strongest evidence for evolution comes in the form of common ancestry, one must understand phylogenies in order to fully appreciate the overwhelming evidence supporting the theory of evolution. Although the morphologies and physiologies of prokaryotes are much simpler than those of eukaryotes, there is a large amount of information in the molecular sequences of their DNA, RNAs and proteins. Thus, it is possible to use molecular similarities to infer the relationships of genes, and by extension, to learn the relationships of the organisms themselves. To infer to relationships that span the diversity of known life, it is necessary to look at genes conserved through the billions of years of evolutionary divergence. An example of genes in this category is of those that define the ribosomal RNAs (rRNAs). The aim of the current study is to construct a phylogenetic tree in MEGA 5 using partial sequences of 16S rRNA gene. By comparing the inferred rRNA sequences (or those of any other appropriate molecule) it is possible to estimate the historical branching order of the species, and also the total amount of sequence change.

Reading Trees

A phylogeny, or evolutionary tree, represents the evolutionary relationships among a set of organisms or groups of organisms, called taxa (singular: taxon). The tips of the tree represent groups of descendent taxa (often species) and the nodes on the tree represent the common ancestors of those descendants. Two descendants that split from the same node are called sister groups. In the tree (Fig. 5.12), species A and B are sister groups—they are each other's closest relatives. Many phylogenies also include an out-group—a taxon outside the group of interest. All the members of the group of interest are more closely related to each other than they are to the out-group. Hence, the out-group stems from the base of the tree. An out-group can give you a sense of where, on the bigger tree of life, the main group of organisms falls. It is also useful when constructing evolutionary trees.

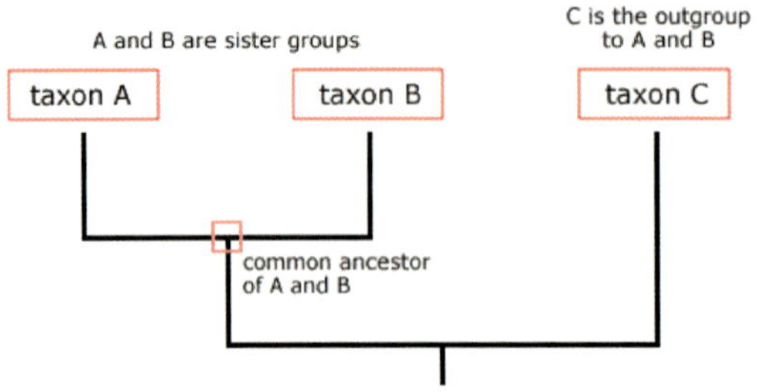

Fig. 5.12 A model tree

Phylogenetic Tree Software

There are many commercially available software programs that can be used for the construction and analysis of phylogenetic trees using their corresponding nucleotide sequences.

- PHYLIP: http://evolution.genetics.washington.edu/phylip.html
- PAUP: http://paup.csit.fsu.edu
- MrBayes: http://mrbayes.csit.fsu.edu
- MEGA (Molecular Evolutionary Genetic Analysis): http://www.megasoftware.net
- BioEdit: http://www.megasoftware.net

Principle

Any phylogenetic tree consists of nodes that are connected by branches where each branch represents the persistence of a genetic lineage through time and each node represents the birth of a lineage. The phylogenetic construction methods are either distance- or character-based methods. In the distance-based methods, the distance between every pair of sequence is calculated and the resultant distance matrix is used for tree construction. The character-based technique explores maximum parsimony, maximum likelihood and Bayesian inference methods. This approach compares simultaneously all the sequences in the alignment considering one character at a time to calculate the score for each tree.

The maximum parsimony technique minimises the number of changes on a phylogenetic tree by assigning character sites to interior nodes of the tree. The character length is the minimum number of changes required for that site, whereas the tree score is the sum of character lengths over all sites. In this case, the maximum parsimony tree is the tree that minimises the tree score. In another approach of maximum likelihood, the likelihood is the function of the parameters with the data observed and fixed. It represents all the information in the data about the parameters. The maximum likelihood estimates of parameters are the parameter values that maximise the likelihood. The MLEs have desirable asymptotic properties as they are

Table 5.4 Functionalities of a few commonly used phylogenetic programs

Name	Brief description	Link
Bayesian evolutionary analysis sampling trees (BEAST)	A Bayesian MCMC program for inferring rooted trees under the clock or relaxed-clock models. It can be used to analyse nucleotide and amino acid sequences, as well as morphological data. A suite of programs, such as Tracer and FigTree, are also provided to diagnose, summarise and visualise results	http://beast.bio.ed.ac.uk/
Genetic algorithm for rapid likelihood inference (GARLI)	A program that uses genetic algorithms to search for maximum likelihood trees. It includes the GTR + Γ model and special cases and can analyse nucleotide, amino acid and codon sequences. A parallel version is also available	http://code.google.com/p/garli
Hypothesis testing using phylogenies (HYPHY)	A maximum likelihood program for fitting models of molecular evolution. It implements a high-level language that the user can use to specify models and to set up likelihood ratio tests	http://www.hyphy.org
Molecular evolutionary genetic analysis (MEGA)	A Windows-based program with a full graphical user interface that can be run under Mac OS X or Linux using Windows emulators. It includes distance, parsimony and likelihood methods of phylogeny reconstruction, although its strength lies in the distance methods. It incorporates the alignment program ClustalW and can retrieve data from GenBank	http://www.megasoftware.net
MrBayes	A Bayesian MCMC program for phylogenetic inference. It includes all of the models of nucleotide, amino acid and codon substitution developed for likelihood analysis	http://mrbayes.net
Phylogenetic analysis by maximum likelihood (PAML)	A collection of programs for estimating parameters and testing hypotheses using likelihood. It is mostly used for tests of positive selection, ancestral reconstruction and molecular clock dating. It is not appropriate for tree searches	http://abacus.gene.ucl.ac.uk/software
Phylogenetic analysis using parsimony* and other methods (PAUP* 4.0)	PAUP* 4.0 is still a beta version (at the time of writing). It implements parsimony, distance and likelihood methods of phylogeny reconstruction	http://www.sinauer.com/detail.php?id=8060
PHYLIP	A package of programs for phylogenetic inference by distance, parsimony and likelihood methods	http://evolution.gs.washington.edu/phylip.html
PhyML	A fast program for searching for the maximum likelihood trees using nucleotide or protein sequence data	http://www.atgc-montpellier.fr/phyml/binaries.php
RAxML	A fast program for searching for the maximum likelihood trees under the GTR model using nucleotide or amino acid sequences. The parallel versions are particularly powerful	http://scoh-its.org/exelixis/software.html
Tree analysis using new technology (TNT)	A fast parsimony program intended for very large data sets	http://www.zmuc.dk/public/phylogeny/TNT

* signifies no aminoacid coded by the nucleotide sequences, rendering to the end during protein synthesis.

unbiased, consistent and efficient. However, in Bayesian interference the general methodology for statistical interference is different. It differs from maximum likelihood in those parameters of the model that are considered to be random variables with statistical distributions, whereas in maximum likelihood they are unknown fixed constants. The most recent Bayesian implementation is in the program BEAST72 where relaxed-clock models were used to infer rooted trees even though the model allows substitution rates to vary across lineage. Table 5.4 depicts the

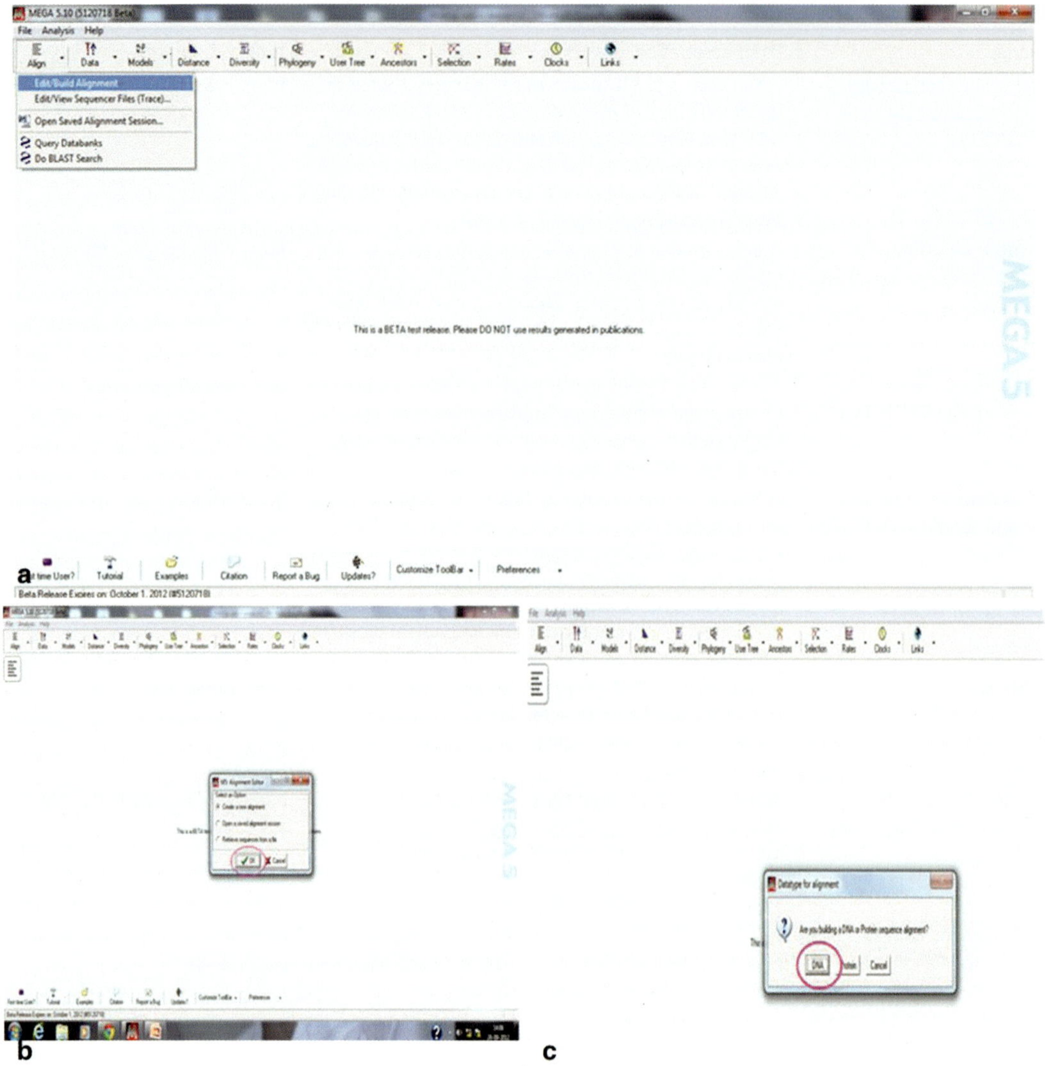

Fig. 5.13 Initial steps of tree drawing in MEGA5

extensive list of commonly used phylogenetic programs with their functions.

Procedure

1. Put all the 16S rDNA sequences you want to use to construct a tree in the same FASTA file.
2. Launch MEGA5. Select Alignment explorer and click on edit/build alignment. A MEGA5 alignment memo box will appear select create new alignment.
3. Align, a memo box will appear, questioning about DNA or Protein sequence. Click onto DNA sequence (Fig. 5.13).
4. Now open all sequence file (Fig. 5.14).

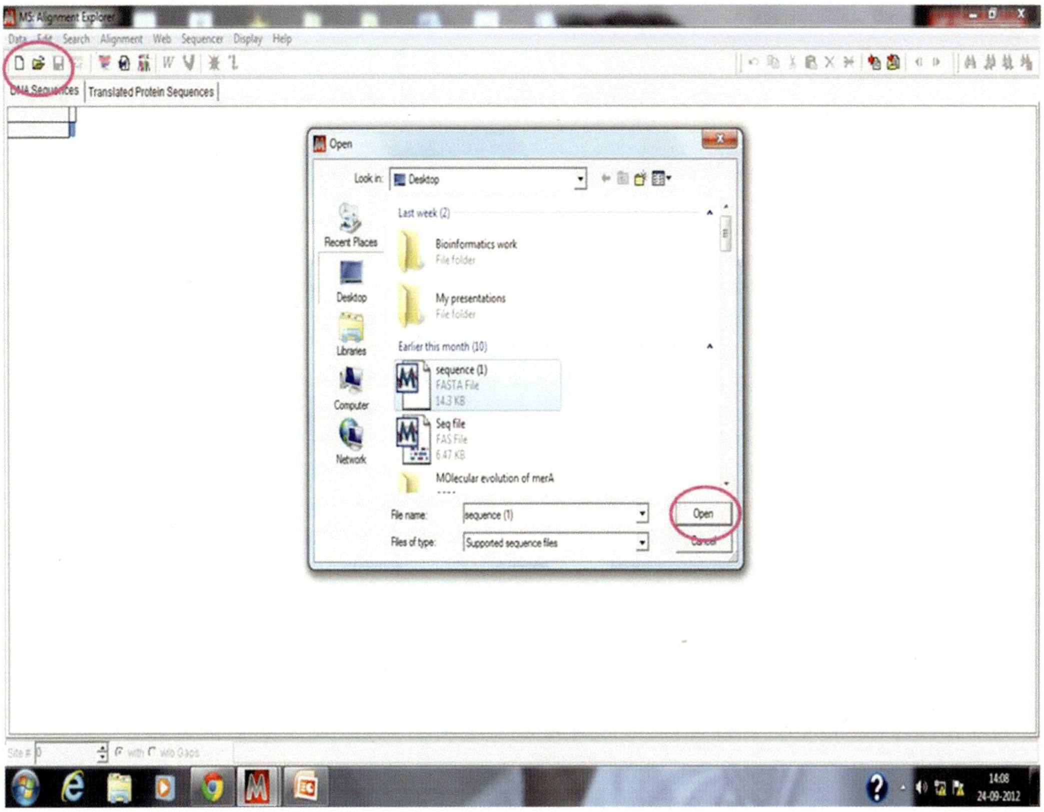

Fig. 5.14 Opening of the sequence files in MEGA5

5. Select all the sequences by clicking on them (using 'up arrow' on your keyboard and your mouse—they will be highlighted in dark blue) then select 'Alignment > Align by ClustalW' > OK.
6. Visually inspect whether the alignment seems to make sense. Add or remove sequences if required or else save it (Fig. 5.15).
7. In MEGA select by click onto Phylogeny Explorer and select construct test minimum likelihood trees (Fig. 5.16).
8. Analysis window will appear, make bootstrap to 100/500/1000 and click on to compute.
9. A phylogenetic tree will appear. Save it in desired format (PNG or PDF) (Fig. 5.17).

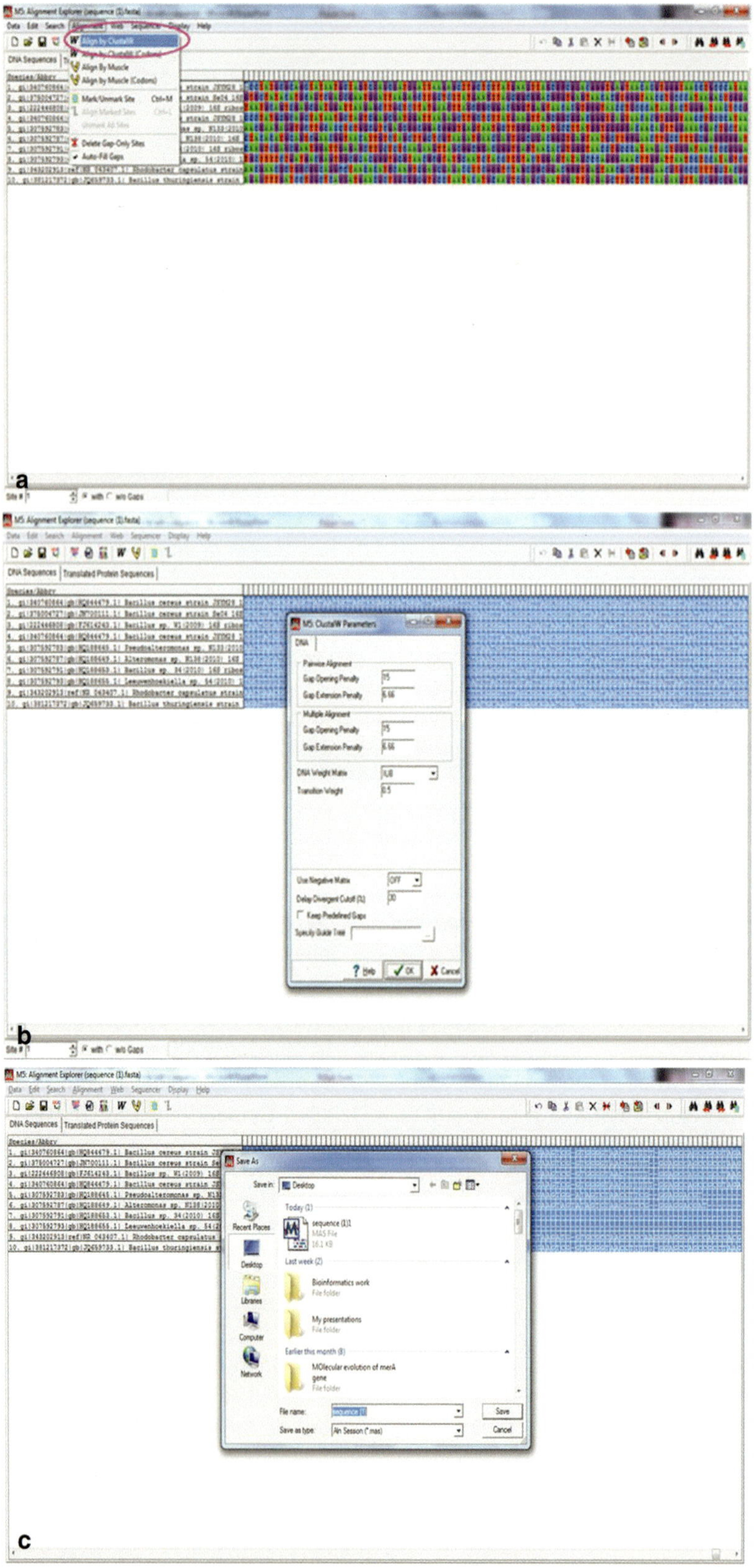

Fig. 5.15 Steps of constructing phylogenetic tree using MEGA5

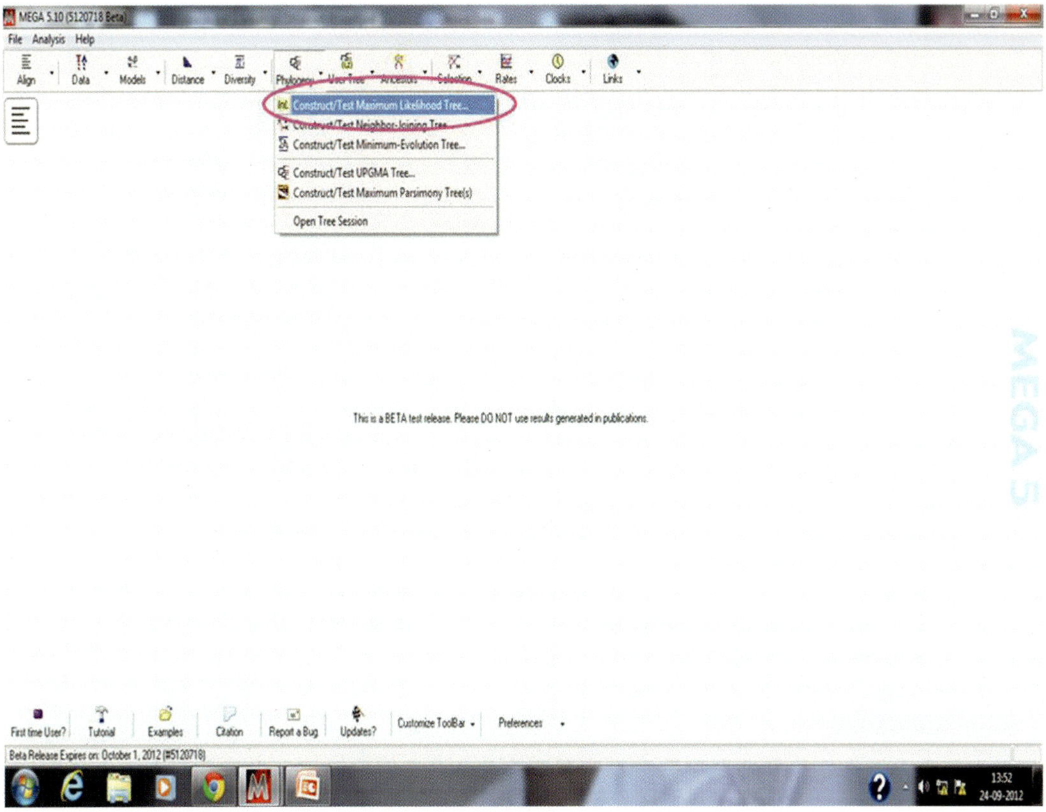

Fig. 5.16 Constructing tree by maximum likelihood method

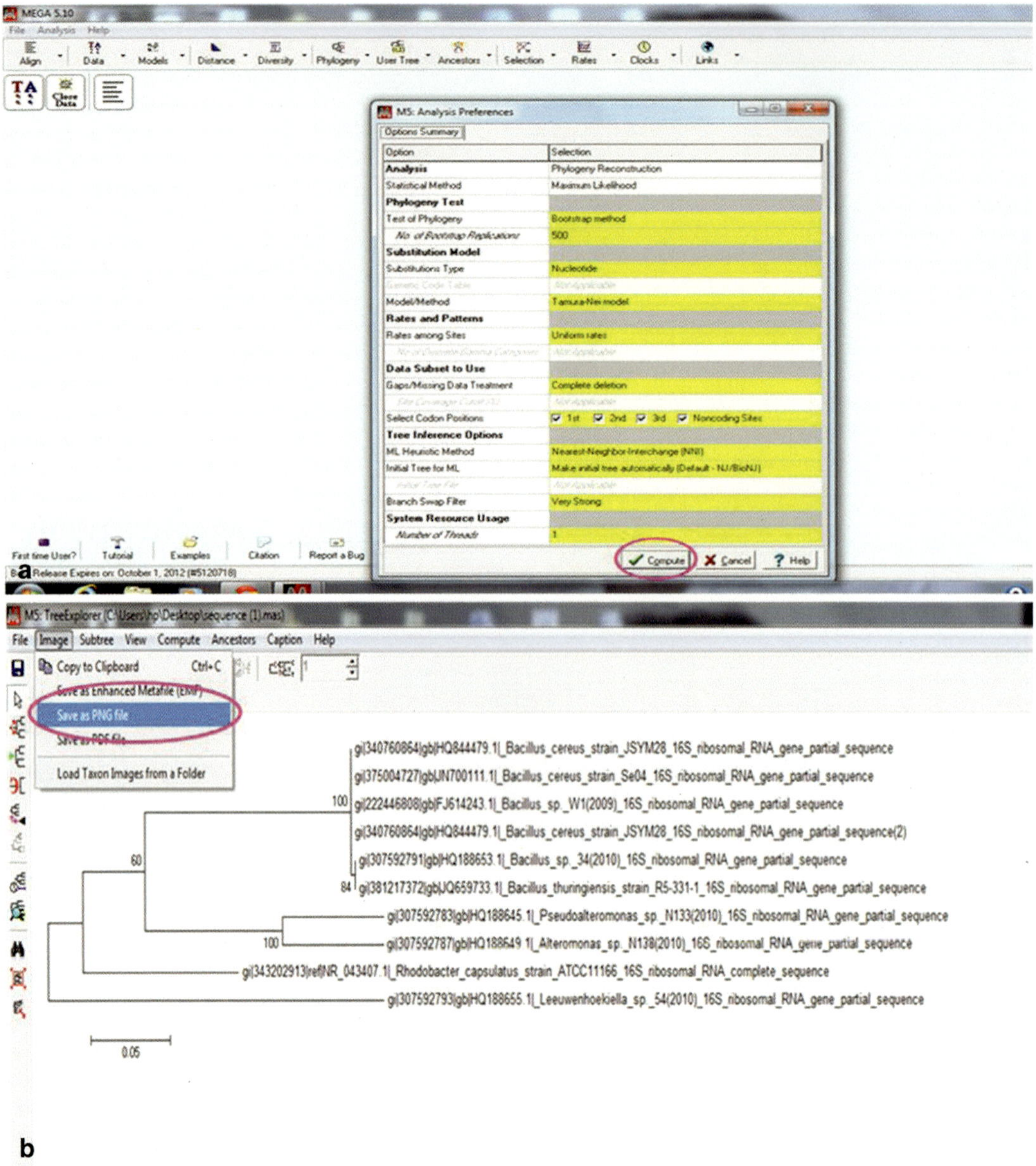

Fig. 5.17 Final tree draw and saving the tree in any format

FLOW CHART

Put all 16S rDNA sequences in FASTA format

↓

Launch MEGA5 and open all sequences

↓

Select all sequence>alignment>align by ClustalW

↓

Save the alignment

↓

MEGA5>Phylogeny>Construct/test minimum likelihood trees

↓

Compute to get tree

↓

Save the phylogenetic tree in desire format

Exp. 5.4 Primer Design

Objective To design a set of primers for the specific amplification of the target gene.

Introduction

PCR is a commonly used method to amplify DNA of interest in many fields such as biomedical research and diagnostic and forensic testing. While the outcome of PCR can be influenced by many other conditions such as the template DNA preparation and reaction conditions, designing a good pair of primers is a critical factor. A general requirement is that the primers should have similar melting temperatures (T_m) and a balanced G/C content, but should avoid self-complementarity and hairpin structure. Additional requirements may also apply in certain cases. For example, to avoid unwanted amplification of genomic DNA in reverse transcription PCR (RT-PCR), it is recommended that a primer pair span an intron, or that one of the primers be located at an exon–exon junction. Another concern is the possible impact of SNPs in the primer regions. Since an SNP may act as a mismatch in some cases, one should consider picking primers outside of such regions. One critical primer property is the target specificity. Ideally, a primer pair should only amplify the intended target, but not any unintended targets. This is especially important for real time quantitative PCR (qPCR) where in many cases the amount of PCR product is represented by the total intensity of fluorescence incorporated into amplified DNA and any amplification of unintended targets can affect the measurement.

Primer Designing Using Software

A number of primer designing tools are available that can assist in PCR primer design for new and experienced users alike. These tools may reduce the cost and time involved in experimentation by lowering the chances of failed experimentation. Currently, there are number of user friendly software available on the web. The tools that are available online have been listed in Table 5.5.

Table 5.5 List of primer designing software for personal computer

Software name	Description	Address
PrimerSelect	Analyses a template DNA sequence and chooses primer pairs for PCR and primer for DNA sequencing	http://www.dnastar.com
DNASIS Max	DNASIS Max is a fully integrated program that includes a wide range of standard sequence analysis features	http://www.medprobe.com/no/dnasis.html
Primer Premier 5	Primer design for Windows and Power Macintosh	http://www.premierbiosoft.com/primerdesign/primerdesign.html
Primer Premier	Comprehensive primer design for Windows and Power Macintosh	http://www.premierbiosoft.com
NetPrimer	Comprehensive analysis of individual primers and primer pairs	http://www.premierbiosoft.com/NetPrimer.html
Array Designer 2	For fast, effective design of specific oligos or PCR primer pairs for microarrays	http://www.premierbiosoft.com/dnamicroarray/dynamicroarray.html
Beacon Designer 2.1	Design molecular beacons and Taqman probes for robust amplification and fluorescence in real time PCR	http://www.premierbiosoft.com/molecular_beacons/taqman_molecular_beacons.html
Genome PRID E 1.0	Primer design for DNA array/chips	http://pride.molgen.mpg.de/genomepride.html
Fast PCR	Software for Microsoft Windows has specific, ready to use templates for many PCR and sequencing applications: standard and long PCR, inverse PCR, degenerate PCR directly on amino acid sequence, multiplex PCR	http://www.biocenter.helsinki.fi/bare-1_html/manual.html
OLIGO 6	Primer Analysis Software for Mac and Windows	http://www.oligo.net/
Primer designer 4	Will find optimal primers in target regions of DNA or protein molecules, amplify features in a molecule, or create products of a specified length	http://www.scied.com/ses_pd5.html
GPRIME	Software for primer design	http://life.anu.edu.au/molecular/software/gprime.html
Sarani Gold	Genome Oligo Designer is software for automatic large-scale design of optimal oligonucleotide probes for microarray experiments	http://mail.strandgenomics.com/products/sarani/
PCR Help	Primer and template design and analysis	http://www.techne.com/CatMol/perhelp.html
Genorama chip Design Software	Genorama Chip Design Software is complete set of programs required for genotyping chip design. The programs can also be bought separately	http://www.asperbio.com/Chip_design_soft.html
Primer Designer	The Primer designer features a powerful, yet extremely simple, real time interface to allow the rapid identification of theoretical ideal primers for your PCR reactions	http://genamics.com/expression/primer.html
Primer Premier	Automatic design tools for PCR, sequencing or hybridization probes, degenerate primer design, nested/multiplex primer design, restriction enzyme analysis and more	http://www.biotechniques.com/freesamples/itembtn21.html
Primer Design	DOS-program to choose primer for PCR or oligonucleotide probes	http://www.chemie.unimarburg.de/%7Ebecker/pdhome.html

Guidelines for Primer Design

When choosing two PCR amplification primers, the following guidelines should be considered:

Primer Length It is accepted that the optimal length of PCR primer is 18–22 bp. This length is long enough for adequate specificity and short enough for primers to bind easily to the template at the annealing temperature.

Melting Temperature (Tm) It can be calculated using the formula of Wallace et al. (1979), T_m (°C) = 2(A+T) + 4(G+C). The optimal melting temperature for primers is in the range 52–58 °C. Primers with melting temperatures above 65 °C should also be avoided because of potential for secondary annealing. The GC content of the sequence gives a fair indication of the primer T_m. All our products calculate it using the nearest neighbour thermodynamic theory, accepted as a much superior method for estimating it, which is considered the most recent and best available.

Primer Annealing Temperature The two primers of a primer pair should have closely matched melting temperatures for maximizing PCR product yield. The difference of 5 °C or more can lead to no amplification.

$$Ta = 0.3 \times T_m(\text{primer}) + 0.7\ T_m(\text{product}) - 14.9$$

Where,

T_m (primer) melting temperature of the primers,
T_m (product) Melting temperature of the product.

GC Content Primers should have GC content between 45 and 60 % (Dieffenbach et al. 1995). GC content, melting temperature and annealing temperature are strictly dependent on one another (Rychlik et al. 1990).

Dimers and False Priming Causes Misleading Results Presence of the secondary structures such as hairpins, self-dimer produced by intermolecular or intramolecular interactions in primers can lead to poor or no yield of the product.

Avoid Cross Homology To improve specificity of the primers it is necessary to avoid regions of homology.

Procedure for Using NETPRIMER Software for Primer Designing

Netprimer is the software used to design and analyse the parameters of designed primer sequences.

1. To start with Netprimer go to the following link http://www.premierbiosoft.com/crm/jsp/com/pbi/crm/clientside/EligibleForDiscountLoginForm.jsp?LoginForFreeTool=true&PID=3
2. After you click on to the above webpage, Netprimer will ask for your email id.
3. New users click to New User? Sign Up register for using Netprimer. For, future access, give your email address and Login as shown in the figure. After logging in, you can see the following webpage on your screen.
4. Click on to 

Note: Netprimer software needs JAVA to be installed on the system. You can download Java 1.4 plug-in directly using the following link.

5. To design the primer for analysing the gene expression of gene *merA*, design primers for an amplicon size of approximately 500 bp.
6. Manually select the sequence of around 20–22 bp from the whole gene sequence, and give the sequence to netprimer software in 5'–3' orientation.
7. Netprimer primer designing and analysis window will appear as below (Fig. 5.18).
8. Netprimer has already set some default values, which can be changed according to requirements (Fig. 5.19).
9. Click on the 'Analyze' tab.

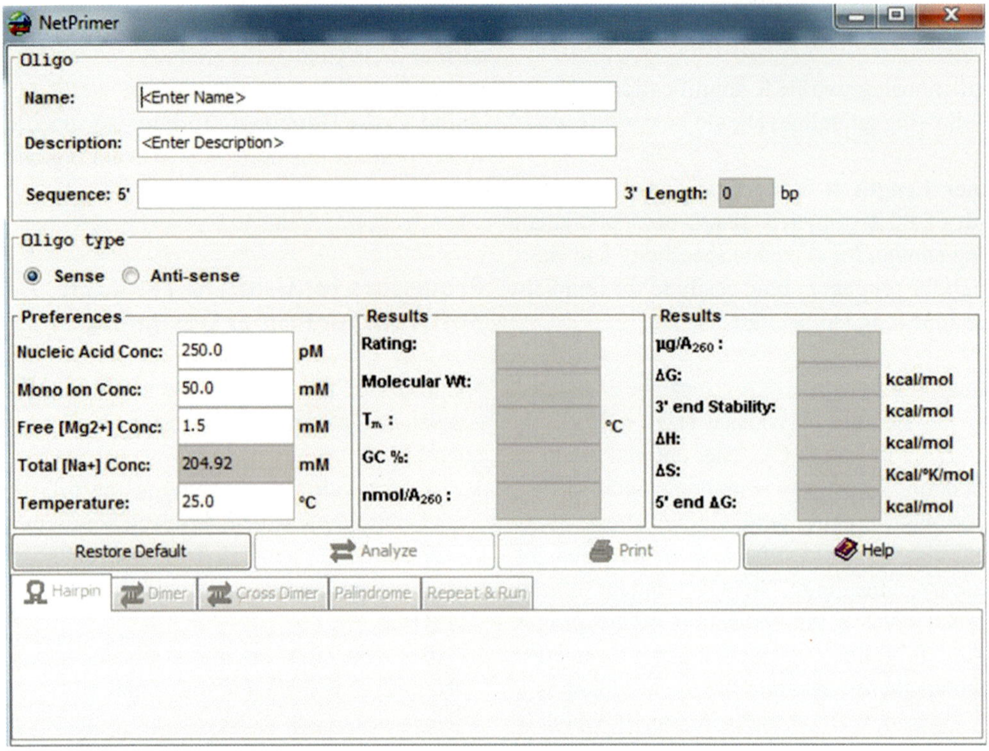

Fig. 5.18 Window of Netprimer for primer designing

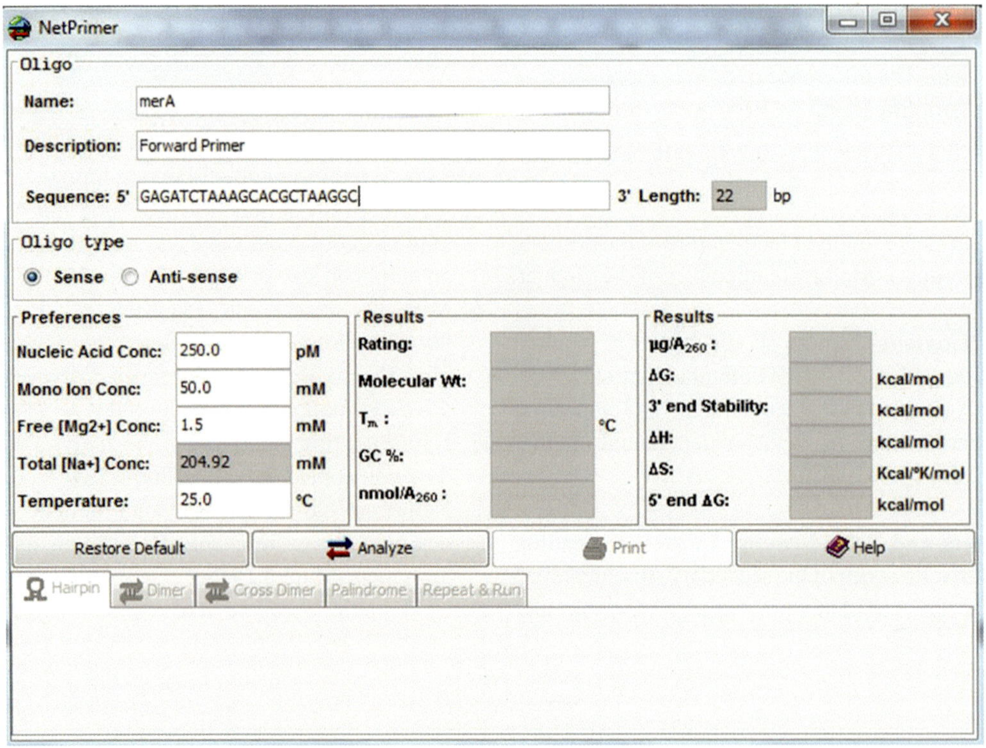

Fig. 5.19 Netprimer window after entering the values and the sequences

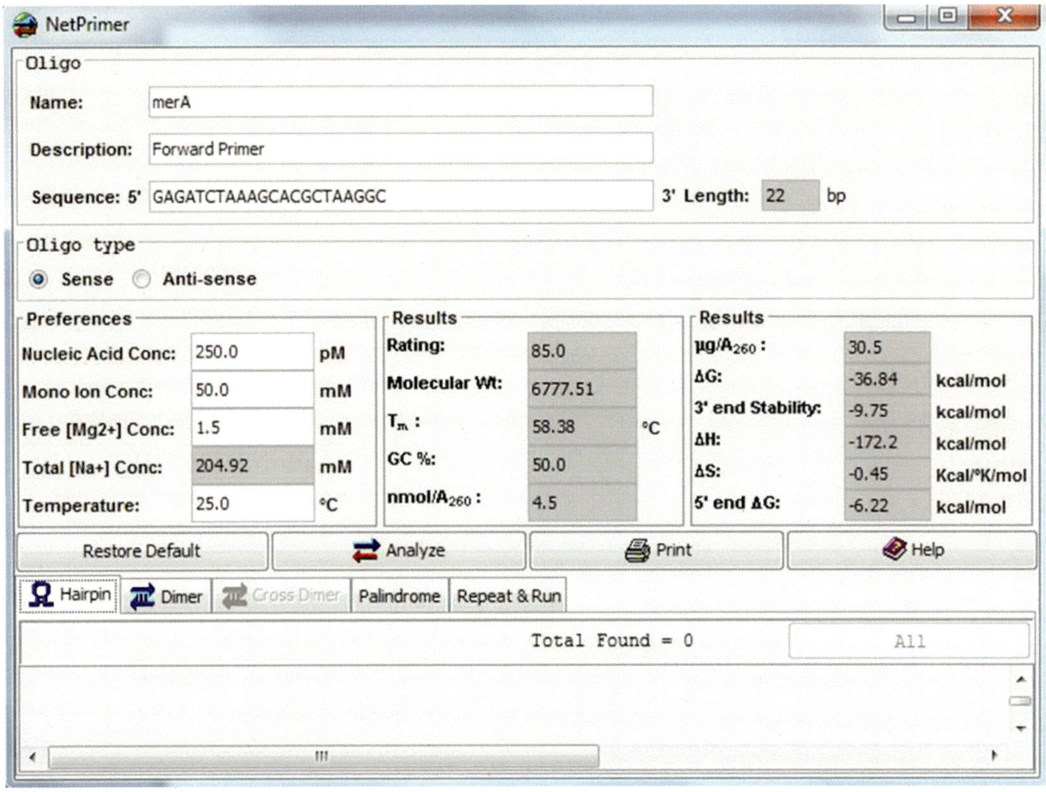

Fig. 5.20 Results of the analysed parameters

10. Following screen shows all the analysed parameters, including T_m, GC content, Hairpin, Dimers etc.
11. Netprimer will show the results of all analysed parameters as shown in Fig. 5.20.
12. Click on to Hairpin, Dimer, Palindrome, Repeat and Run Tabs to check presence of hairpins, dimers, palindromes and repeats.
13. Repeat the same for reverse primer sequence. Select the 20–22 bp oligonucleotides from the region such that you will get amplification of 1 kb.
14. Change the oligo type option to antisense, reverse complement the sequence and then give it to the software, e.g. if the sequence selected for reverse primer is 5'CCACCGAAACTCCAGGCTTTG3'; reverse complement and the sequence of the reverse primer will be 5'CAAAGCCTGGAGTTTCGGTGG3'.
15. Check the difference between T_m of both the primers. It should not be more than 5 °C
16. Rating indicates the quality or efficiency of primer to give desired amplicon. The rating value 100.0 indicates that the primer pair designed is good and more likely to produce desired amplification.

FLOW CHART

Go to
http://www.premierbiosoft.com/crm/jsp/com/pbi/crm/clientside/EligibleForDiscountLoginForm.jsp?LoginForFreeTool=true&PID=3

↓

Click on to Launch Netprimer

↓

Manually select around 20-22 bp from the whole gene sequence and give that sequence to Netprimer in 5'-3' direction

↓

If required change the default values of the set parameters and click on Analyze tab

↓

Click on to Hairpin, Dimer, Palindrome, Repeat and Run Tabs to check presence of hairpins, dimers, palindromes and repeats

↓

Repeat the same for reverse primer sequence. Select the 20-22 bp oligonucleotides from the region such that you will get amplification of 1 kb

↓

Change the oligo type option to antisense, reverse complement the sequence and then give it to the software

↓

Check the difference between Tm of both the primers. It should not be more than $5^{0}C$

Rating indicates the quality or efficiency of primer to give desired amplicon. The rating value 100.0 indicates the primer pair designed is good and more likely to produce desired amplification.

Application of Molecular Microbiology

6

Exp. 6.1 Biofilm Formation in Glass Tubes

Objective Screening of biofilm-forming bacteria by a glass tube assay.

Introduction

Biofilm is an aggregate of microorganisms with a distinct architecture where cells stick to a static surface. These cells are embedded with the self-produced matrix that consists of extracellular polymeric substances (EPS). These EPS consist of extracellular deoxyribonucleic acid (DNA), proteins and polysaccharides. Biofilm may be formed on both living and nonliving substances, which is of wide concern both from the environmental as well as clinical point of view. The biofilm-forming bacterial community is distinct from planktonic cells as the planktonic cells are mostly single cells that float or swim in the liquid media. Microbial biofilm may be the result of many factors including cellular recognition for different attachment sites on a surface, nutrition deficiency or exposure of planktonic cells to sub-inhibitory concentration of antibiotics.

Biofilm is the irreversible attachment between the surface and the matrix of a primarily polysaccharide material of bacterial origin. In certain instances, the biofilm-forming cells differ from the planktonic counterparts with respect to the genes that are transcribed. Biofilms can be formed on a variety of surfaces including living tissues, medical devices, industrial or portable water system piping or in aquatic natural ecosystem. When biofilm is formed in the water system, it is highly complex as it is formed from the corrosion products, clay material, fresh water diatoms as well as filamentous bacteria. However, biofilm on a medical device consists of single coccids organisms that are associated with EPS. In these cases, they cause both useful and harmful impacts on the ecosystem.

The formation of biofilm requires the coordinated chemical signalling between the participating cells. Unless an adequate amount of neighbouring cells are present, the cost of biofilm formation overtakes the benefit for the bacterium. In this regard, a signalling system comes into play for allowing the bacteria to sense the presence of other bacterial systems around them and to respond to the variety of conditions. This process is popularly called quorum sensing. Quorum sensing utilizes the signal molecules secreted by the bacteria called autoinducers. Autoinducer molecules are secreted by the bacteria and are continuously produced and diffused through the cell membrane to interact with the specific repressor or activator sequences in the DNA. The presence or absence of these autoinducer molecules regulate the production of messenger ribonucleic acid (mRNA) and the protein molecules which are coded for biofilm formation.

This intracellular signalling system exhibits many advantages to the bacterial entities. For instance, certain species of bacteria secrete antibiotics that inhibit the growth of other bacteria;

S. Das, H. R. Dash, *Microbial Biotechnology- A Laboratory Manual for Bacterial Systems*,
DOI 10.1007/978-81-322-2095-4_6, © Springer India 2015

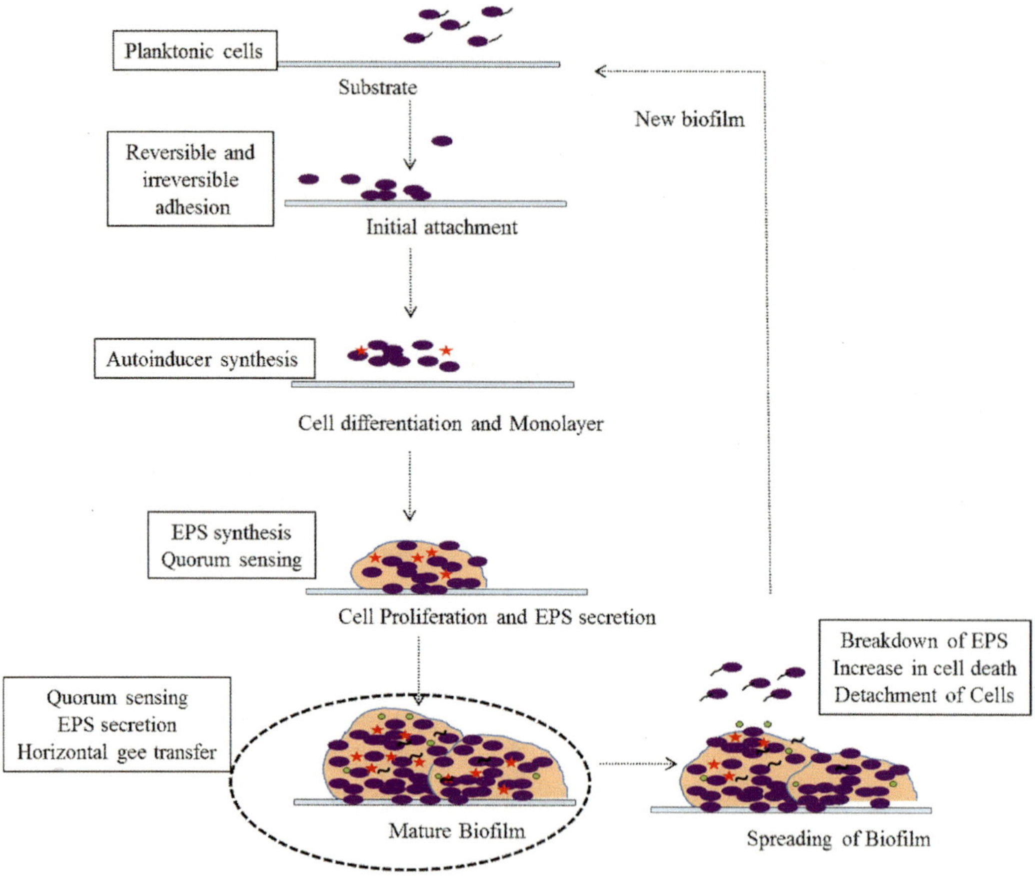

Fig. 6.1 Steps of biofilm formation in natural environmental conditions. *EPS* extracellular polymeric substances

however, in quorum-sensing-mediated biofilm formation, bacterial species come together to form biofilm. It helps bacteria to coordinately release the virulence factors and overcome the animal or plant's immune system. Signal transduction among bacterial proximity present in a biofilm enhances bacterial mating and thus the acquisition of novel DNA by transformation which is increased and enriches bacterial diversity.

Principle

There are five stages of bacterial biofilm formation, i.e. initial attachment, irreversible attachment, maturation I, maturation II and dispersion. Each step involves the firm attachment of microorganisms with the surface, and thus they become protected from the action of cleaners and sanitizers. The first step of biofilm formation involves the loose collection of cells that starts within seconds of exposures to the bacterial surface. This step is followed by the formation of electric charge on the surface, and the bacteria become firmly attached with the surface for its opposite charge. However, the electrostatic force is weak and reversible at this stage, and the microorganisms can be removed easily. This step is followed by the attachment one, where the polysaccharides entrap the cells and debris in a glue-like matrix that renders firm attachment of the cells with the substratum. At this time, the biofilm environment becomes rich in nutrient content and thus is capable of supporting rapid growth in thickness (Fig. 6.1).

The basic structural unit of any biofilm is the micro-colony. Proximity of cells within the micro-colony provides an ideal environment for the creation of nutrient gradients, exchange of genes and quorum sensing. Since micro-colonies are composed of multiple species, the cycling of various nutrients through redox reactions can readily occur in aquatic and soil biofilms. Biofilm formation by a bacterial community has an enhanced role in bioremediation of toxic substances from the environment. Biofilm formation increases the biosorption capability of the organism and hence their subsequent removal from the environment. In certain instances, a higher amount of pollutants is entrapped in the EPS matrix synthesized by bacteria and hence increases the expression level of certain genes in bacteria that enhances bioremediation capability.

In a glass tube assay, biofilm is formed at the air–water interface of the grown bacterial culture under static condition. When the biofilm is stained with crystal violet, being a basic dye, it ionizes in water to form anions and cations. In a basic dye, it has the chromophore, i.e. the coloured portion and the anion, i.e. the proton acceptor. As the bacterial cell wall and the cytoplasmic membranes are acidic in nature, the cell envelope becomes negatively charged. In this regard, the crystal violet being positively charged readily binds with the cell wall and colour of crystal violet.

Reagents Required and Their Role

Luria Bertani Broth

Luria bertani (LB) broth is a rich medium that permits fast growth and good growth yields for many species including *Escherichia coli*. It is the most commonly used medium in microbiology studies. Easy to make, fast growth of most bacterial, readily available and simple compositions contribute to the popularity of LB broth. LB can support *E. coli* growth OD_{600} 2–3 under normal shaking incubation conditions.

Crystal Violet

Crystal violet is a triarylmethane dye that is mostly used in Gram's staining of bacteria. It has an absorption maximum at 590 nm with an extinction coefficient of 87,000 M^{-1} cm^{-1}. It has been reported to be a potential antibacterial, antifungal and anthelmintic agent. The dye being positively charged can bind readily with the negatively charged cell envelope to give rise to its coloration at the air–water interface where biofilm grows and gives proper staining.

Procedure

1. Streak pure culture of bacterial strain on LB plates and incubate at 37 °C for 24 h.
2. Inoculate 2–3 colonies in 1 ml of LB tubes and incubate the tube for 48 h at 37 °C under static condition.
3. After 48 h decant the media from the tube and remove all the media from the tube.
4. Wash the tube with autoclaved milli-Q to remove any suspended media from the tube.
5. Add 5 ml of methanol to the tube and incubate at room temperature for 15 min.
6. After 15 min discard methanol from the tube carefully so that no more methanol remains in the tube.
7. Add 5 ml of 1 % crystal violet solution to the tube so that the air–water interface, where biofilm growth is present, dips inside the crystal violet.
8. Incubate the tubes for 5 min at room temperature.
9. Decant all the crystal violet from the tube after incubation and take care so that no more crystal violet remains inside the tube.
10. Wash the tubes with water twice by adding water to it and remove crystal violet by decantation.
11. A ring develops at the air–water interface confirming the biofilm formation, and a missing ring confirms the non-biofilm-forming nature of the isolate.

Observation

Observe the development of a ring at the air–water interface of the tube. A strong biofilm former gives a distinct thick ring at the interface where weak biofilm formers develop thin structures at the interface. Hence, qualitatively the isolates can be screened for the presence or absence of a biofilm formation. Optionally, at the final stage, 33 % acetic acid is added to each tube, and the absorbance can be measured at 570 nm with acetic acid keeping as blank. Thus, the biofilm-forming ability can be quantified and compared among the isolates for a strong or weak biofilm formation.

Result Table

Isolate	Intensity of the ring at the air–water interface	Absorbance at 570 nm	Inference[a]

[a] Biofilm-forming ability of the isolates can be determined based upon the thickness of the ring formed at the air–water interface after staining with crystal violet (Fig. 6.2)

Troubleshooting

Problems	Possible cause	Possible solution
No ring developed	Incubation time with methanol is more	Never incubate the isolate for more than 10 min with methanol as it may degrade the biofilm structure
	Crystal violet incubation time is not proper	Crystal violet only gives the stain to the ring, and it should be properly stained with crystal violet; the biofilm structure developed at the air–water interface should dip inside the crystal violet solution
	Disintegrated biofilms during washing	Wash the tubes with utmost care so that the biofilm structure should not disintegrate by washing
	Incubation under shaking conditions	Never incubate the broth under shaking condition as the firm attachment is required between the culture and the substratum for the development of a strong biofilm
Poor rings developed	Incubation time with methanol is more	Never incubate the isolate for more than 10 min with methanol as it may degrade the biofilm structure
	Crystal violet incubation time is not proper	Crystal violet only gives the stain to the ring, and it should be properly stained with crystal violet; the biofilm structure developed at the air–water interface should dip inside the crystal violet solution
	Disintegrated biofilms during washing	Wash the tubes with utmost care so that the biofilm structure should not disintegrate by washing
	Incubation under shaking conditions	Never incubate the broth under shaking condition as the firm attachment is required between the culture and the substratum for the development of a strong biofilm
Rings with debris	Washing with water is not proper	If the debris of the culture and media remains in the tube, the visualization of a clear ring will not be appropriate, hence wash the tubes carefully with water so that there will be no more debris remaining in the tube
	Washing with methanol is not proper	Washing with methanol removes all the remaining debris except the bound biofilm and crystal violet. Hence, washing with methanol should be carried out with proper care so that no debris remains in the tube except the biofilm structure

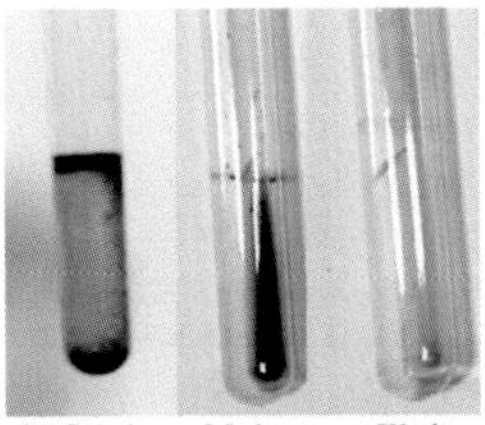

Fig. 6.2 Interpretation of biofilm-forming ability of the bacterial strains by a glass tube assay

Precaution

1. Do not incubate the tubes under static conditions.
2. Wash the tubes carefully with subsequent incubation at room temperature at a proper time for the optimum removal of the cell debris except the biofilm structure.
3. While pipetting, do not dispense the solutions through the walls of the tubes as there is the chance of disintegration of biofilm structure that develops at the glass surface.
4. Do not incubate the tubes for a longer time after adding methanol as overwashing may disintegrate the biofilm structure.
5. Always use gloves while performing this experiment.
6. Use autoclaved pipettes and tips while performing this experiment.
7. Take utmost care to remove all the crystal violet from the tube except at the biofilm structure developed at the air–water interface that gives a clear image of the biofilm structure.
8. Note the results as +, ++ or +++ for weak biofilm formers, biofilm formers and strong biofilm structures, respectively.

FLOW CHART

Streak pure culture of bacterial strain on LB plates at 37°C for 24 h

↓

Inoculate 2-3 colonies in 1 ml of LB tubes and incubate for 48 h at 37°C under static conditions

↓

After 48 h decant the media from the tube and remove all the media from the tube

↓

Wash the tubes with autoclaved milli-Q to remove any suspended media from the tube

↓

Add 5 ml of methanol to the tube and incubate at room temperature for 15 min

↓

After 15 min discard methanol from the tube carefully so that no more methanol remains in the tube

↓

Add 1% crystal violet solution to the tube so that the air water interface where biofilm growth is there should dip inside crystal violet

↓

Incubate the tubes for 5 min at room temperature

↓

Decant all the crystal violet from the tube after incubation and take care so that no more crystal violet remains inside the tube

↓

Wash the tubes with water twice by adding water to it and removing by decanting

↓

A ring develops at the air water interface confirming the biofilm formation and the no ring confirms the non-biofilm forming nature of the isolate

Exp. 6.2 Screening of Biofilm Formation in Micro-Titre Plates

Objective Quantitative estimation of biofilm formation by micro-titre plate assay.

Introduction

Biofilm is a group of microorganisms where cells stick to a static surface. These cells are embedded with the self-produced matrix that consists of extracellular polymeric substances. These extracellular polymeric substances (EPS) consist of extracellular DNA, proteins and polysaccharides. Biofilm may be formed on both living and nonliving substances which is of wide concern both from the environmental as well as clinical point of view. The biofilm-forming bacterial community is distinct from planktonic cells as the planktonic cells are mostly single cells that float or swim in the liquid media. Microbial biofilm may be the result of many factors including cellular recognition for different attachment sites on a surface, nutrition deficiency or exposure of planktonic cells to subinhibitory concentration of antibiotics.

Biofilm is the irreversible attachment between the surface and the matrix of a primarily polysaccharide material of bacterial origin. In certain instances, the biofilm-forming cells differ from the planktonic counterparts with respect to the genes that are transcribed. Biofilms can be formed on a variety of surfaces including living tissues, medical devices, industrial or portable water system piping or in aquatic natural ecosystem. When biofilm is formed in the water system, it is highly complex as it is formed from the corrosion products, clay material, fresh water diatoms as well as filamentous bacteria. However, biofilm on a medical device consists of single coccids organisms that are associated with EPS. In these cases, they cause both useful and harmful impacts on the ecosystem.

The formation of biofilm requires the coordinated chemical signalling between the participating cells. Unless an adequate amount of neighbouring cells are present, the cost of biofilm formation overtakes the benefit for the bacterium. In this regard, a signalling system comes into play for allowing the bacteria to sense the presence of other bacterial systems present around them and to respond to the variety of conditions. This process is popularly called quorum sensing. Quorum sensing utilizes the signal molecules secreted by the bacteria called autoinducers. Autoinducer molecules are secreted by the bacteria that are continuously produced and diffused through the cell membrane to interact with the specific repressor or activator sequences in the DNA. The presence or absence of these autoinducer molecules regulate the production of mRNA and the protein molecules which are coded for biofilm formation.

This intracellular signalling system exhibits many advantages to the bacterial entities. For instance, certain species of bacteria secrete antibiotics that inhibit the growth of other bacteria; however, in quorum-sensing-mediated biofilm formation, bacterial species come together to form biofilm. It helps bacteria to co-ordinately release the virulence factors and overcome the animal or plant's immune system. Signal transduction among bacterial proximity present in a biofilm enhances bacterial mating, and thus the acquisition of novel DNA by transformation which is increased and enriches bacterial diversity.

Principle

There are five stages of bacterial biofilm formation, i.e. initial attachment, irreversible attachment, maturation I, maturation II and dispersion. Each step involves the firm attachment of microorganisms with the surface, and thus they become protected from the action of cleaners and sanitizers. The first step of biofilm formation involves the loose collection of cells that starts within seconds of exposures to the bacterial surface. This step is followed by the formation of electric charge on the surface, and the bacteria become firmly attached with the surface for its opposite charge. However, the electrostatic force is weak and reversible at this stage, and the microorganisms can be removed easily. This step is followed by the attachment one, where the polysaccharides entrap the cells and debris in a glue-like matrix that renders firm attachment of the cells with the substratum. At this time, the biofilm environment becomes rich in nutrient content and thus is capable of supporting rapid growth in thickness.

The basic structural unit of any biofilm is the micro-colony. Proximity of cells within the micro-colony provides an ideal environment for the creation of nutrient gradients, exchange of genes and quorum sensing. Since micro-colonies are composed of multiple species, the cycling of various nutrients through redox reactions can readily occur in aquatic and soil biofilms. Biofilm formation by a bacterial community has an enhanced role in bioremediation of toxic substances from the environment. Biofilm formation increases the biosorption capability of the organism and hence their subsequent removal from the environment. In certain instances, higher amount of pollutants are entrapped in the EPS matrix synthesized by bacteria and hence increases the expression level of certain genes in bacteria that enhances bioremediation capability.

Biofilm formation may be determined in several different ways; however, the most frequently used technique is the glass tube assay. However, this technique only allows the qualitative screening of biofilm positive isolates by the development of rings at the air–water interface after staining with the cationic dye. In this regard, micro-titre plate assay determines the qualitative value of the biofilm formation by the isolates with respect to the respective controls. The optical density of the stained bacterial biofilm can be determined by spectrophotometer.

Reagents Required and Their Role

Luria Bertani Broth

Luria Bertani (LB) broth is a rich medium that permits fast growth and good growth yields for many species including *E. coli*. It is the most commonly used medium in microbiology studies for many bacterial cultures. Easy to make, fast growth of most bacterial strains, readily available and simple compositions contribute to the popularity of LB broth. LB can support *E. coli* growth OD_{600} 2–3 under normal shaking incubation conditions.

Crystal Violet

Crystal violet is a triarylmethane dye that is mostly used for histological staining during Gram's staining of bacteria. It has an absorption maximum at 590 nm with an extinction coefficient of 87,000 M^{-1} cm^{-1}. It has been reported to be a potential antibacterial, antifungal and anthelmintic agent. The dye being positively charged can bind readily with the negatively charged cell envelope to give rise to its coloration at the air–water interface where biofilm grows and gives proper staining.

Procedure

1. Inoculate 2–3 colonies of pure culture of bacteria into 2 ml of LB medium. Incubate at 37 °C for 24 h at shaking with 180 rpm.
2. Dilute the overnight grown culture to 1:100 into fresh medium of LB.
3. Add 100 μl of the diluted medium to each well of the micro-titre plate. Use three replicates for the standard biofilm assay of each culture.
4. Incubate the micro-titre plate for 24, 48 and 72 h at 37 °C under static condition.
5. After incubation, dump out the cells by turning the plate over and shocking out the remaining media.
6. Gently submerge the plate in a small tub of water. Shake out water. Repeat this process a second time. This step helps to remove the unattached cells and media components to lower the background staining.
7. Add 125 μl of a 0.1 % solution of crystal violet in water to each well of the micro-titre plate.
8. Incubate the plate at room temperature for 10–15 min.
9. Rinse the plate 3–4 times with water by submerging in a tub of water, shake out and blot vigorously on a stack of paper towels to rid the plate of all excess cells and dye.
10. Turn the micro-titre plate upside down and dry for a few hours or overnight.
11. For qualitative assay, take photograph of the wells when they are dry.
12. Add 125 μl of 30 % acetic acid in water to each well of the micro-titre plate to solubilize the crystal violet.
13. Incubate the micro-titre plate at room temperature for 10–15 min.
14. Transfer 125 μl of the solubilized crystal violet to a new flat bottomed micro-titre plate.
15. Take absorbance at 550 nm using 30 % acetic acid in water as blank.

Observation

Compare the Optical Density (OD) values at 550 nm to get a comparative idea about the biofilm formation among the bacterial culture. Bacterial culture having the higher OD value at 550 nm corresponds to its higher biofilm-forming capability.

Fig. 6.3 **a** Aerial and **b** transverse view of biofilm-forming bacterial isolate *Bacillus thuringiensis* PW-05 in micro-titre plate after staining with crystal violet

Result Table

Bacterial culture	$OD_{550\ nm}$			Inference[a]
	24 h	48 h	72 h	
Biofilm +ve control				
Test strain 1				
Test strain 2				

[a] The higher the $OD_{550\ nm}$ value, the higher is the biofilm-forming ability of the bacterial isolate. After staining with crystal violet, the wells of the micro-titre plate show adherent stains along the walls. It signifies the biofilm attachment on the plate surface (Fig. 6.3)

Troubleshooting

Problems	Possible cause	Possible solution
No ring developed	Incubation time with methanol is more	Never incubate the isolate for more than 10 min with methanol as it may degrade the biofilm structure
	Crystal violet incubation time is not proper	Crystal violet only gives the stain to the ring, and it should be properly stained with crystal violet; the biofilm structure developed at the air–water interface should dip inside the crystal violet solution
	Disintegrated biofilms during washing	Wash the tubes with utmost care so that the biofilm structure should not disintegrate by washing
	Incubation under shaking conditions	Never incubate the broth under shaking condition as the firm attachment is required between the culture and the substratum for the development of a strong biofilm
Poor rings developed	Incubation time with methanol is more	Never incubate the isolate for more than 10 min with methanol as it may degrade the biofilm structure
	Crystal violet incubation time is not proper	Crystal violet only gives the stain to the ring, and it should be properly stained with crystal violet; the biofilm structure developed at the air–water interface should dip inside the crystal violet solution
	Disintegrated biofilms during washing	Wash the tubes with utmost care so that the biofilm structure should not disintegrate by washing
	Incubation under shaking conditions	Never incubate the broth under shaking condition as the firm attachment is required between the culture and the substratum for the development of a strong biofilm
Rings with debris	Washing with water is not proper	If the debris of the culture and media remains in the tube, the visualization of a clear ring will not be appropriate, hence wash the tubes carefully with water so that there will be no more debris remaining in the tube
	Washing with methanol is not proper	Washing with methanol removes all the remaining debris except the bound biofilm and crystal violet. Hence, washing with methanol should be carried out with proper care so that no debris remains in the tube except the biofilm structure

Precaution

1. Do not incubate the tubes under static conditions.
2. Wash the tubes carefully with subsequent incubation at room temperature for proper time for the optimum removal of the cell debris except the biofilm structure.
3. While pipetting, do not dispense the solutions through the walls of the tubes as there is the chance of disintegration of biofilm structure that develops at the glass surface.
4. Do not incubate the tubes for a longer time after adding methanol as overwashing may disintegrate the biofilm structure.
5. Always use gloves while performing this experiment.
6. Use autoclaved pipettes and tips while performing this experiment.
7. Take utmost care to remove all the crystal violet from the tube except at the biofilm structure developed at the air–water interface that gives a clear image of the biofilm structure.
8. Note the results as +, ++ or +++ for weak biofilm formers, biofilm formers and strong biofilm structures, respectively.

FLOW CHART

Inoculate 2-3 colonies of pure culture of bacteria into 2 ml of LB medium. Incubate at 37°C for 24 h under shaking at 180 rpm

↓

Dilute the over-night grown culture to 1:100 into fresh medium of LB

↓

Add 100 µl of the diluted medium to each well of the micro-titre plate. Use three replicates for the standard biofilm assay of each culture

↓

Incubate the micro-titre plate for 24, 48 and 72 h at 37°C under static condition

↓

After incubation, dump out the cells by turning the plate over and shaking out the liquid

↓

Gently submerge the plate in a small tub of water. Shake out water, repeat this process a second time

↓

Add 125 µl of a 0.1% solution of crystal violet in water to each well of the micro-titre plate

↓

Incubate the plate at room temperature for 10-15 min

↓

Rinse the plate 3-4 times with water by submerging in a tub of water, shake out and blot vigorously on a stack of paper towels to rid the plate of all excess cells and dye

↓

Turn the micro-titre plate upside down and dry for few h or overnight

↓

For qualitative assay, take photograph of the wells when they are dry

↓

Add 125 µl of 30% acetic acid in water to each well of the micro-titre plate to solubilize the crystal violet

↓

Incubate the micro-titre plate at room temperature for 10-15 min

↓

Transfer 125 µl of the solubilized crystal violet to a new flat bottomed micro-titre plate

↓

Take absorbance at 550 nm using 30% acetic acid in water as blank

Exp. 6.3 Confocal Laser Scanning Microscopy for Biofilm Analysis

Objective To analyse biofilm architecture of bacterial strain by confocal laser scanning microscopy.

Introduction

Confocal laser scanning microscopy (CLSM) is the most versatile and effective approach for analysis of bacterial biofilms. The advantage of using this technique is its nondestructive nature imparted to the biofilm architecture. This technique reduces greatly the pretreatments such as disruption and fixation and thus maintains the evidence for microbial relationships, complex structures and organization of bacterial cells in biofilm. In addition to CLSM, other microscopic techniques such as light microscopy, scanning electron microscopy can also be used for the analysis of biofilm architecture. As a result of its noninvasive and nondestructive nature, CLSM provides a huge advantage over other microscopic techniques to reconstitute the 3D structure of bacterial biofilms in their naturally hydrated form. In addition to that, a set of computational tools can be applied to analyse the data to obtain a clear image on microbial architecture and its nature.

CLSM is the microscopic technique to obtain high-resolution optical images with deep selectivity. The advantage of this microscopic technique is the acquiring of in-focus images at selected depths by the process of optical sectioning. During this process, images are acquired point by point and reconstructed using software to form a three dimensional image of topologically complex objects. For the analysis of biological samples, CLSM can be combined with laser scanning method for detection of biological samples labelled with fluorescent markers. In addition to that, statistical programmes such as Computer Statistics Programme (COMSTAT) can be used to quantify the biofilm architecture in terms of average colony size, average thickness, and roughness coefficient and other parameters.

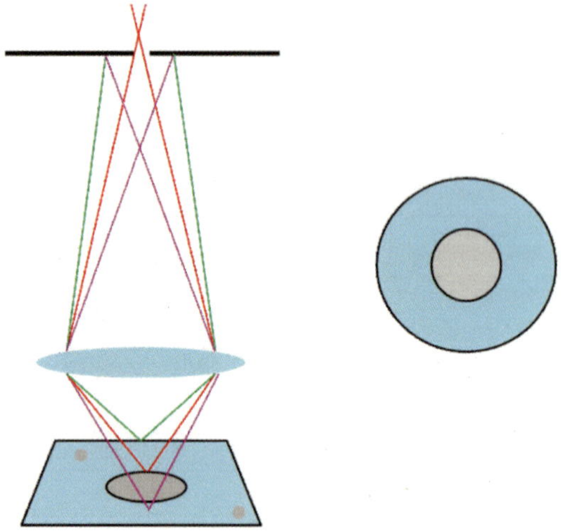

Fig. 6.4 Development of different focus levels in the specimen to generate three dimensional datasets

Principle

Confocal microscopy is the optical imaging technique that increases the optical resolution and contrast of the micrograph by using point illumination that enables the construction of three dimensional structures from the obtained image. In standard microscopy, areas of above and below of the focal planes of thick samples contribute to the image as 'out of focus blur'. In confocal microscopy, the pinhole between specimen and detector is used to select information from a single focal plane. This in turn produces a sharply focussed optical slice through the specimen (Fig. 6.4). Finally, after taking a series of optical slices from different focus levels a three dimensional data set can be generated.

The advanced version of confocal microscopy is the laser scanning confocal microscopy. In this case, a pair of oscillating mirrors scans a point of laser light across the specimen via objective. The fluorescent light emitted by the specimen passes back through the mirror system to a beam splitter that rejects any reflected excitation wavelengths and then passes it through the pinhole to generate the optical slice. The detector is a photomultiplier tube that simply records the brightness of fluorescence at each raster point and maps into a 2D (XY) image. The advanced versions of confocal microscopes use acousto optical beam splitter

(AOBS) that acts as a dichroic beam splitter to separate shorter wavelength from longer wavelength fluorescent emission. AOBS shows a very clean, sharp cut off between signal transmission and rejection compared to standard optical beam splitters. The other advantages of using AOBS system includes the fast switching between channels, less bleed through with multiple labelling, high signal transmission, less bleaching by optimising excitation intensity and easier region of interest scanning. In addition to that, it can detect any emission wavelength in the visible and near infrared and there is no need of buying new filters when new dyes are introduced.

Biofilm is a group of microorganisms where the cells stick to each other in a substratum. In order to analyse the biofilm-forming efficacy of an organism, many factors are to be taken care of including their biomass, numbers in terms of Colony Forming Units (CFU) per millilitre of square centimetre, viability of cells and their activity. Other factors regarding their biodiversity constituents, biofilm architecture, and growth rate, spectrum of utilized substrates, products or intermediates formed and impact on the environment can also be studied. Quantification of biofilm formation by the microorganisms can be assessed in many ways, either by determining the maximum number and determination of active organisms or indirect methods of determination of oxygen consumption, increase in conductivity, CO_2 production and production of other cellular metabolites. The majority of the biofilm constituents can be determined by using nucleic acid stains including acridine orange, 4',6-diamidino-2-phenylindole (DAPI), but this staining practice cannot distinguish between live and dead cells. Similarly, active organisms can be determined by cultivation methods on various media, determination of enzymatic activity, determination of elongation and turbidity measurement.

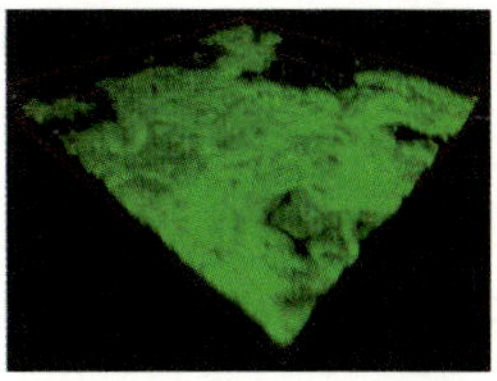

Fig. 6.5 Three dimensional image of bacterial biofilm (*Pseudomonas mendocina* NR802) under confocal laser scanning microscope after Syto-9 staining

Many microscopic tools can also be employed to analyse the bacterial biofilm samples that include light microscopy, epifluorescent, scanning electron microscopy and CLSM (Fig. 6.5). However, CLSM is the most advanced technique of biofilm visualization. CLSM with fluorescent dyes makes it possible to visualize bacteria stained with propidium iodide and matrix polysaccharide bound to ConA lectin conjugated with fluorescein isothiocyanate (FITC). Lectins of different carbohydrate specificities conjugated with fluorochromes are generally used for the analysis of biofilm composition. In an advanced technique for cell visualization in CSLM bacteria are labelled by insertion into the chromosome of a sequence encoding a fluorescent label, i.e. green fluorescent protein (GFP). The various dyes used for analysis of biofilm samples have been listed in Table 6.1.

Table 6.1 List of dyes and their specifications used for the analysis of bacterial biofilm samples by confocal laser scanning microscope

Dye	Nature of the dye/binding affinity	Excitation/emission
Syto-9	Green fluorescent nucleic acid stain that stains Gram-positive and Gram-negative bacteria	486/501 nm (RNA) 485/498 nm (DNA)
ConcavlinA-Tetramethylrhodamine (ConA-TRITC)	Binding affinity for glucose and dextran, lectin derivative	555/580 nm
Propidium Iodide (PI)	Nucleic acid stain	535/617 nm
Acridine orange	Cell permeable nucleic acid stain, green fluoresce upon binding with double-stranded DNA and red fluoresce with single stranded RNA	500/526 nm (DNA) 460/650 nm (RNA)
Syto17	Red fluorescent nucleic acid stain that exhibits bright red fluorescence	621/634 nm
ConA-fluorescein	Binding affinity for glucose and dextran, lectin derivative	494/518 nm

RNA ribonucleic acid, *DNA* deoxyribonucleic acid, *ConA* conacanavalin A

Reagents Required and Their Role

Luria Bertani Broth

Luria bertani (LB) broth is a rich medium that permits fast growth and good growth yields for many bacterial species. It is the most commonly used medium in microbiology studies for many bacterial cultures. Easy to make, fast growth of most bacterial strains, readily available and simple compositions contribute to the popularity of LB broth. LB can support bacterial growth OD_{600} 2–3 under normal shaking incubation conditions.

Biofilm-Forming Bacteria

Biofilm-forming characteristic is the de novo property of certain bacterial species that can form a mat-like structure. The adherent cells are embedded in the self-produced matrix of extracellular polymeric substances. The organisms may form biofilms in response to many factors including attachment sites on a surface, nutrient deficiency, exposure to antibiotic, presence of toxic metals and other stress conditions. Many bacterial species including *Bacillus* sp., *Pseudomonas* sp., *E. coli, Lactobacillus* sp. have been reported to form biofilms under environmental conditions.

Syto-9

Syto-9 is a green fluorescent nucleic acid dye that stains live and dead Gram-positive and Gram-negative bacteria as well as the eukaryotes. It can be used to stain both DNA and RNA. There are many advantages of using this strain as it is permeable to virtually all cell membranes. It possesses high molar absorptivity with extinction coefficient of >50,000 cm^{-1} M^{-1} at visible absorption maxima. Syto-9 has extremely low intrinsic fluorescence with quantum yields <0.01 when not bound to nucleic acids. It is soluble in dimethyl sulfoxide (DMSO) and possesses the excitation/emission maximum at 486/501 nm for single-stranded RNA and 485/498 nm for double-stranded DNA.

ConA-TRITC

Concanavalin A (ConA) is the most widely used lectins in staining practices. When conjugated with tetramethylrhodamine exhibits the bright orange-red fluorescence of TRITC with absorption/emission maxima at 555/580 nm. In neutral and alkaline solutions, concanavalin A exists as a tetramer with a molecular weight of approximately 104,000 Da. In acidic solutions (pH below 5.0), concanavalin A exists as a dimer. ConA-TRITC selectively binds to α-mannopyranosyl and α-glucopyranosyl residues.

Phosphate Buffer Saline

Phosphate buffer saline (PBS) is used to wash the unstained bacterial cells as well as the excess stain to avoid noise during the acquisition of image under fluorescence microscopy. There are many advantages of using PBS for washing over distilled water or milli-Q water for washing of bacterial cells. Due to the low level of salinity difference, there are less chances of cell bursting followed by cell death due to washing with phosphate buffer saline. However, milli-Q water may serve the purpose of washing; however, due to high difference in salinity, there is a huge chance of bursting of cells and subsequently killing of cells. 1X PBS buffer can be prepared using the following components: 4.3 mM sodium phosphate dibasic (Na_2HPO_4), 137 mM sodium chloride, 2.7 mM potassium chloride and 1.4 mM potassium phosphate monobasic in water. To prepare 1000 ml of 1X PBS, add 8 g of NaCl, 1.44 g of Na_2HPO_4, 0.25 g of KH_2PO_4 in 800 ml of milli-Q water, allow the solutes to dissolve for 3–5 min, slowly add 1 M HCl to adjust the pH to 7.4, make up the volume to 1000 ml and autoclave. Autoclaved PBS can be stored at room temperature till further use.

Protocol

Sample Preparation

1. Inoculate 2–3 biofilm-forming bacterial colonies in 100 ml of LB broth and incubate at 37 °C for 24 h with vigorous shaking at 180 rpm.
2. Mark a six-welled plate as per the requirement with a permanent marker and carefully transfer 2.5 ml of LB broth to the wells of the plate.
3. Put unbreakable coverslips in the wells so that, a layer of medium will remain at the top of the coverslip.
4. Carefully inoculate 10 μl of bacteria from the grown culture and incubate at 37 °C for 48 h under static conditions in a moist chamber.
5. After incubation, take out the coverslip from the media with the help of forceps and keep it on tissue paper.

Staining

1. Wash the coverslips with PBS to remove the planktonic cells that are attached to it, allow to dry at room temperature.
2. Float the coverslip with 100 μl of Syto-9 solution for 30 min by incubating at room temperature on a gel rocker.
3. Wash the stained coverslip two times with PBS.
4. Allow to air dry at room temperature.
5. Float the coverslip with 100 μl of ConA-TRITC for 30 min by incubating at room temperature on a gel rocker.
6. Drain out the remaining stain and wash with PBS two times.
7. Dry the slides at room temperature.

Confocal Imaging

1. Fix coverslip upside down on a glass slide and put a drop of oil over it.
2. Place the slide upside down over the objective lens of CLSM.
3. Use 63X objective lens with 1.2 numerical apertures.
4. Take all the images by scanning a frame of 512 × 512 pixels with the laser beam in the x, y plane.
5. Collect randomly ten stacks from different points in order to get the significant data.
6. Collect each image with 1.33 pinhole size and take each optical slice at 1 μm z-interval to construct the 3D image of stacks.
7. Gather the biofilm parameters of the samples by using COMSTAT such as average thickness, maximum thickness, total biomass, average colony size, average colony volume, average fractional dimension, average surface area of biomass in each image stack, surface to volume ratio and roughness coefficient.

Observation

Observe the green and orange red coloured sections of the image. The green colour portion signifies the bacterial biomass, whereas orange red colour signifies the production of exopolysaccharides. A higher amount of green fluorescence compared to red fluorescence implies less EPS production by the bacterial biomass, whereas a higher amount of red fluorescence may be due to the higher biofilm-forming ability of the test isolate.

Observation Table

Isolate	Average thickness	Average colony size	Surface area to volume ratio	Roughness coefficient	Total biomass	Inferences
Biofilm positive control						
Test isolate 1						
Test isolate 2						

Precautions

1. Always operate the laser products within controlled access areas. At the entrance of the controlled access areas, post laser warning signals.
2. When you want to observe the direct light or reflected light, protect your eyes by wearing protected glasses.
3. All the dyes used during the confocal microscopy are carcinogenic in nature. Hence, be careful while dealing with these dyes.
4. Make sure the work area to be cleaned regularly for the dust particles.
5. Always put on gloves, lab coat and eye protection while working in confocal microscope.

Troubleshooting

Problems	Possible cause	Possible solution
No image when microscope light path is at camera port	Light path switching knob is pulled	Push the knob and switch to first port
	Motor not rotating	Turn on the key switch
	Too weak illumination	Increase illumination power of the light
Visible dust	Dust at the specimen	Clean the slides and coverslips
	Unclean objective lens	Clean the objective lens following manufacturer's guidelines
	Dust on camera port	Remove dusts from the camera port using air blow
Background noise	Dust on specimen	Clean the slides and coverslips
	Washing not proper	Wash the samples repeatedly in between staining procedure
	Higher incubation time with the dye	Strictly follow the incubation time with each dye
Images cannot be distinguished with multiple dyes	Overlapping emission/excitation maximum	Before using multiple dyes in a single experiment, check their emission/excitation maximum so that they should not overlap each other
	Improper sample preparation	Strictly follow the aforementioned procedure for sample preparation of the biofilm samples
	More washing	Vigorous washing may decrease the cell number as well as the other components to be stained giving rise to improper results. Follow the required amount of washing at regular intervals

FLOW CHART

Inoculate 10 μl of overnight grown bacterial culture in a 6 welled plate containing 2.5 ml LB broth medium and the coverslip

↓

Incubate the plate at 37°C for 48 h under static conditions

↓

Take out the coverslip and wash with PBS, stain with 100 μl of Syto9 solution for 30 min at room temperature, wash the coverslip two times with PBS

↓

Float the coverslip with 100 μl of ConA-TRITC for 30 min by incubating at room temperature on a gel rocker, drain out the remaining stain and wash two times with PBS

↓

Observe the stained coverslip under oil immersion confocal microscope with 63X objective and 1.2 numerical aperture

↓

Gather biofilm parameters such as average thickness, total biomass, average colony size, surface area to volume ratio, roughness coefficient etc. by using COMSTAT software

Exp. 6.4 Fluorescence Microscopy of Bacterial Biofilm and Image Analysis

Objective Analysis of bacterial biofilm by fluorescence microscope and analysis of image by IMAGE J software.

Introduction

Biofilms are complex communities of microorganism that develop on surface in diverse environment. Biofilms are formed when millions of microorganisms accumulate on a solid surface in moist environment, creating a complex structure that functions as a community. Bacterial biofilms are three dimensional sessile layered structures encapsulated in hydrated extracellular polymeric substances (EPS) on the substratum. Microbial EPS are biopolymers consisting of polysaccharides, proteins and nucleic acids. EPS are involved in establishment of stable arrangement of microorganisms in biofilm. EPS are mainly composed of high molecular weight compound including polysaccharide protein and amphiphilic polymers. A thorough understanding of biofilm matrix ultrastructure is critical for biofilm-related studies. Biofilm formation by the organisms used in the study was analysed by fluorescence and confocal laser scanning microscopy (CLSM). The fluorescence microscopy and the more versatile CLSM allow the nondestructive in situ study of biofilms. When combined with the application of fluorescent dyes, fluorescence microscopy and CLSM can be effectively used for the visualization and quantification of biofilm components. Concerning biofilm, a number of fluorescent dye specific to the component of biofilm can be used to stain biofilm components like EPS, live cell and dead cells. The aim of the present study is to grow and visualize bacterial biofilm stained with acridine orange using fluorescence microscope. The biofilm images will be analysed using IMAGE J software to calculate fluorescent intensity and draw 3D structure of biofilm.

Many biofilm-forming bacteria can be isolated from the natural environment and the most abundant of them include *Pseudomonas aeruginosa, Bacillus subtilis, Escherichia coli* and *Staphylococcus aureus*. However, in order to obtain a clear image on the biofilm architecture of the isolates, it needs to be stained with a proper dye. The most common dyes that are used for this purpose are acridine orange (binds with nucleic acid), Syto-9 (binds to nucleic acid), propidium iodide (binds to DNA), tetramethylrhodamine isothiocyanate concanacalin A (binds to glycoprotein).

Images acquired in the process of staining followed by fluorescence microscopy needs to be analysed to obtain a clear image and a comparison of biofilm formation by the bacterial isolates. In this regard, a Java-based image processing software developed by the National Institute of Health comes into use. The biofilm architecture as well as the comparative account of biofilm-forming ability of the isolates can be determined by analysing the raw integrated density of the biofilm matrix using this software tool.

Principle

A fluorescence microscope uses fluorescence to generate an image. Biofilm either stained with fluorescent dye or tagged with fluorescent protein is illuminated with the light of a specific wavelength (or wavelengths) which is absorbed by the fluorophores, causing them to emit light of longer wavelengths (i.e. of a different colour than the absorbed light). The filters and the dichroic are chosen to match the spectral excitation and emission characteristics of the fluorophore used to label the specimen. In this manner, the distribution of a single fluorophore (colour) is imaged at a time. Multicolour images of several types of fluorophores must be composed by combining several single-colour images. Each dye has its own emission and excitation spectra which is crucial in selection of filters to visualize the specimen under microscope.

Reagents Required and Their Role

Biofilm-Forming Bacterial Isolates

There exists a certain group of bacteria harbouring the de novo potential of biofilm formation. They differ from their planktonic counterparts in terms of the formation of biofilm matrix under stress conditions which are of huge importance from bioremediation point of view. In this experiment, two potential biofilm formers *P. aeruginosa* and *Pseudomonas mendocina* can be used. Optionally, you can use any isolate showing positive result towards biofilm formation by glass tube and micro-titre plate assay.

Luria Bertani Broth

Luria bertani (LB) broth is the routinely used growth medium in microbiological experiments. Easy to make, fast growth of most bacterial strains, readily available and simple compositions contribute to the popularity of LB broth. LB can support bacterial growth OD_{600} 2–3 under normal shaking incubation conditions in 12–24 h.

Acridine Orange

Acridine orange (AO) is a nucleic acid selective fluorescent dye. It is cell permeable. When interacting with DNA, it gives an excitation maximum at 502 nm and an emission maximum at 525 nm (green). However, upon interaction with RNA, the excitation maximum shifts to 460 nm and the emission maximum shifts to 650 nm (red). Thus, the biofilm attached to a surface can be stained with acridine orange and can be visualised at both above excitation wave lengths after staining. For proper staining of biofilm samples, 0.02 % of acridine orange solution is used. It can be prepared by dissolving 20 mg of acridine orange powder in 100 ml of milli-Q water. The prepared dye solution should be stored at room temperature under dark conditions.

Phosphate Buffer Saline

Phosphate buffer saline (PBS) is used to wash the unstained bacterial cells as well as the excess stain to avoid noise during the acquision of image under fluorescence microscopy. There are many advantages of using PBS for washing over distilled water or milli-Q water for washing of bacterial cells. Due to the low level of salinity difference, there are less chances of cell bursting followed by cell death due to washing with phosphate buffer saline. However, milli-Q water may serve the purpose of washing; but, due to high difference in salinity, there is a huge chance of bursting of cells and subsequently killing of cells. 1X PBS buffer can be prepared using the following components: 4.3 mM sodium phosphate dibasic (Na_2HPO_4), 137 mM sodium chloride, 2.7 mM potassium chloride and 1.4 mM potassium phosphate monobasic in water. To prepare 1000 ml of 1X PBS, add 8 g of NaCl, 1.44 g of Na_2HPO_4, 0.25 g of KH_2PO_4 in 800 ml of milli-Q water, allow the solutes to dissolve for 3–5 min, slowly add 1 M HCl to adjust the pH to 7.4, make up the volume to 1000 ml and autoclave. Autoclaved PBS can be stored at room temperature till further use.

IMAGE J (Version 1.46)

The computer programme used for the analysis of biofilm images obtained by fluorescent microscopy is IMAGE J version 1.46. The software is written in Java and is freely available in public domain. It can support different data types in the form of TIFF, GIF, JPEG, BMP, PNG and many more. It is user-friendly and can be operated easily using a single command. It can be downloaded freely using the link http://imagej.nih.gov/ij/index.html.

Protocol

Biofilm Growth and Microscopic Study

1. Inoculate a loop full of *P. aeruginosa* and *P. mendocina* to 5 ml LB broth and incubate at 37 °C overnight.
2. Dilute the above culture to 1:100 in LB broth (i.e. 1 ml of culture to 99 ml of LB broth).
3. To grow biofilm at liquid air interphase, transfer 5 ml of above-diluted culture to a test tube with a glass slide so that half of the slide is immersed in the media.
4. To grow submerged biofilm, transfer 15 ml of above-diluted culture to a petri-plate with a glass slide so that the slide gets completely immersed in the media.
5. Incubate at 37 °C for 24–48 h at static conditions.
6. After sufficient incubation, remove the glass slide and wash it with PBS 2–3 times, gently vortex to remove planktonic cells.
7. Stain the slides with 0.02 % aqueous solution of AO and leave it for 5 min in dark.
8. Wash gently with 1X PBS; allow drying and put a cover slip over the stained area.
9. Observe under a fluorescent microscope.

IMAGE J Analysis

1. Launch IMAGE J and open an acquired biofilm image (Fig. 6.6).
2. Go to analyse and click on it to set the measurement and select the parameters, click ok (Fig. 6.7).
3. Select an area from the image biofilm area and click on to measure. Select at least ten different images or fields to get statically significant data (Fig. 6.8).
4. A result window will appear. Take the average of the raw integrated density to quantify the biofilm. Raw integrated density can be used to quantify the biofilm growth at a different time interval or to compare the biofilm growth of different bacterial species (Fig. 6.9).
5. To plot a 3D-biofilm structure, select plugin and interactive 3D-surface plot. A 3D-plot window will appear. Adjust different image parameters to get a superior 3D plot (Fig. 6.10).
6. If required, plot a graph of the raw integrated density in a Microsoft Excel sheet.

Observation Table

Strain	Raw integrated density of AO	Average raw integrated density	Biofilm-forming ability
Control			
Test isolate 1			
Test isolate 2			

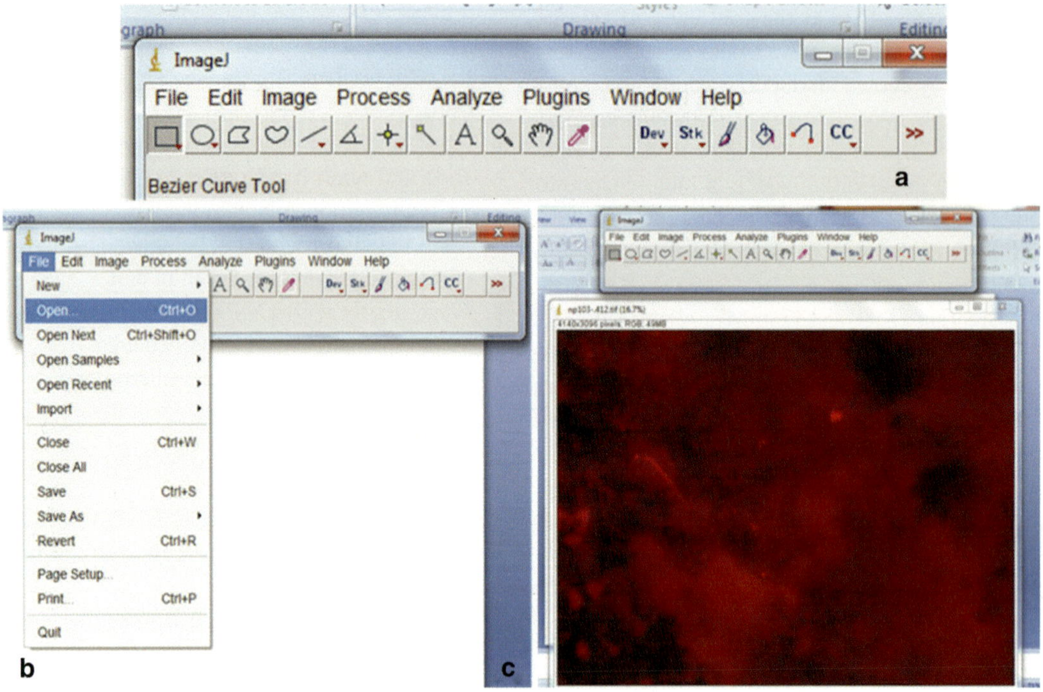

Fig. 6.6 Analysis of a biofilm matrix by IMAGE J software

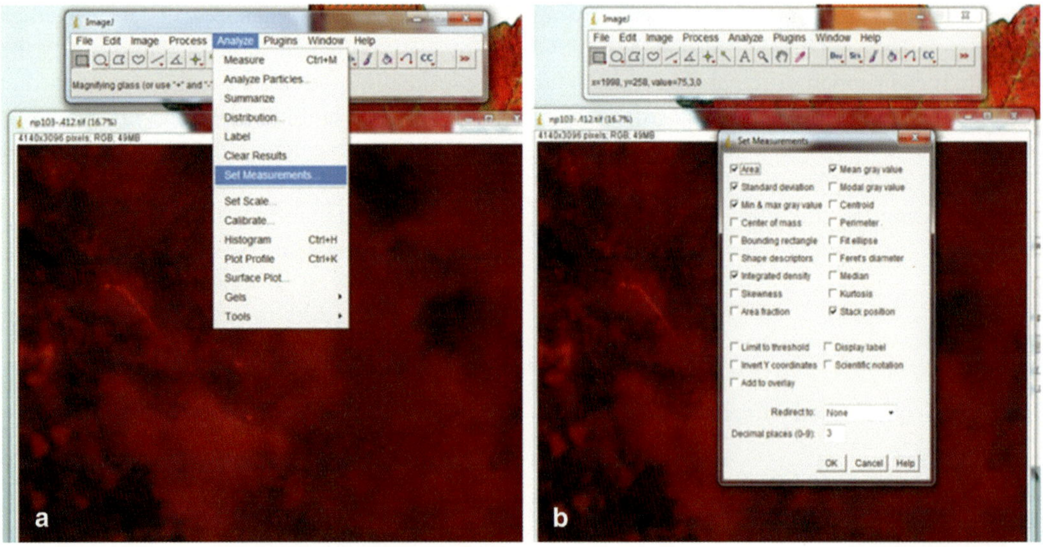

Fig. 6.7 Selection of parameters for the biofilm analysis

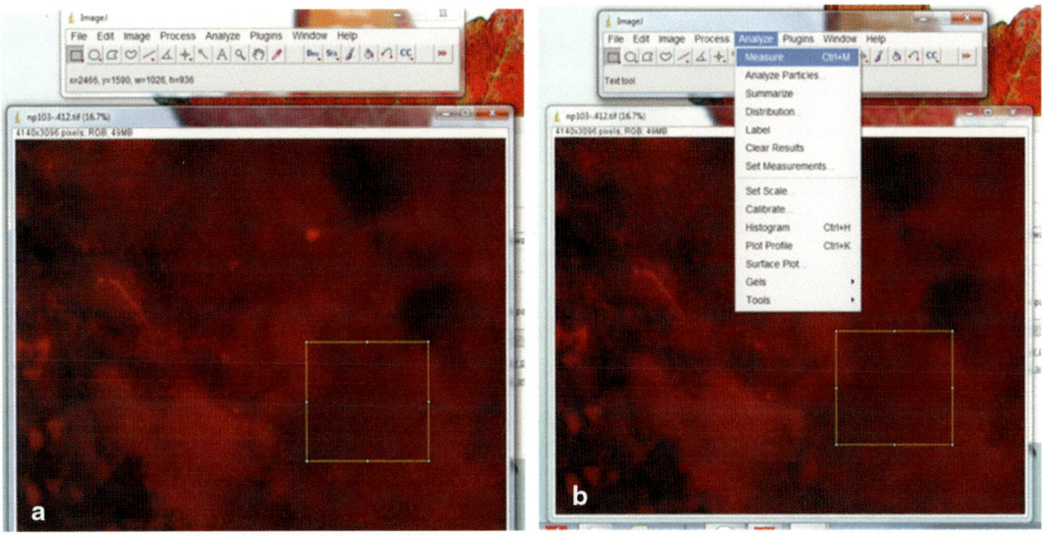

Fig. 6.8 Selection of an area for the image analysis

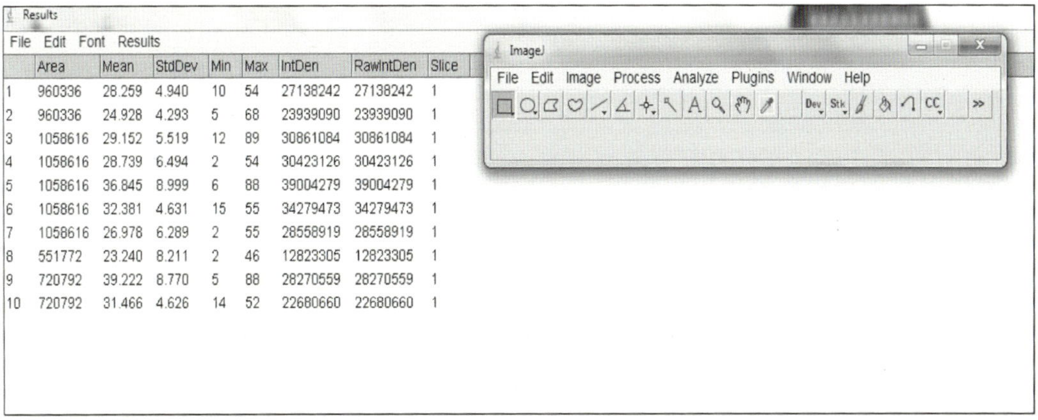

	Area	Mean	StdDev	Min	Max	IntDen	RawIntDen	Slice
1	960336	28.259	4.940	10	54	27138242	27138242	1
2	960336	24.928	4.293	5	68	23939090	23939090	1
3	1058616	29.152	5.519	12	89	30861084	30861084	1
4	1058616	28.739	6.494	2	54	30423126	30423126	1
5	1058616	36.845	8.999	6	88	39004279	39004279	1
6	1058616	32.381	4.631	15	55	34279473	34279473	1
7	1058616	26.978	6.289	2	55	28558919	28558919	1
8	551772	23.240	8.211	2	46	12823305	12823305	1
9	720792	39.222	8.770	5	88	28270559	28270559	1
10	720792	31.466	4.626	14	52	22680660	22680660	1

Fig. 6.9 Result window for different parameters

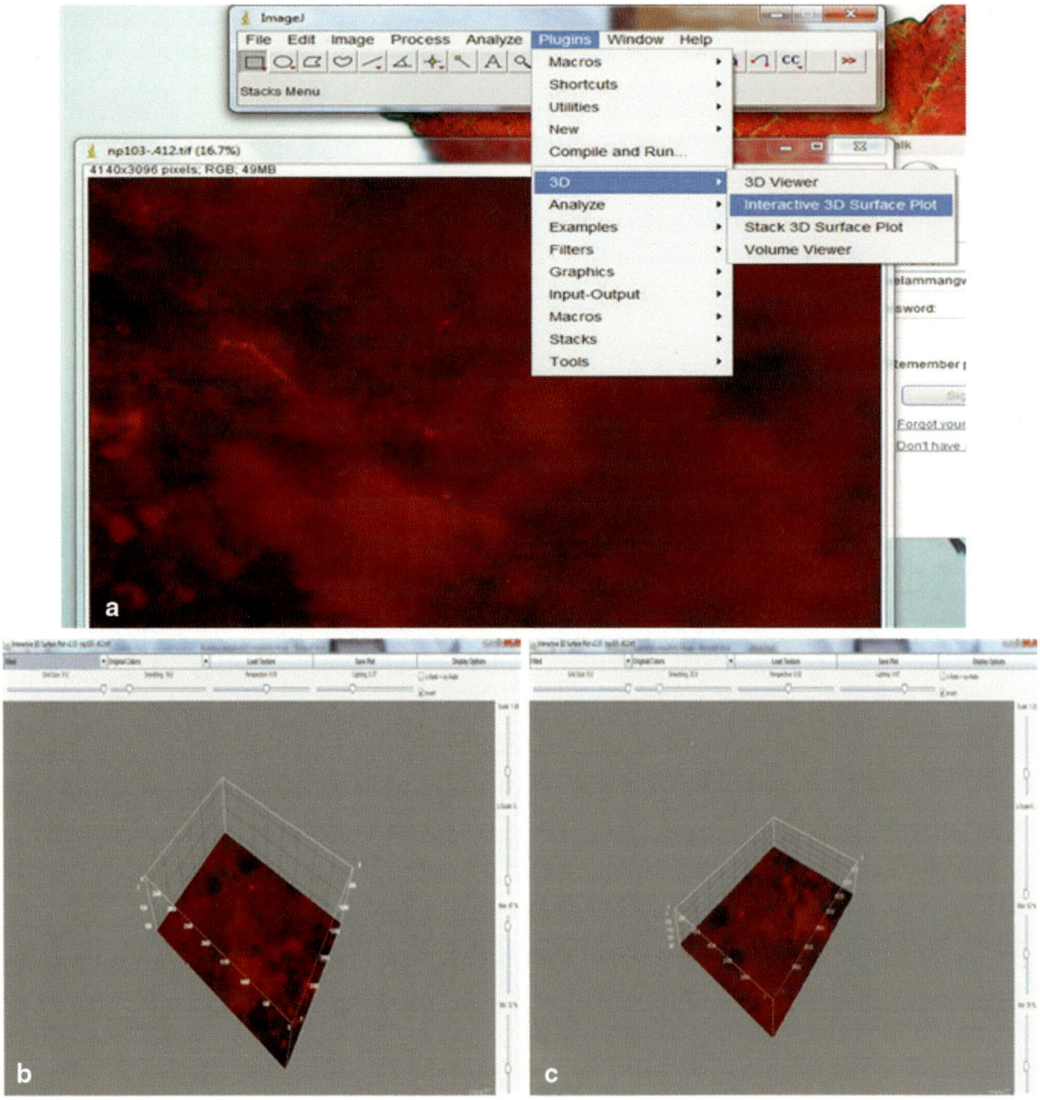

Fig. 6.10 Plotting of a 3D-surface plot of the biofilm matrix

Precautions

1. Always wear gloves.
2. PI and AO are potential mutagens and should be handled with care. The dye must be disposed safely and in accordance with applicable local regulations.
3. Carefully handle the dyes. Avoid spilling, skin and eye contact. Wash hands after handling.

FLOW CHART

Inoculate bacterial culture in LB broth

↓

Incubate overnight at 37°C

↓

Dilute in 1:100 dilution in LB

↓

Transfer it to either glass tube or petri-plate with glass slide

↓

Incubate required time

↓

Wash and stain with fluorescent dye (0.02% AO)

↓

Fluorescent microscopic studies (Ex $_{460nm}$, Em $_{650nm}$)

↓

Image J analysis

↓

Raw integrated density, 3D plot and Surface plot

Exp. 6.5 Screening for Biosurfactants

Objective Screening of biosurfactant-producing bacteria among the environmental isolates by drop collapse assay and oil spreading assay.

Introduction

Biosurfactants are the amphiphilic compounds synthesized on the living surfaces, mostly on the microbial cell surfaces. It may also be excreted extracellular, harbouring both hydrophilic and hydrophobic moieties that are known to be capable of reducing surface tension and interfacial tension between the individual molecules. Most of the biosurfactants contain any of the following components, i.e. mycolic acid, glycolipids, polysaccharide-lipid complex, lipoprotein or lipopeptide, phospholipid or the microbial cell surface itself. Biosurfactant is known to be impeded by the lack of available economic and versatile products. Surfactin, sophorolipids and rhamnolipids are among the limited number of commercially available biosurfactants. Though the type and amount of microbial surfactants produced depend on the producer organism, entities such as carbon and nitrogen source, trace element, temperature and aeration also play an important role for the efficient production of surfactants by the organism.

There exist many hydrophobic pollutants in the environment that needs to be solubilized before subjecting to degradation by microbial cells. In this regard, mineralization is governed by desorption of hydrocarbons from the soil. Hence, during the past few decades, there is an increasing demand for the biological surface active compounds or biosurfactants that are produced by large varieties of microorganisms for their wide application in biodegradation, low toxicity and widespread application in comparison to chemical surfactants. They are of much use such as emulsifiers, de-emulsifiers, wetting agents, spreading agents, foaming agents, functional food ingredients and detergents. Although the conventional chemical surfactants are inexpensive as well as highly efficient, they impart high adverse effect on the environment causing pollution. In this regard, the use of biosurfactant is the lower level of pollution, low toxicity, biocompatibility and digestibility thus allowing their use in cosmetics, pharmaceuticals and food additives.

Biosurfactants show huge level of compatibility with chemical products leading to the formation of novel formulations. There are many bacterial strains reported to produce biosurfactants, i.e. *Aeromonas* sp. (Glycolipid), *Bacillus subtilis* (Lipopeptide), *Klebsiella oxitoca* (Lipopolysaccharide), *Pseudomonas aeruginosa* (Rhamnolipid), *Pseudomonas fluorescence* (Glycolipid) and many others. There are many simplest criteria for screening of biosurfactant-producing bacterial strains that include haemolysis on blood agar, determination of emulsion index value, drop collapse assay and others. These biosurfactant-producing strains impart various physiological advantages to them that include increasing surface area for water insoluble substrates by emulsification, increased bioavailability of hydrophobic substrates, binding to heavy metals, involvement in pathogenesis, possessing antibacterial activity and regulation of attachment or detachment of microorganisms to and from the surface.

Principle

Structurally, biosurfactants are a diverse group of biomolecules, i.e. glycolipids, lipopeptides, lipoproteins, lipopolysaccharides or phospholipids. Most of the techniques for screening of biosurfactant producing bacteria are based on the interfacial or surface activity of them. In another approach of biosurfactant screening, their interference with hydrophobic interfaces is explored. There are also many other specific screening techniques such as the colorimetric Cetyl trimethylammonium bromide (CTAB) agar assay which are applied successfully to a certain group of biosurfactants. These screening techniques provide qualitative as well as quantitative results. However, for the initial screening of biosurfactant-producing microorganisms, qualitative screening techniques are sufficient. In some cases, addition of a little amount of biosurfactant increases the growth of microorganisms (Fig. 6.11).

The majority of biosurfactant-producing bacterial screening techniques involves the measurement of interfacial or surface activity. In this regard, the direct measurement of the interfacial or surface activity of the culture supernatant renders a straight forward screening technique for biosurfactant producing strains. The results of this technique give a clear idea of the indication of strong biosurfactant production. Another approach of measurement of interfacial activity includes the drop collapse assay which relies on the destabilization of liquid droplets by surfactants. When drops of cell suspension/culture supernatants are placed on oil coated, solid surface, the result can be visualized with naked eye. When the liquid does not contain surfactants, the polar water molecules are repelled from the hydrophobic surface and drops remain stable. When the liquid contains surfactants, the drops spread or collapse because of the reduction of interfacial tension between the liquid drop and the hydrophobic surface. The stability of drops is dependent on surfactant concentration and correlates with surface and interfacial tension.

There are many other screening techniques for biosurfactant production including measurement of cell-surface hydrophobicity by bacterial adhesion to hydrocarbons assay (BATH), hydrophobic interaction chromatography (HIC), replica plate assay, salt aggregation assay or CTAB agar plate assay and blood agar haemolysis assay. In this experiment, we discuss the various

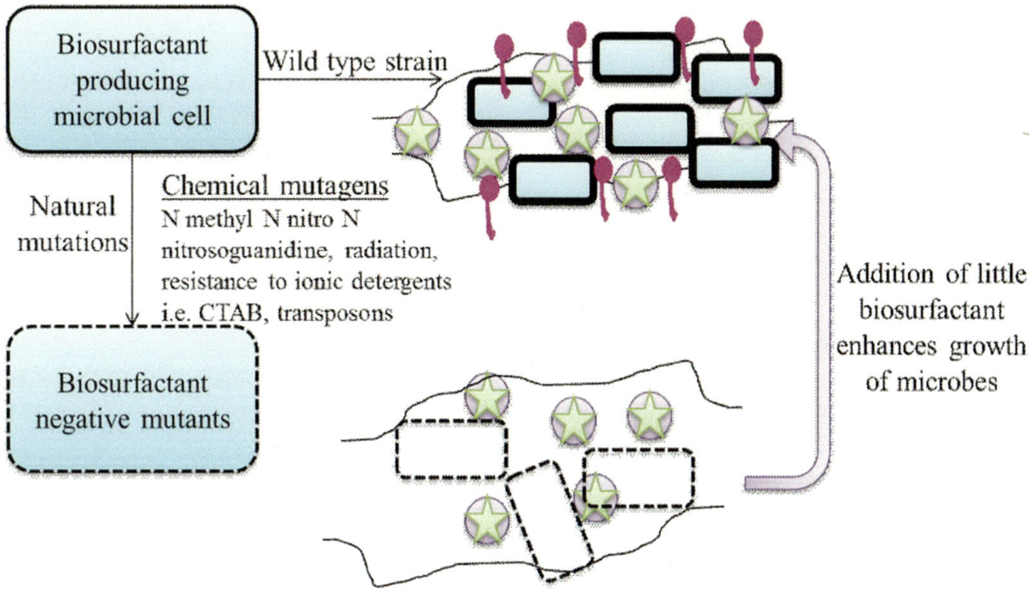

Fig. 6.11 Addition of a biosurfactant enhancing the growth of microorganisms and further enhancement of biosurfactant production

techniques involving biosurfactant screening, i.e. drop collapse assay and oil-spreading assay.

Reagents Required and Their Role

Luria Bertani Broth

Luria Bertani (LB) broth is a rich medium that permits fast growth and good growth yields for many bacterial species. It is the most commonly used medium in microbiological studies. Easy to make, fast growth of most bacterial strains, readily available and simple compositions contribute to the popularity of LB broth. LB can support bacterial growth OD_{600} 2–3 under normal shaking incubation conditions.

Frying Oil

Oil is the neutral, nonpolar chemical substance with the characteristic of viscous liquid at room temperature and pressure. It is both hydrophobic and lipophilic in nature. Most of the oils contain a high amount of carbon and hydrogen. Frying oil may be derived from vegetable or animal oils, or fats that are used for various processes of cooking and food preparation. Cooking oils are generally derived from animal fat or from plant oils of olive, maize, sunflower and many other species.

Procedure

1. Inoculate 2–3 colonies from the plates containing pure culture of bacterial strains to be tested into 5 ml LB tubes.
2. Incubate the tubes for 24 h at 37 °C with shaking at 180 rpm.
3. Transfer the cell into the 1.5 ml of micro-centrifuge tube and centrifuge at 6000 rpm for 5 min at room temperature.
4. Collect the supernatant and transfer it to a fresh micro-centrifuge tube for further screening of biosurfactant production.

Oil-Spreading Technique

1. Take 30 ml of distilled water in a glass petriplate.
2. Add 1 ml of used frying oil at the centre of the plate containing distilled water.

3. Add 20 μl of the culture supernatant at the centre of the plate containing water and the fried oil.
4. Carefully observe the displacement of oil and its subsequent spreading on the water.
5. If the culture supernatant is capable of displacing the oil to spread, it can be considered to be biosurfactant positive.

Drop Collapse Method

1. This assay relies on the destabilization of liquid droplets by the surfactants.
2. Place a drop of the culture supernatant on the oil-coated solid surface.
3. Carefully observe the repelling of the water molecules from the hydrophobic surface.
4. The liquid contains biosurfactant; the drop spreads or collapses because of the reduction of force or interfacial tension between the liquid drop and the hydrophobic surface.
5. The stability of the drop depends on the surfactant concentration, and it can be correlated with the surface and interfacial tension.

Observation

Observe the displacement of oil molecules after the addition of the culture supernatant. If the liquid contains biosurfactant, the drop spreads or collapses because of the reduction of force or interfacial tension between the liquid drop and the hydrophobic surface. The stability of the drop depends on the surfactant concentration, and it can be correlated with the surface and interfacial tension.

Result Table

Organism	Emulsification		% of emulsification activity
	Cell free culture	Intracellular	
Control			
Test strain			

FLOW CHART

Grow bacterial culture in LB broth for 24 h at 37°C with shaking at 180 rpm, after suitable growth centrifuge the grown culture to collect the supernatant at 6,000 rpm for 5 min at room temperature

↓

Take 30 ml of distilled water in a glass petriplate, add 1 ml of used frying oil at the centre of the plate and add 20 μl of culture supernatant at the centre of the petriplate

↓

Carefully observe for the displacement of oil to spread and its subsequent spreading on water

↓

For drop collapse technique place a drop of the culture supernatant on the oil coated solid surface

↓

Carefully observe for the repelling of water molecules from the hydrophobic surface

↓

When the liquid contains biosurfactant the drop spreads or collapses because of the reduction of force or interfacial tension between the liquid drop and the hydrophobic surface.

Exp. 6.6 Spectrophotometric Analysis of Bioremediation of Polycyclic Aromatic Hydrocarbons by Bacteria

Objective To analyse biodegradation of PAH by spectrophotometric measurement.

Introduction

Contamination of natural water bodies by highly hydrophobic, toxic and low-availability organic pollutant has become an issue of considerable environment apprehensions. Polycyclic aromatic hydrocarbons (PAHs) are among the prior list of organic persistent pollutants in marine sediment and water because of their toxic, mutagenic and carcinogenic effects. PAHs with two or more benzene ring are widely present in environment. The removal of PAHs from contaminated environments is of great concern. Microbial degradation is believed to be one of principal means of successfully removing PAHs from natural environments. Therefore, the biodegradation of PAHs has been studied extensively. Molecular biology techniques may allow the detection of specific genes, but the presence of these genes does not guarantee that the bacteria possessing them are viable or that the genes are being expressed in situ. Microbial activity has been deemed the most influential and significant cause of PAH removal. Numerous studies have been conducted on microbial consortia and enrichment, and several diverse genera of bacteria have been isolated. The advancement in technology has given many instruments such as high performance liquid chromatography (HPLC), gas chromatography (GC), mass spectrometry (MS), etc. that are being used for rapid and straightforward PAHs' degradation. However, a simple ultraviolet-visible (UV-VIS) spectrophotometric analysis can be a cost-effective method that can be used for biodegradation studies of pure compound. The aim of the current study is a quantitative analysis of biodegradation of PAH using the spectrophotometric method.

Biodegradation is the variable bioremediation technology for organic pollutants. It has been known for a long time that microorganisms degrade environmental pollutants in various matrices and environments. The bioremediation utilizes the metabolic versatility of the microorganisms to degrade hazardous pollutants. The feasible remedial technology requires microorganisms capable of quick adaptation to the environmental conditions and remediation of toxic substances in a reasonable period of time. Therefore, screening of a potential microorganism for efficient PAH removal capability is of huge importance from bioremediation point of view. Prior to going for bioremediation studies of microorganisms, they should be tested for their bioremediation potential using commonly available tools such as UV-VIS spectrophotometer.

Principle

Many molecules absorb ultraviolet or visible light. The absorbance of a solution increases as the attenuation of the beam increases. Absorbance is directly proportional to the path length, b, and the concentration, c, of the absorbing species. Beer's Law states that

$$A = ebc,$$

where e is a constant of proportionality, called the absorptivity.

Different molecules absorb radiation of different wavelengths. An absorption spectrum will show a number of absorption bands corresponding to structural groups within the molecule. Absorbance is a characteristic of molecules useful many times in characterization of a compound. Every molecule has maximum absorbance at a particular λ known as λ_{max}. In this regard, PAHs generally absorb light in the 200–400 nm range as well as strongly fluoresce. UV-VIS absorption and fluorescence spectroscopic techniques are sensitive for PAHs detection at an order of 0.1–1.0 μg/l and hence are widely used for the analysis of PAHs. There are some PAHs with their reported absorption maximum under UV light. They include Benzo [α] anthracene (288 nm), Benzo [α] pyrene (297 nm), Benzo [κ] fluoranthracene (307 nm), Chrysene (268 nm) and phenanthrene (251 nm) (Rivera-Figueroa et al. 2004).

Phenanthrene is a polycyclic aromatic hydrocarbon which is composed of three fused benzene rings. It is found mostly in cigarette smoke and is a potential irritant. The pure form of phenanthrene appears as a white powder with blue fluorescence. It is nearly insoluble in water but is soluble in most low polarity organic solvents such as toluene, carbon tetrachloride, ether, chloroform, acetic acid and benzene.

Reagents Required and Their Role

Luria Bertani Broth

Luria Bertani (LB) broth is a rich medium that permits fast growth and good growth yields for many species including *Escherichia coli*. It is the most commonly used medium in microbiology studies. Easy to make, fast growth of most bacterial, readily available and simple compositions contribute to the popularity of LB broth. LB can support *E. coli* growth OD_{600} 2–3 under normal shaking incubation conditions.

Phenanthrene

Phenanthrene is a potential PAH which is a colourless crystalline solid that also looks yellow. Most of the phenanthrene pollution in the environment comes due to burning of coal, oil, gas and garbage. The stock concentration of phenanthrene to be used in this study is 10 mg/ml. As it is sparingly soluble in water, the stock concentration should be prepared in absolute acetone.

Basal Salt Medium

Minimal salt medium is a highly referenced microbial growth medium that is used for the cultivation of microorganisms. The composition of minimal salt medium is KH_2PO_4—0.8 g, K_2HPO_4—1.2 g, NH_4NO_3—1.0 g, $MgSO_4.7H_2O$—0.2 g, $FeCl_3$—50 mg, $CaCl_2$—20 mg, $MnSO_4$—1.0 mg, Na_2MoO_4—0.2 mg, pH 7.2, distilled water—1000 ml. Thus, this buffered medium contains only salt and nitrogen and the microorganisms specifically degrading the PAHs can be grown in this media when it is supplemented with phenanthrene. Basalt salt medium (BSM) agar plate can be prepared by adding 1.5 % agar.

n-Hexane

Physical properties of PAH includes their low solubility/sparingly solubility in water. In contrast to that, they are highly soluble in most of the low polarity organic solvents such as toluene, carbon tetrachloride, ether, chloroform, acetic acid, hexane and benzene. Thus, in order to extract residual phenanthrene from bacterial culture n-hexane is used.

Procedure

Growth of Bacteria

1. Transfer an isolated bacterial colony to LB broth and incubate at 37 °C overnight with constant shaking at 160 rpm.
2. Dilute the above culture in a ratio of 1:100 in LB broth and incubate at 37 °C with constant shaking at 160 rpm.
3. Allow it to grow up to log phase. At log phase, harvest cells by centrifugation at 6000 rpm at room temperature for 10 min.
4. Discard the supernatant and resuspend the cell pellet with 1X phosphate buffer saline.
5. Resuspend the cell pellet to basal mineral media (about 100 ml), and adjust OD 595 nm to 0.1.
6. From stock solution of phenanthrene transfer 1 ml to a 250 ml flask containing 100 ml basal mineral medium and bacterial culture.
7. Leave it inside the laminar hood till acetone gets evaporated completely.
8. Incubate the flask at 37 °C with shaking at 160 rpm till 7 days.
9. At regular time interval (24 h) extract the residual phenanthrene with equal volume of n-hexane twice.
10. Perform the extraction step in triplicates.

Extraction and Quantification of Phenanthrene

1. To 100 ml of culture add 100 ml of n-hexane and vortex it for 10–20 min so that residual phenathrene get transferred to n-hexane.
2. Centrifuge the above mixture at 6000 rpm at 4 °C for 10 min.
3. Separate the upper organic phase and repeat the above step with aqueous phase again.
4. Combine both organic phases and allow it to dry.
5. Add fresh n-hexane to it equal to the volume of n-hexane extracted.
6. Use above extract for spectrophotometric analysis.

Preparation of Standard Curve

1. Prepare a stock solution (5 mg/ml) of phenanthrene in n-heaxne.
2. From the stock solution, prepare 5 ml working stock of 0.5, 1, 10, 50 and 100 µg/ml in n-hexane.
3. Take the lowest concentration of working stock solution and scan it in between 200–400 nm to get λ_{max}.
4. Take the absorbance of stocks at λ_{max} and prepare a standard curve. Perform in triplicates.
5. From the standard curve calculate the amount of residual phenanthrene in the media to find out the degradation amount.

Observation

Observe the absorption maximum of phenanthrene after scanning through the UV range in the spectrophotometer. Use the following equation to calculate the percentage of degradation of phenanthrene by the test bacterial isolate: % degradation $= R/50 \times 100$

Observation Table

Sl. No.	Concentration of phenanthrene (in µg/ml)	Volume of phenanthrene stock (in ml)	Volume of n-hexane (in ml)	OD at λ_{max}
1	0.5			
2	1			
3	5			
4	10			
5	25			
6	50			
7	100			
8	Residual (R)			

Precautions

1. Avoid direct contact with organic solvents and phenanthrene as they are flammable.
2. Properly mix n-hexane to the degradation culture to get maximum extraction efficiency.

FLOW CHART

Inoculate bacterial culture in LB broth and Incubate overnight at 37°C

↓

Centrifuge at 6000 rpm/ RT/ 10 min

↓

Adjust OD at 595 nm to 0.1 in BSM

↓

100 ml above culture + phenanthrene (50 mg/l)

↓

Incubate 37°C/160 rpm

↓

Extraction of residual phenanthrene with equal volume of n-hexane

↓

Preparation of standard curve for phenanthrene by UV-VIS spectrophotometer and quantification of residual phenanthrene

Exp. 6.7 H_2S Assay to Screen Metal-Accumulating Bacteria

Objective To study metal accumulation in bacteria by H_2S assay.

Introduction

Bioremediation is the process utilizing microorganisms or their enzymes for promoting degradation or removal of contaminants from the environment. In this regard, the use of microbial metabolic potential to remove environmental pollutants provides an economic and safe alternative compared to other physicochemical methodologies. Bacteria and other microorganisms exhibit a number of metabolism dependent and independent processes for uptake and accumulation of toxic metals. Both living and dead cells as well as their products are considered to be effective metal accumulators, and there are many evidences for biomass-based clean-up processes to be viable economically. However, till now, many aspects of metal–microbe interactions have been unexplored for biotechnological applications and further development and application before their release to the environment.

Microorganisms including bacteria, cyanobacteria, algae, fungi and yeasts are capable of removing heavy metals and other pollutants from the environment by various physicochemical mechanisms, i.e. adsorption, metabolism, transformation or transport. Microbial metabolism is also responsible for the synthesis of certain cell components or metabolites that are responsible for the creation of particular environmental condition facilitating deposition or precipitation of toxic substances. In this regard, living or dead microbial biomass is responsible for metal accumulation as well as the products derived from microbial cells. This process of metal accumulation is of particular interest as they are of wide applications for varieties of industries including those concerned with the provision for nuclear power. There are many studies depicting the natural capability of microorganisms for biosorption of toxic heavy metal ions that render different degrees of intrinsic resistance.

Metal bioremediation by bioaccumulation or bioadsorption is having huge advantage over other resistant mechanisms as this mechanism does not lead to the formation of other subsequent harmful by-products during processing and the contaminants are localized. Another major advantage of bioaccumulating microorganisms is that the metals can be recovered from them using standard procedures and can be reused subsequently for the synthesis of other substances and hence the chance of elimination to the environment by contaminating can be limited. Thus, a rapid technique for the screening of metal accumulating bacteria by biosorption by H_2S assay has been described here.

Principle

The complex structure of microorganisms is responsible for exploring the many ways for metal accumulation by microbial cell. However, there are various biosorption mechanisms which have not been explored so far. The transportation of metal across the cell membrane yields intracellular accumulation that depends on the metabolism of the cells and in this technique biosorption is carried out only within the viable cells. This mechanism is associated with the normal defence system of the bacterial cell upon exposure to the toxic substances. In another approach, during non-metabolism-dependent biosorption, metal accumulation occurs by means of physicochemical interaction between the metals and the functional groups present on the microbial cell surface. In this case, the involved mechanisms include physical adsorption, ion exchange and chemical sorption which are not dependent on cell metabolism. As the cell wall of microorganisms consists of mostly polysaccharides, proteins and lipids, they possess many metal-binding groups such as carboxyl, sulphate, phosphate and amino groups. However, these types of metabolism independent biosorption mechanisms are relatively rapid and this process is highly reversible.

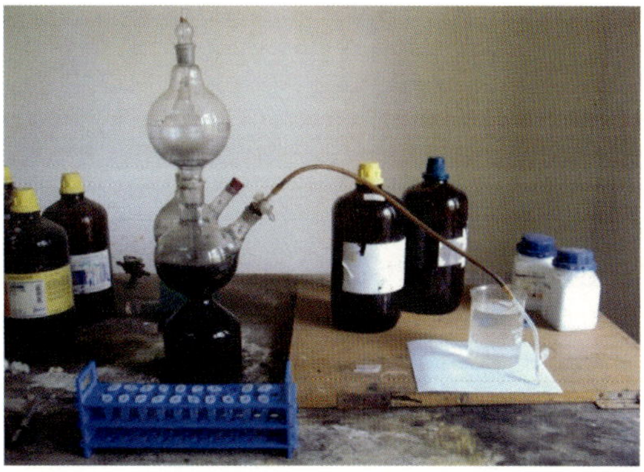

Fig. 6.12 Experimental set up for the determination of metal bioaccumulation in the isolates at H_2S gas exposure

In another approach of bioaccumulation involving precipitation where metal uptake takes place both in solutions and on the cell surface, this process is dependent on the cellular metabolism. In this case, in the presence of the toxic metals, microorganisms produce certain compounds that are responsible for favouring the precipitation process. In certain instances, precipitation was also found to be independent of cellular metabolism and is found to be the mere interaction between metals and the cell surface. Metals are having higher affinity towards the sulphur molecules and can readily form the corresponding metal sulphides. Metal sulphides precipitate and give rise to black colour precipitate which can be distinguished visibly. H_2S gas containing sulphur molecules can readily penetrate the bacterial cells accumulating mercury or other metals and bind with those metal ions to form metal sulphides. When bacterial colonies accumulate metals and are grown on plates and are exposed to H_2S gas, bacterial colonies develop black colorization for the positive screening of bioaccumulation of bacterial colonies. Alternatively, bacterial colonies grown in broth culture in the presence of metal solutions and the cell mass can be harvested by centrifugation (Fig. 6.12). The cell pellet collected in the micro-centrifuge tube on exposure to H_2S gas may or may not give rise to black colour precipitation depending on the nature of the isolate.

Reagents Required and Their Role

Luria Bertani Broth

Luria bertani (LB) broth is a rich medium that permits fast growth and good growth yields for many bacterial species. It is the most commonly used medium in microbiological studies. Easy to make, fast growth of most bacterial strains, readily available and simple compositions contribute to the popularity of LB broth. LB can support bacterial growth OD_{600} 2–3 under normal shaking incubation conditions.

H_2S Gas

Hydrogen sulphide is a colourless gas with characteristic foul odour of rotten eggs. It possesses high-penetrating power inside the bacterial cell. H_2S binds with metal compounds to form the metal sulphides which readily form the black precipitate. In normal laboratory conditions, ferrous sulphide is used to generate H_2S gas with a strong acid in a Kipp generator. The reaction involves: $FeS + 2HCl \rightarrow FeCl_2 + H_2S$.

Mercuric Chloride

Mercuric chloride ($HgCl_2$) is the crystalline solid salt of mercury which is highly toxic in nature. In earlier practices, it was used for the treatment of syphilis, but nowadays it is no longer in use due to its high toxicity level and availability of superior treatments. There are many bacterial strains resistant to mercury, and the resistance mechanisms can be studied by the indirect resistance mechanism to the corresponding salt solution. It should be added to the medium at the sublethal concentration to the microorganism.

Phosphate Buffer Saline

Phosphate buffer saline (PBS) is the buffer solution most commonly used in any biological research. The water-based salt solution contains sodium phosphate, and in some instances potassium phosphate or potassium chloride is also used. It is mostly used for its isotonic nature; it is non-toxic to the bacterial cultures. Ingredients for preparation of PBS are shown in Table 6.2.

Table 6.2 Ingredients and concentrations to prepare phosphate buffer saline

Salt	Concentration (mM/l)	Concentration (g/l)
NaCl	137	8.01
KCl	2.7	0.20
pH 7.4		

Procedure

1. Prepare 5 ml of LB broth.
2. Add concentration of mercury at a level of sub-minimum inhibitory concentration of the organism tested.
3. Inoculate 2–3 bacterial colonies from the pure culture of bacteria streaked after incubation for 24 h.
4. Incubate the tubes at 37 °C for 24 h with shaking at 180 rpm.
5. After incubation, transfer the grown culture into 1.5 ml micro-centrifuge tube and centrifuge at 6000 rpm for 5 min at 4 °C.
6. Collect the cell pellet and discard the supernatant.
7. Add the remaining culture to the tube and repeat centrifugation to collect the cell pellet.
8. Wash the cell pellet twice with autoclaved PBS.
9. Prepare H_2S in a Kipp generator by adding FeS with concentrated HCl solution.
10. Expose the tubes containing bacterial cell pellets to H_2S gas for 10 min.
11. Observe the development of black coloration of the pellets.

Observation

Observe carefully the change in colour of the cell pellets from colourless to dark black. The black colouration of the cell pellets is due to the accumulated mercury compounds in the bacterial cell where H_2S molecules bind with them to generate the corresponding mercuric sulphide precipitation. The level of mercury biosorption can be estimated by the development of coloration at an interval of time.

Result Table

Isolate	Development of black coloration with time (min)										Inference
	1	2	3	4	5	6	7	8	9	10	
Control											
Test 1											
Test 2											

Troubleshooting

Problem	Possible cause	Possible solution
Nondevelopment of black coloration	Check with the +ve control	If positive control did not give the positive result, then the bacterial strain showing negative result is confirmed to be negative for mercury biosorption
	Metal did not dissolve properly in media	If the metal did not dissolve in the media properly, the bacteria may not be able to take it up proficiently from the media to accumulate inside it
	Intensity of H_2S gas production is poor	If the intensity of the H_2S gas production is not at a high intensity level, it does not possess the proper penetrating power to go inside the bacterial cell for the development of black precipitate
False +ve results	Washing step not appropriate	If washing will not be appropriate, there is the chance of getting false positive results for bioaccumulation; carefully repeat the washing step with PBS
	Media remaining along with the pellet	The media should be removed completely after centrifugation that remains in the supernatant. If the media will not be removed properly, mercury remaining in the media may bind with the H_2S gas to give rise to false positive results

PBS phosphate buffer saline

Precautions

1. Always wear gloves while performing this experiment.
2. Use mask to cover your face while exposing to H_2S gas.
3. Never let mercury salts to come in contact with your body parts.
4. Never inhale mercury salts or any other reagents used during this experiment as they may cause potential health hazards.
5. Try to expose the cell pellets with H_2S gas in sterile conditions.

FLOW CHART

Prepare 5 ml of LB broth

↓

Add concentration of mercury at a level of sub minimum inhibitory concentration of the organism tested

↓

Inoculate 2-3 bacterial colonies from the pure culture of bacteria streaked after incubation for 24 h

↓

Incubate the tubes at 37°C for 24 h with shaking at 180 rpm

↓

After incubation transfer the grown culture into 1.5 ml micro-centrifuge tube and centrifuge at 6,000 rpm for 5 min at 4°C

↓

Collect the cell pellet and discard the supernatant

↓

Add the remaining culture to the tube and repeat centrifugation to collect the cell pellet

↓

Wash the cell pellet twice with autoclaved PBS

↓

Prepare H_2S in a Kipp generator by adding FeS with concentrated HCl solution

↓

Expose the tubes containing bacterial cell pellets to H_2S gas for 10 min

↓

Observe the development of black coloration of the pellets

References

Birnboim HC, Doly J (1979) A rapid alkaline extraction procedure for screening recombinant plasmid DNA. Nucleic Acids Res 7:1513–1523

Brock, TD (1997) The value of basic research: discovery of Thermus aquaticus' and other extreme thermophiles. Genetics. 146:1207–1210

Chomczynski P, Sacchi N (1987) Single step method of RNA isolation by acid guanidinium thiocyanate-phenol-chloroform extraction. Anal Biochem 162:156–159

Espinosa L, Borowsky R (1998) Evolutionary divergence of AP-PCR (RAPD) patterns. Mol Biol Evol 15:408–414

Healy M, Huong J, Bittner T, Lising M, Frye S, Raza S, Schrock R, Manry J, Renwick A, Nieto R, Woods C, Versalovic J, Lupski JR (2005) Microbial DNA typing by automated repetitive-sequence-based PCR. J Clin Microbiol 43:199–207

Lederberg J, Cavalli-Sforza LL, Lederberg EM (1952) Sex compatibility in *Escherichia coli*. Genetics 37:720–730

Mantri CK, Mohapatra SS, Ramamurthy T, Ghosh R, Colwell RR, Singh DV (2006) Septaplex PCR assay for rapid identification of *Vibrio cholera* including detection of virulence and *int* SXT genes. FEMS Microbiol Lett 265:208–214

Olive DM, Bean P (1999) Principles and applications of methods for DNA-based typing of microbial organisms. J Clin Microbiol 37:1661–1669

Rivera-Figueroa AM, Ramazan KA, Finlayson-Pitts BJ (2004) Fluorescence, absorption, and excitation spectra of polycyclic aromatic hydrocarbons as a tool for quantitative analysis. J Chem Educ 81:242–245

Saiki R, Gelfand D, Stoffel S, Scharf S, Higuchi R, Horn G, Mullis K, Erlich H (1988) Primer-directed enzymatic amplification of DNA with a thermostable DNA polymerase. Science 239:487–491

Versalovic J, Schneider M, De Bruijn FJ, Lupski JR (1994) Genomic fingerprinting of bacteria using repetitive sequence based polymerase chain reaction. Met Mol Cell Biol 5:25–40

Woese CR (1987) Bacterial evolution. Microbiol Rev 51:221–271

Zhu YY, Machleder EM, Chenchik A, Li R, Siebert PD (2001) Reverse transcriptase template switching: a SMART^{TM} approach for full-length cDNA library construction. Bio Techniques 30:892–897

S. Das, H. R. Dash, *Microbial Biotechnology- A Laboratory Manual for Bacterial Systems,*
DOI 10.1007/978-81-322-2095-4, © Springer India 2015

Further Readings

Arora DK, Das S, Sukumar Mesapogu (2013) Analyzing microbes: manual of molecular biology techniques. Springer Protocols Handbooks. Springer Verlag. ISBN: 978-3-642-34409-1

Atlas RM (1997) Principles of microbiology. Mosby-Year Book, WmC Brown Publ, USA

Birge EA (2000) Bacterial and bacteriophage genetics. Springer-Verlag, USA

Caldwell DR (2000) Microbial physiology and metabolism, 2nd edn. Star Publ Comm, Belmont Calif

Cappuccino JG, Sherman N (1999) Microbiology–a laboratory manual, 4th edn. Addison-Wesley Longman, USA

Das S, Dash HR, Mangwani N, Chakraborty J, Kumari S (2014) Understanding molecular identification approaches for genetic relatedness and phylogenetic relationships of microorganisms. J Microbiol Methods (in press). doi:10.1016/j.mimet.2014.05.013

Glazer AN, Nikaido H (2007) Microbial biotechnology: fundamentals of applied microbiology. 2nd edn. Cambridge University Press, Cambridge. ISBN: 13-978-0-521-84210-5

Green MR, Sambrook J (2012) Molecular cloning: a laboratory manual. 4th edn. Cold Spring Harbor Laboratory Press, Cold Spring Harbor

Hurst GH, Crawford RL, Knudsen GR (2002) Manual of environmental microbiology, 2nd edn. ASM Press, USA

Prescott LM, Harley JP, Klein DA (2002) Microbiology, 5th edn. McGraw Hill, USA

Primrose SB, Twyman R, Old RW (2001) Principles of gene manipulation and genomics, 7th edn. Wiley, USA

Verma AS, Das S, Singh A (2014) Laboratory manual for biotechnology. S. Chand and Co. Pvt. Ltd, New Delhi. ISBN: 978-93-83746-22-4

Wilson K, Walker J (2010) Principles and techniques of biochemistry and molecular biology. 7th edn. Cambridge university press, Cambridge. ISBN: 978-0-521-51635-8

S. Das, H. R. Dash, *Microbial Biotechnology- A Laboratory Manual for Bacterial Systems*,
DOI 10.1007/978-81-322-2095-4, © Springer India 2015

Printed by Printforce, the Netherlands